PROGRESS IN BRAIN RESEARCH

VOLUME 180

NANONEUROSCIENCE AND NANONEUROPHARMACOLOGY

Other volumes in PROGRESS IN BRAIN RESEARCH

Volume 145: Acetylcholine in the Cerebral Cortex, by L. Descarries, K. Krnjevićand M. Steriade (Eds.) – 2004, ISBN 0-444-51125-3.
Volume 146: NGF and Related Molecules in Health and Disease, by L. Aloe and L. Calzà (Eds.) – 2004, ISBN 0-444-51472-4.
Volume 147: Development, Dynamics and Pathology of Neuronal Networks: From Molecules to Functional Circuits, by J. Van Pelt, M. Kamermans, C.N. Levelt, A. Van Ooyen, G.J.A. Ramakers and P.R. Roelfsema (Eds.) – 2005, ISBN 0-444-51663-8.
Volume 148: Creating Coordination in the Cerebellum, by C.I. De Zeeuw and F. Cicirata (Eds.) – 2005, ISBN 0-444-51754-5.
Volume 149: Cortical Function: A View from the Thalamus, by V.A. Casagrande, R.W. Guillery and S.M. Sherman (Eds.) – 2005 ISBN 0-444-51679-4.
Volume 150: The Boundaries of Consciousness: Neurobiology and Neuropathology, by Steven Laureys (Ed.) – 2005, ISBN 0-444-51851-7.
Volume 151: Neuroanatomy of the Oculomotor System, by J.A. Büttner-Ennever (Ed.) – 2006, ISBN 0-444-51696-4.
Volume 152: Autonomic Dysfunction after Spinal Cord Injury, by L.C. Weaver and C. Polosa (Eds.) – 2006, ISBN 0-444-51925-4.
Volume 153: Hypothalamic Integration of Energy Metabolism, by A. Kalsbeek, E. Fliers, M.A. Hofman, D.F. Swaab, E.J.W. Van Someren and R.M. Buijs (Eds.) – 2006, ISBN 978-0-444-52261-0.
Volume 154: Visual Perception, Part 1, Fundamentals of Vision: Low and Mid-Level Processes in Perception, by S. Martinez-Conde, S.L. Macknik, L.M. Martinez, J.M. Alonso and P.U. Tse (Eds.) – 2006, ISBN 978-0-444-52966-4.
Volume 155: Visual Perception, Part 2, Fundamentals of Awareness, Multi-Sensory Integration and High-Order Perception, by S. Martinez-Conde, S.L. Macknik, L.M. Martinez, J.M. Alonso and P.U. Tse (Eds.) – 2006, ISBN 978-0-444-51927-6.
Volume 156: Understanding Emotions, by S. Anders, G. Ende, M. Junghofer, J. Kissler and D. Wildgruber (Eds.) – 2006, ISBN 978-0-444-52182-8.
Volume 157: Reprogramming of the Brain, by A.R. Møller (Ed.) – 2006, ISBN 978-0-444-51602-2.
Volume 158: Functional Genomics and Proteomics in the Clinical Neurosciences, by S.E. Hemby and S. Bahn (Eds.) – 2006, ISBN 978-0-444-51853-8.
Volume 159: Event-Related Dynamics of Brain Oscillations, by C. Neuper and W. Klimesch (Eds.) – 2006, ISBN 978-0-444-52183-5.
Volume 160: GABA and the Basal Ganglia: From Molecules to Systems, by J.M. Tepper, E.D. Abercrombie and J.P. Bolam (Eds.) – 2007, ISBN 978-0-444-52184-2.
Volume 161: Neurotrauma: New Insights into Pathology and Treatment, by J.T. Weber and A.I.R. Maas (Eds.) – 2007, ISBN 978-0-444-53017-2.
Volume 162: Neurobiology of Hyperthermia, by H.S. Sharma (Ed.) – 2007, ISBN 978-0-444-51926-9.
Volume 163: The Dentate Gyrus: A Comprehensive Guide to Structure, Function, and Clinical Implications, by H.E. Scharfman (Ed.) – 2007, ISBN 978-0-444-53015-8.
Volume 164: From Action to Cognition, by C. von Hofsten and K. Rosander (Eds.) – 2007, ISBN 978-0-444-53016-5.
Volume 165: Computational Neuroscience: Theoretical Insights into Brain Function, by P. Cisek, T. Drew and J.F. Kalaska (Eds.) – 2007, ISBN 978-0-444-52823-0.
Volume 166: Tinnitus: Pathophysiology and Treatment, by B. Langguth, G. Hajak, T. Kleinjung, A. Cacace and A.R. Møller (Eds.) – 2007, ISBN 978-0-444-53167-4.
Volume 167: Stress Hormones and Post Traumatic Stress Disorder: Basic Studies and Clinical Perspectives, by E.R. de Kloet, M.S. Oitzl and E. Vermetten (Eds.) – 2008, ISBN 978-0-444-53140-7.
Volume 168: Models of Brain and Mind: Physical, Computational and Psychological Approaches, by R. Banerjee and B.K. Chakrabarti (Eds.) – 2008, ISBN 978-0-444-53050-9.
Volume 169: Essence of Memory, by W.S. Sossin, J.-C. Lacaille, V.F. Castellucci and S. Belleville (Eds.) – 2008, ISBN 978-0-444-53164-3.
Volume 170: Advances in Vasopressin and Oxytocin – From Genes to Behaviour to Disease, by I.D. Neumann and R. Landgraf (Eds.) – 2008, ISBN 978-0-444-53201-5.
Volume 171: Using Eye Movements as an Experimental Probe of Brain Function—A Symposium in Honor of Jean Büttner-Ennever, by Christopher Kennard and R. John Leigh (Eds.) – 2008, ISBN 978-0-444-53163-6.
Volume 172: Serotonin–Dopamine Interaction: Experimental Evidence and Therapeutic Relevance, by Giuseppe Di Giovanni, Vincenzo Di Matteo and Ennio Esposito (Eds.) – 2008, ISBN 978-0-444-53235-0.
Volume 173: Glaucoma: An Open Window to Neurodegeneration and Neuroprotection, by Carlo Nucci, Neville N. Osborne, Giacinto Bagetta and Luciano Cerulli (Eds.) – 2008, ISBN 978-0-444-53256-5.
Volume 174: Mind and Motion: The Bidirectional Link Between Thought and Action, by Markus Raab, Joseph G. Johnson and Hauke R. Heekeren (Eds.) – 2009, 978-0-444-53356-2.
Volume 175: Neurotherapy: Progress in Restorative Neuroscience and Neurology — Proceedings of the 25th International Summer School of Brain Research, held at the Royal Netherlands Academy of Arts and Sciences, Amsterdam, The Netherlands, August 25–28, 2008, by J. Verhaagen, E.M. Hol, I. Huitinga, J. Wijnholds, A.A. Bergen, G.J. Boer and D.F. Swaab (Eds.) – 2009, ISBN 978-0-12-374511-8.
Volume 176: Attention, by Narayanan Srinivasan (Ed.) – 2009, ISBN 978-0-444-53426-2.
Volume 177: Coma Science: Clinical and Ethical Implications, by Steven Laureys, Nicholas D. Schiff and Adrian M. Owen (Eds.) – 2009, 978-0-444-53432-3.
Volume 178: Cultural Neuroscience: Cultural Influences On Brain Function, by Joan Y. Chiao (Ed.) – 2009, 978-0-444-53361-6.
Volume 179: Genetic models of schizophrenia, by Akira Sawa (Ed.) – 2009, 978-0-444-53430-9.

PROGRESS IN BRAIN RESEARCH

VOLUME 180

NANONEUROSCIENCE AND NANONEUROPHARMACOLOGY

EDITED BY

HARI SHANKER SHARMA
Laboratory of Cerebrovascular Research, Department of Surgical Sciences, Anesthesiology and Intensive Care Medicine, University Hospital, Uppsala University, SE-75185 Uppsala, Sweden

ELSEVIER

AMSTERDAM – BOSTON – HEIDELBERG – LONDON – NEW YORK – OXFORD
PARIS – SAN DIEGO – SAN FRANCISCO – SINGAPORE – SYDNEY – TOKYO

Elsevier
Radarweg 29, PO Box 211, 1000 AE Amsterdam, The Netherlands
Linacre House, Jordan Hill, Oxford OX2 8DP, UK
360 Park Avenue South, New York, NY 10010-1710

First edition 2009

Library of Congress Cataloging-in-Publication Data
A catalog record for this book is available from the Library of Congress

British Library Cataloguing in Publication Data
A catalogue record for this book is available from the British Library

ISBN: 978-0-444-53431-6
ISSN: 0079-6123

Transferred to Digital Printing in 2010.

List of Contributors

R.J. Andrews, Smart Systems and Nanotechnology, NASA Ames Research Center, Moffett Field, CA, USA
M. Aschner, Department of Pediatrics, Pharmacology and The Kennedy Center for Research on Human Development, Vanderbilt University School of Medicine, Nashville, TN, USA
A. Barras, Laboratory of Chemistry and MicroNanotechnology for Therapy, UMR 8161 CNRS-Université de Lille 2-Université de Lille 1-Institut Pasteur de Lille, Lille Cedex, France
D. Betbeder, Laboratory of Physiology, University of Lille, Faculté de Médecine, Lille, France
L. Bondioli, Department of Pharmaceutical Sciences, University of Modena and Reggio Emilia, Modena, Italy
A.M. Brioschi, Department of Neurology and Laboratory of Clinical Neurobiology, Ospedale S. Giuseppe, Istituto Auxologico Italiano, IRCCS, Verbania, Italy
S. Calderoni, Department of Neurology and Laboratory of Clinical Neurobiology, Ospedale S. Giuseppe, Istituto Auxologico Italiano, IRCCS, Verbania, Italy
J. Chang, Laboratory of Physiology, University of Lille, Faculté de Médecine, Lille, France and School of Materials Science and Engineering, Tianjin University, Tianjin, China and Laboratory of Blood-Brain Barrier, University of Artois, Faculté des Sciences Jean Perrin, Lens, France
R.P. Choudhury, Department of Cardiovascular Medicine, John Radcliffe Hospital, Headington, Oxford, UK
L. Costantino, Department of Pharmaceutical Sciences, University of Modena and Reggio Emilia, Modena, Italy
C.J. Destache, Department of Pharmacy Practice, Creighton University School of Pharmacy & Health Professions, Omaha, NE, USA
N. Dupont, Laboratory of Chemistry and MicroNanotechnology for Therapy, UMR 8161 CNRS-Université de Lille 2-Université de Lille 1-Institut Pasteur de Lille, Lille Cedex, France
W. Feng, School of Materials Science and Engineering, Tianjin University, Tianjin, China
F. Forni, Department of Pharmaceutical Sciences, University of Modena and Reggio Emilia, Modena, Italy
M.R. Gasco, Nanovector s.r.l, Torino, Italy
Y. Jallouli, Laboratory of Blood-Brain Barrier, University of Artois, Faculté des Sciences Jean Perrin, Lens, France
J.V. Lafuente, Lab Neurociencias Clínicas y Experimentales (LaNCE), Dpt. de Neurociencias, Universidad del País Vasco – EuskalHerriko Unibertsitatea, Bilbao, España
W. Lee, Department of Neurobiology, Center for Glial Biology in Medicine, Atomic Force Microscopy & Nanotechnology Laboratories, Civitan International Research Center, Evelyn F. McKnight Brain Institute, University of Alabama, Birmingham, AL, USA
G. Liu, Department of Radiology, University of Utah, Salt Lake City, UT, USA
A. Mauro, Department of Neurology and Laboratory of Clinical Neurobiology, Ospedale S. Giuseppe, Istituto Auxologico Italiano, IRCCS, Verbania, Italy and Departement of Neurosciences, University of Torino, Torino, Italy
M.A. McAteer, Department of Cardiovascular Medicine, John Radcliffe Hospital, Headington, Oxford, UK

P. Men, Department of Radiology, University of Utah, Salt Lake City, UT, USA
D.F. Muresanu, Department of Neurology, University of Medicine and Pharmacy "Iuliu Hatieganu", Cluj-Napoca, Romania
V. Parpura, Department of Neurobiology, Center for Glial Biology in Medicine, Atomic Force Microscopy & Nanotechnology Laboratories, Civitan International Research Center, Evelyn F. McKnight Brain Institute, University of Alabama, Birmingham, AL, USA
R. Patnaik, Department of Biomaterials, School of Biomedical Engineering, Institute of Technology, Banaras Hindu University, Varanasi, India
G. Perry, College of Sciences, University of Texas at San Antonio, San Antonio, TX, USA and Department of Pathology, Case Western Reserve University, Cleveland, OH, USA
L. Priano, Department of Neurosciences, University of Turin, Torino, Italy and IRCCS – Istituto Auxologico Italiano, Ospedale S. Giuseppe – Piancavallo, Verbania, Italy
B. Ruozi, Department of Pharmaceutical Sciences, University of Modena and Reggio Emilia, Modena, Italy
A. Sharma, Laboratory of Cerebrovascular and Pain Research, Department of Surgical Sciences, Anesthesiology and Intensive Care Medicine, University Hospital, Uppsala University, SE-75185 Uppsala, Sweden
H.S. Sharma, Laboratory of Cerebrovascular and Pain Research, Department of Surgical Sciences, Anesthesiology and Intensive Care Medicine, University Hospital, Uppsala University, SE-75185 Uppsala, Sweden
G.A. Silva, Departments of Bioengineering and Ophthalmology and Neurosciences Program, University of California, San Diego, CA, USA
M.A. Smith, Department of Pathology, Case Western Reserve University, Cleveland, OH, USA
G. Tosi, Department of Pharmaceutical Sciences, University of Modena and Reggio Emilia, Modena, Italy
M.A. Vandelli, Department of Pharmaceutical Sciences, University of Modena and Reggio Emilia, Modena, Italy
X.-B. Yuan, School of Materials Science and Engineering, Tianjin University, Tianjin, China
G.P. Zara, Department of Anatomy Pharmacology and Forensic Medicine, University of Turin, Torino, Italy

Preface

Recent advancements in our knowledge about Nanoscale materials and their possible effects on the biological system have resulted in an increased awareness about their modulatory role on the human health system (Sharma, 2009a, 2009b; Sharma & Sharma, 2009). However, the effects of these nanoscale materials comprising "microfine particles" normally present in the environment, or "engineered materials from metals" emanating from some industrial sources at certain work places on our central nervous system (CNS) are still not well known (Sharma, 2007). On the other hand, some attempts have been made to use nanodrug delivery to achieve better therapeutic value in clinical situations (Hekmatara, Gelperina, Vogel, Yang, & Kreuter, 2009). In addition, nanoparticles are also used now for neurodiagnostic purposes (Pathak et al., 2009). Based on these studies from the past 3–4 years led to the development of a new discipline, "Nanoneuroscience," that deals with the effects of nanoparticles on the CNS related to their both beneficial and harmful effects. This book is the first to be solely directed to understand the new developments in the field of Nanoneuroscience and Nanoneuropharmacology that is currently needed by researchers and clinicians alike to follow the rapid growth of this newly emerging field.

Research on nanoparticles has attracted the attention of scientists in the past 5 years to find out whether these nanomaterials could affect our vital organs, for example, lung, liver, kidney, and heart adversely when they enter into our body fluid environments possibly through inhalation (Oberdörster, Elder, & Rinderknecht, 2009). However, effects of these nanoparticles on the CNS toxicity *in vivo* are still not examined in detail (Sharma, 2007; Sharma & Sharma, 2007). Using *in vitro* models few studies demonstrated neurotoxicity of nanoparticles to neurons and glial cells (Andrews, 2009). This indicates a possible adverse effect on our brain function following exposure to these microfine particles. However, it is still unclear whether human population living in areas that are heavily polluted with carbon nanoparticles due to motor vehicle exhausts, or with silica dust in desert environment are more vulnerable to CNS injuries or combat stress resulting in an exacerbation of their cognitive, sensory–motor disturbances, or brain pathology as compared to normal populations (Andrews, 2009; Campbell, Araujo, Li, Sioutas, & Kleinman, 2009; Sharma, Ali, Hussain, Schlager, & Sharma, 2009; Sharma, Patnaik, Sharma, Sjöquist, & Lafuente, 2009; Sharma & Sharma, 2009; Sharma, Ali, Tian, Hussain, et al., 2009).

In addition, whether the neuroprotective effects of drugs are also evident with CNS injuries occurring in nanoparticles inoculated subjects is still unclear (Sharma, Ali, Tian, Hussain, et al., 2009). Thus, the need of the hour is to find out whether our military personnel working in desert environments, where they are exposed to silica dust, are more vulnerable to CNS injuries or heat stress during their combat or peacekeeping operations (Sharma & Sharma, 2007; Sharma, Ali, Hussain, et al., 2009; Sharma, Ali, Tian, Hussain, et al., 2009). In such situations, pharmacological use of nanoparticles, to enhance drug delivery to the brain, could be a great opportunity to enhance the neurotherapeutic capabilities of the known neuroprotective agents (Sharma, Ali, Tian, Patnaik, et al., 2009; Sharma et al., 2007). Based on these investigations, it appears that use of nanodrug delivery could be useful in attaining superior neuroprotection in CNS injuries (Sharma, 2007; Sharma & Sharma, 2007; Sharma et al., 2007). However, it is still uncertain whether nanoparticles used to enhance drug delivery as such may have some neurotoxic effects.

Recent developments in this field show that nanoparticles derived from metals can induce profound neurotoxicity probably by inducing breakdown of the blood–brain barrier (Sharma, Ali, Hussain, et al., 2009; Sharma, Patnaik, et al., 2009; Sharma & Sharma, 2007). Furthermore, nanoparticles are also able to enhance the perception and symptoms of heat stress and lead to exacerbation of brain pathology and cognitive dysfunctions in laboratory conditions (Sharma, Ali, Tian, Hussain, et al., 2009; Sharma & Sharma, 2007). Under such circumstances, normal drugs that are able to induce neuroprotection in normal animals failed to protect brain pathology in nanoparticles-treated animals (Sharma, Ali, Tian, Hussain, et al., 2009). This indicates that nanoparticles exposure alters the physiological response of the organisms following CNS injury or stress and results in aggravation of cellular and molecular reactions within the CNS. Thus, new investigations are needed to further expand our knowledge in the field of Nanoneurosciences and Nanoneuropharmacology.

This volume is the first to summarize recent developments in this newly emerging discipline of Nanoneuroscience by leading experts across the world. The volume represents carefully selected and refereed collections of papers by top nanoneuroscientists engaged in this cutting edge research who present their newly available data in the field of Nanoneuroscience and Nanoneuropharmacology. All the chapters included in this maiden volume by leading experts provide "state-of-the-art" knowledge of Nanoneuroscience either dealing with nanoneurotoxicity, nanoneuropharmacology, and/or nanoneurodiagnostic aspect. This book will thus further expand our understanding and can serve as a reference book in the rapidly expanding field of Neurotherapeutics and related disciplines, for example, neuropharmacology, neuropsychiatry, neurotraumatology, neuropathology, neurorehabilitation, neurodiagnostics, neurophysiology, and neurobiology.

I strongly hope that this novel volume in this emerging area of Nanoneuroscience will help in an increased understanding on the roles of nanoparticles in neuroscience with regard to their neurotoxicity or neuroprotective capabilities. The volume is indispensable to clinicians and basic researchers alike to find new avenues of Nanoneuroscience research to reduce nanoneurotoxicity and/or to enhance nanodrug delivery to achieve better human health-care effectively in the near future.

Hari Shanker Sharma (Uppsala)

References

Andrews, R. J. (2009). Nanotechnology and neurosurgery. *Journal of Nanoscience and Nanotechnology, 9*, 5008–5013.

Campbell, A., Araujo, J. A., Li, H., Sioutas, C., & Kleinman, M. (2009). Particulate matter induced enhancement of inflammatory markers in the brains of Apolipoprotein E knockout mice. *Journal of Nanoscience and Nanotechnology, 9*, 5099–5104.

Hekmatara, T., Gelperina, S., Vogel, V., Yang, S.-R., & Kreuter, J. (2009). Encapsulation of water-insoluble drugs in poly(butyl cyanoacrylate) nanoparticles. *Journal of Nanoscience and Nanotechnology, 9*, 5091–5098.

Oberdörster, G., Elder, A., & Rinderknecht, A. (2009). Nanoparticles and the brain: Cause for concern? *Journal of Nanoscience and Nanotechnology, 9*, 4996–5007.

Pathak, S., Tolentino, R., Nguyen, K., D'Amico, L., Barron, E., Cheng, L., et al. (2009). Quantum dot labeling and imaging of glial fibrillary acidic protein intermediate filaments and gliosis in the rat neural retina and dissociated astrocytes. *Journal Nanoscience and Nanotechnology, 9*, 5047–5054.

Sharma, H. S. (2007, December). Nanoneuroscience: Emerging concepts on nanoneurotoxicity and nanoneuroprotection. *Nanomedicine, 2*(6), 753–758. Review.

Sharma, H. S. (2009a, June). Birth of a new journal. *Journal of Nanoneuroscience.*

Sharma, H. S. (2009b). Nanoneuroscience: Nanoneurotoxicity and nanoneuroprotection. *Journal of Nanoscience and Nanotechnology, 9*, 4992–4995.

Sharma, H. S., Ali, S. F., Dong, W., Tian, Z. R., Patnaik, R., Patnaik, S., et al. (2007, December). Drug delivery to the spinal cord tagged with nanowire enhances neuroprotective efficacy and functional recovery following trauma to the rat spinal cord. *Annals of the New York Academy of Sciences, 1122*, 197–218.

Sharma, S., Ali, S. F., Hussain, S. M., Schlager, J. J., & Sharma, A. (2009). Influence of engineered nanoparticles from metals on the blood–brain barrier permeability, cerebral blood flow, brain edema and neurotoxicity. An experimental study in the rat and mice using biochemical and morphological approaches. *Journal of Nanoscience and Nanotechnology, 9*, 5055–5072.

Sharma, H. S., Ali, S. F., Tian, Z. R., Hussain, S. M., Schlager, J. J., Sjöquist, P.-O., et al. (2009). Chronic treatment with nanoparticles exacerbate hyperthermia induced blood–brain barrier breakdown, cognitive dysfunction and brain pathology in the rat. Neuroprotective effects of nanowired-antioxidant compound H-290/51. *Journal of Nanoscience and Nanotechnology, 9*, 5073–5090.

Sharma, H. S., Ali, S., Tian, Z. R., Patnaik, R., Patnaik, S., Lek, P., et al. (2009). Nano-drug delivery and neuroprotection in spinal cord injury. *Journal of Nanoscience and Nanotechnology, 9*, 5014–5037. Review.

Sharma, H. S., Patnaik, R., Sharma, A., Sjöquist, P.-O., & Lafuente, L. V. (2009). Silicon dioxide nanoparticles (SiO_2, 40–50 nm) exacerbate pathophysiology of traumatic spinal cord injury and deteriorate functional outcome in the rat. An experimental study using pharmacological and morphological approaches. *Journal of Nanoscience And Nanotechnology, 9*, 4970–4980.

Sharma, H. S., & Sharma, A. (2007). Nanoparticles aggravate heat stress induced cognitive deficits, blood–brain barrier disruption, edema formation and brain pathology. *Progress in Brain Research, 162*, 245–273. Review.

Sharma, H. S., & Sharma, A. (2009, July). Conference scene: New perspectives on nanoneuroscience, nanoneuropharmacology and nanoneurotoxicology. *Nanomedicine, 4*(5), 509–513.

Acknowledgments

I wish to express my sincere gratitude to Hilary Rowe, Cindy Minor, Susan Lee (USA), and Maureen Twaig (The Netherlands) during the initial development of this book project. I am indebted to Johannes Menzel, Lisa Tickner, and Lyndse Dixon (UK) for their untiring help through the incubation period of this book project on various aspects. My special thanks are due to Aruna Sharma (Sweden); Gayathri Venkatasamy (India), and Clare Caruana (UK) for all necessary help during preparation and editing of this volume.

Contents

Section IV. Nanoneurotoxicity and Nanoneuroprotection

SECTION I

Nanodrug delivery and Imaging Techniques

H.S. Sharma (Ed.)
Progress in Brain Research, Vol. 180
ISSN: 0079-6123

CHAPTER 1

Drug delivery to the brain using colloidal carriers

Jiang Chang[1], Youssef Jallouli[2], Alexandre Barras[3], Nicole Dupont[1] and Didier Betbeder[1,4,*]

[1]*EA 2689, Laboratory of Physiology, IMPRT, University of Lille 2, 1 place Verdun, 59045 Lille, France*
[2]*EA 2648, Laboratory of the Blood Brain Barrier, IMPRT, University of Artois, rue Jean Souvraz, 62307 Lens, France*
[3]*Interdisciplinary Research Institut, USR 3078 CNRS Parc Scientifique de la Haute Borne, 50 avenue de Halley – BP 70478, 59 658 Villeneuve d'Ascq, France*
[4]*University of Artois, 62000 Arras, France*

Abstract: Many neurodegenerative diseases, cancer, and infections of the brain become more prevalent as populations become older. Despite major advances in neuroscience, the blood–brain barrier (BBB) ensures that many potential therapeutics cannot reach the central nervous system (CNS). The BBB is formed by the complex tight junctions between the endothelial cells of the brain capillaries and their low endocytic activity. It results in the capillary wall that behaves as a continuous lipid bilayer and prevents the passage of polar substances. Drug delivery to the brain has remained one of the most vexing problems in translational neuroscience research, because of the difficulties posed by the BBB. Several strategies for delivering drugs to the CNS have been developed. This review rationalizes the strategies to target drugs to the brain by using different colloids.

Keywords: nanoparticles; drug delivery; blood–brain barrier; colloid; targeting

Introduction

Most of the intractable central nervous system (CNS) disorders have not been beneficially treated by classical small molecule therapy, including Alzheimer's disease, the neurodegeneration of Parkinson's disease, stroke, cerebral AIDS, brain cancer, the ataxias, the inherited inborn errors of metabolism, and other brain disorders. The diffusion of drugs from blood into the brain depends mainly upon the ability of biologically active molecules to cross lipid membranes. For this reason the design and development of colloids containing bioactive agents for brain therapeutic application may be a resourceful approach to overcome limitations. Moreover, it is important to require a fundamental understanding of the *in vivo* interaction between the nanoparticles (NPs) and the blood–brain barrier (BBB).

The blood–brain barrier

Progress in the brain drug delivery has lagged behind other areas in the molecular neuroscience, because of the difficulties posed by the BBB. An

* Corresponding author.
Tel.: +33-320-626968; Fax: +33-320-626963;
E-mail: dbetbeder@aol.com

DOI: 10.1016/S0079-6123(08)80001-5

overview of the BBB helps to elucidate the problems of drug delivery to the brain and to propose potential target to cross these barriers.

In humans, there are approximately 650 km of capillaries perfusing the brain, and the surface area of the brain microvascular endothelium is approximately 20 m^2 (Pardridge, 2003), which is 1000-fold greater than the surface area of either the blood–cerebrospinal fluid (CSF) barrier or the arachnoid membrane. Therefore, the quantitatively important barrier system within the brain is the BBB at the capillary endothelium. Despite the vast surface area of the human BBB, the thickness of the BBB is very thin (200–300 nm), and the total intracellular volume of the brain capillary endothelium is only 5 mL in the entire human brain. This very thin cellular barrier has some of the most restrictive permeability properties of any biological membrane (Oldendorf, 1971).

The BBB provides the brain with nutrients, prevents the introduction of harmful blood-borne substances, and restricts the movement of ions and fluid to ensure an optimal environment for CNS. Meanwhile, BBB much more represents an insurmountable barrier for the majority of drugs including anticancer agents, antibiotics, peptides, and other oligo- and macromolecular drugs. It is located at the level of the brain capillaries, where there is a convergence of different cell types: endothelial cells, pericytes, astrocytes, microglias (perivascular macrophages), and neurons. The brain microvessel endothelial cells that form the BBB display important morphological characteristics such as the presence of tight junctions between the cells, the absence of fenestrations, enzymes, high level of p-glycoprotein (P-gp) involved in drug efflux mechanisms, and a diminished pinocytic activity that together help to restrict the passage of compounds from the blood into the extracellular environment of the brain (Fig. 1). Thus, the BBB prevents the uptake of all large molecule drugs. Only small (with a molecular weight <400 Da), lipid-soluble, electrically neutral molecules and weak bases are able to diffuse passively across the BBB (Banks & Kastin, 1985).

Because of tight junction, there is no paracellular passage between the brain endothelial cells (Fig. 1a). Only lipid-soluble amphiphilic agents, such as alcohol, could transport BBB by passive

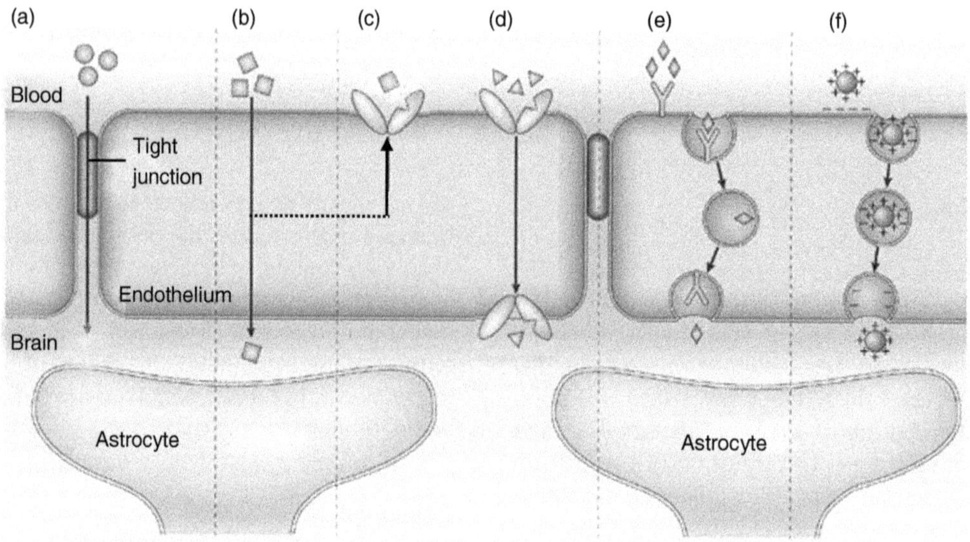

Fig. 1 Pathways across the blood–brain barrier. (a) No paracellular passage; (b) passive diffusion, only for lipid-soluble amphiphilic agents, such as alcohol; (c) active efflux transcytosis (P-glycoprotein acts as efflux transporter); (d) carrier-mediated influx transport; (e) receptor-mediated transcytosis; (f) adsorptive transcytosis. Modified from Abbott, Rönnbäck, and Hansson (2006).

diffusion (Fig. 1b). To allow the passage of polar and macromolecules, there are three broad classes of carriers: carrier-mediated influx transport (CMT) and carrier-mediated active efflux transport (AET) for small molecules (Fig. 1c, d) and receptor-mediated transport (RMT) for large molecules (Fig. 1e). CMT systems exist that enable the entry or the elimination of a variety of compounds including hydrophilic substances such as hexoses and amino acids. The RMT systems include receptors such as the insulin, low-density lipoprotein (LDL), and transferrin receptor (Tf-R). Adsorptive-mediated transcytosis (AMT) also provides a means for brain delivery of medicines across the BBB (Fig. 1f). The BBB provides both the potential for binding and the uptake of cationic molecules to the luminal surface of endothelial cells, and then for exocytosis at the abluminal surface. The transcytotic pathways present at the BBB provide the means for movement of the molecules through the endothelial cytoplasm.

Strategies to overcome BBB

Routes of administration

There have been three major paths proposed for delivery of active molecules to the brain: intracerebral, nasal, and intravascular.

Neurosurgery-based strategies are intracerebral invasive, but neuropharmaceutical agents have been delivered to the brain by intracerebroventricular infusion of drugs or intracerebral implants of either genetically engineered cells or biodegradable polymers loaded with drugs (Bobo et al., 1994; Kroll, Pagel, Muldoon, Roman-Goldstein, & Neuwelt, 1996). They have been the most widely used for circumventing the BBB drug delivery problem. Implantation of intraventricular catheters could allow for chronic drug delivery to the brain. However, this approach is limited by the much faster rates of bulk flow of CSF through the ventricular compartments as compared to the slow rates of drug diffusion down into the brain parenchyma.

Intranasal drug administration is a noninvasive method of bypassing the BBB to deliver neurotrophins and other therapeutic agents to the brain and the spinal cord. The advantages of nasal route for systemic delivery of drug can include rapid onset of action and the preferential drug delivery to the brain via olfactory pathway (Illum, 2000). A number of hydrophilic and lipophilic therapeutic agents have been shown to enter the brain directly from the nasal cavity via olfactory pathway. In humans, intranasal insulin has been shown to improve memory in normal adults and patients with Alzheimer's disease (Stockhorst, Fries, Steingrueber, & Scherbaum, 2004). Delivery from the nose to the CNS occurs within minutes along both the olfactory and trigeminal neural pathways. It would appear that the nasal route can be considered for drugs with the following criteria: ineffective orally, lipophilic, and administered in minute doses (Hussain, 1998).

Intra arterial method involves the temporary disruption of the tight junctions by infusing hypertonic solutions or biologically active agents such as bradykinin (Emerich, Snodgrass, Pink, Bloom, & Bartus, 1998; Neuwelt & Rapoport, 1984). Although these modes of delivery are traditional, the risks of infection and neuropathological changes due to disruption of the BBB emphasize the need to develop new noninvasive delivery strategies.

The most preferred intravascular route is the intravenous (i.v.) route; however, it is necessary to cross BBB. Intravenous injection can treat virtually 100% of the neurons in the brain. Because every neuron is perfused by its own blood vessel, the drug is delivered to the "doorstep" of every neuron following initial transport across the vascular barrier. However, the brain is virtually impenetrable by the majority of drug candidates. Autoradiograph of a mouse taken 30 min after i.v. injection of radiolabeled histamine, a small molecule, shows that the drug is taken up by all organs of the body except the brain and the spinal cord (Pardridge, Oldendorf, Cancilla, & Frank, 1986). The [^{111} In] radiolabeled epidermal growth factor (EGF) injected intravenously into tumor-bearing rats does not label a large brain cancer. The neurotrophins must be injected into the brain because these large molecules do not cross the BBB. Therefore, in the absence of BBB disruption, neuroprotection is not possible following delayed i.v. administration of macromolecular drugs if they are not recognized by BBB receptors (Pardridge, 2001).

Colloids as vectors

Colloidal drug carriers such as liposomes and NPs are able to modify the distribution of an associated substance. They can therefore be used to improve the therapeutic index of drugs by increasing their efficacy and/or reducing their toxicity. If these delivery systems are carefully designed with respect to the target, they may provide one solution to some of the delivery problems posed by new classes of active molecules such as peptides, proteins, genes, and oligonucleotides. They may also extend the therapeutic potential of established drugs such as doxorubicin and amphotericin B. Colloidal drug delivery vehicles have been studied for almost 30 years, but the few liposome-based formulations already on the market are mainly concerned with reducing the side effects of the encapsulated drugs. Now that the interactions between particles and biological milieu are better understood, "stealth" liposomes and NPs which show diminished phagocytosis have been developed and the range of sites which can be reached has been extended. Even without specific targeting technologies, sites of inflammation and infection and solid tumors can be reached by the enhanced permeation and retention (EPR) effect (Rice et al., 2003). If specificity for a particular cell type is required, ligands such as monoclonal antibodies, glucides, lectins, or growth factors can be coupled with these long-circulating systems. Colloidal drug carriers are particularly useful for formulating new drugs derived from biotechnology (peptides, proteins, genes, oligonucleotides) because they can provide protection from degradation in biological fluids and promote their penetration into cells.

The ability to cross the BBB to deliver drugs while targeting a group of cells (e.g., a tumor) requires several things to happen together. For example, in an ideal situation, if the nanocarrier–drug complex were to be administered systemically (e.g., intravenously) it would need to find the CNS while exhibiting minimal systemic effects, be able to cross the BBB, target whatever it needs to target once in the CNS, and only then releases the drug. These are technically very demanding challenges that require multidisciplinary interactions between different fields including engineering, chemistry, cell biology, physiology, pharmacology, medicine, liposomes or NP technology, and others.

Extensive endocytotic brain capillary endothelial cell uptake of colloids was already shown in *in vitro* cell cultures of human, bovine, porcine, rat, and mouse origin (Bickel, Kang, Yoshikawa, & Pardridge, 1994; Fenart et al., 1999; Kratzer et al., 2007; Ramge et al., 2000; Vorbrodt & Dobrogowska, 1999). After the endocytosis two mechanisms are possible: one is transcytosis through the endothelial cell layer, a mechanism suggested for LDL, Tf, and lactoferrin transport (Dehouck et al., 1997, Fillebeen et al., 1999); the other is the simple release of the drug within the endothelial cells and delivery to the brain.

Brain drug delivery strategies via colloid carriers

In general, colloidal drug carriers include micelles, emulsions, liposomes, and NPs (nanospheres and nanocapsules). It is noteworthy that only liposomes and NPs have been largely exploited for brain drug delivery. The aim in using colloidal carriers is generally to increase the specificity toward cells or tissues, to improve the bioavailability of drugs by increasing their diffusion through biological membranes, and/or to protect them against enzyme inactivation. Moreover, the colloidal systems allow access across the BBB of nontransportable drugs by masking their physico-chemical characteristics through their encapsulation in these systems. Active targeting to the brain can be achieved by the attachment of a specific ligand (such as a monoclonal antibody) onto the surface of the colloidal particle, preferentially at the end of the polyethylene glycol (PEG) molecules since the targeted colloidal particles will be much more efficient if they are also sterically stabilized (Maruyama et al., 1995).

The *in vivo* biodistribution and opsonization of nanosystems in blood circulation is governed by their size and surface characteristics. When NPs have reached the blood circulation, for example, after i.v. injection, they are opsonized and rapidly cleared by the reticuloendothelial system (RES), which is predominantly distributed in liver, lungs, spleen, and bone marrow. Unless there is desired drug delivery into these tissues (for hepatic metastases or primary hepatocarcinoma), the surface of the NPs must be modified to escape the RES with a hydrophilic surface to reduce opsonization thereby delaying macrophage clearance. Hydrophilic polymers coating the NPs surface repel plasma proteins and escape opsonization and clearance. This has been described as a "cloud" effect (Brigger et al., 2002). These polymers can be PEG or glucides such as dextrans and chitosan. This technology has yielded the "stealth" particle, which is invisible to the macrophage (Bazile et al., 1995). Paclitaxel-loaded methoxy-PEG-PLA NPs provided approximately threefold longer half-life and effective concentration of the drug in rats, without concentration peaks above toxic level compared to a conventional formulation.

Over the past few decades, pharmaceutical technology has lead to the emergence of different nanosystems or nanoplatforms tailored to deliver drugs to the brain, including liposomes, lipid NPs, and polymeric NPs.

Liposomes

Liposomes are vesicles composed of one (unilamellar) or several (multilamellar) lipid bilayers surrounding internal aqueous compartments. They are composed of biocompatible and biodegradable lipids similar to biological membranes. Their biophysical properties, such as size, surface charge, lipid composition, and amount of cholesterol, are varied and are able to control distribution, tissue uptake, and drug delivery. There are three types of liposomal systems: small unilamellar vesicles (SUV, 10–50 nm), large unilamellar vesicles (LUV, 50–1000 nm), and multilamellar vesicles (MLV, 100–20 μm) (Hoekstra & Martin, 1982).Various therapeutic molecules can be encapsulated inside the vesicles. Hydrophilic and hydrophobic substances are incorporated in the internal aqueous phase and the lipid bilayer, respectively. Moreover, some liposome components can inhibit P-gp efflux and consequently, drug permeability across BBB (Miller et al., 1991). Conventional liposomes are rapidly cleared from circulation by the RES. Extended circulation time can be accomplished by decreasing the particle size (<100 nm), incorporation in their structure of gangliosides (Allen & Chonn, 1987), and by liposome surface modification with PEG or other hydrophilic polymers. In humans, a 200-fold decrease of the systemic plasma clearance was observed for PEGylated liposomes in comparison to conventional liposomes (Allen, 1994). An enhanced transport to the brain of liposome-encapsulated drugs has been observed in several reported studies.

Passive targeting of the brain can be obtained using PEGylated liposome. Fabel et al. (2001) used doxorubicin as PEGylated liposomal encapsulated formulation (Caelyx, Scheringh-Plough, Munich, Germany) in therapy of recurrent malignant glioma. Stabilization of the disease was observed in 54% (7 of 13) of patients. Median time-to-progression was 11 weeks. Progression-free survival at 12 months was 15%.

Experimental autoimmune encephalitis is another brain disease in which PEGylated liposomes have been found useful for drug delivery. In inflammatory conditions, it is believed that the disruption of BBB allows the free diffusion of colloids such as liposomes. Thus, prednisolone entrapped into PEGylated liposomes has demonstrated an effective restoration of the BBB integrity; macrophage infiltration was diminished in the treated animals. Additionally, the use of liposomes may reduce systemic side effects and could be employed for the treatment of multiple sclerosis (Schmidt et al., 2003).

Active targeting can be achieved by complexing the liposomes with an antibody or a ligand that will be recognized by cell surface receptor in the targeted tissue. PEGylated liposomes can be chemically modified (vectorized) with ligands such as Tf, insulin, or antibodies directed to BBB receptors inducing the transcytose of the whole vector (Chekhonin et al., 2005; Pardridge, 1999). This approach may be the most striking advance in BBB targeting and translocation. Such immunoliposome constructs were successfully used to deliver small drugs, daunomycin and digoxin, as well as DNA to the brain (Shi, Zhang, Zhu, Boado, & Pardridge, 2001). Huwyler et al. have shown that a specific antibody to the Tf-R called OX26 could mediate the targeting of daunomycin to the brain by the use of PEGylated MAb-liposomes (Huwyler, Yang, & Pardridge, 1997).

Omori et al. (2003) used PEGylated liposome conjugated with Tf (Tf–PEG-liposome) and treated animals having transient middle cerebral occlusion. The Tf–PEG fluorescence was marginally detectable in sham control brain, but remarkably increased with a peak at 2 days, showing about 70% of Tf-R positive vascular endothelium double-labeled with Tf–PEG. These results indicate that the Tf–PEG-liposome could be utilized as an efficient drug delivery tool to the brain after stroke.

Indirect targeting by macrophages can also be used to deliver liposomal drugs to the brain. Since it is now known that the brain is under immunological surveillance, the authors hypothesized that phagocytic cells of the innate immune system, mainly neutrophils and monocytes, can be exploited as transporters of drugs to the brain. To target circulating mononuclear phagocytic cells, negatively-charged nanosized liposomes were formulated encapsulating serotonin, a BBB-impermeable neurological drug. The brain uptake of liposomal serotonin was significantly higher (0.138% $\pm$0.034 and 0.097% $\pm$0.011 vs. 0.068% $\pm$0.02 and 0.057% $\pm$0.01, 4 h and 24 h after i.v. administration in rats, in serotonin liposomes, and in solution, respectively) (Afergan et al., 2008).

In conclusion, liposomes have been extensively investigated for the brain delivery of molecules, showing increased drug efficacy and reduced drug toxicity.

Lipid nanoparticles

Solid lipid nanoparticles (SLNs) and lipid nanocapsules (LNCs) are under investigation as drug delivery systems for brain targeting.

Solid lipid nanoparticles

SLNs represent an attractive colloidal drug carrier system (Müller & Lucks, 1996). SLNs are spherical particles in the nanometer range made from solid lipid(s), emulsifier(s), and water. The solid hydrophobic core is made up of solid lipids including triglycerides, partial glycerides, fatty acids, steroids, and waxes and they are stabilized by all classes of emulsifiers including lecithin, polymers, and their mixtures (Mehnert & Mäder, 2001).

The three basic production techniques for SLN are the high shear homogenization and ultrasound, the high pressure homogenization (HPH), and the microemulsion technique. High shear homogenization and ultrasound are prevalent and simple nevertheless low dispersion quality is often the disadvantage. Another more efficient method to produce submicron-sized SLN is HPH (Mehnert & Mäder, 2001).

SLNs are considered particularly useful for the administration of lipophilic drugs but they also have the potential to carry hydrophilic drugs (Fundaro et al., 2000), peptides (Chattopadhyay, Zastre, Wong, Wu, & Bendayan, 2008), or contrast agents (Peira et al., 2003).

Clozapine (CLZ), a lipophilic antipsychotic drug, has very poor oral bioavailability (<27%) due to first-pass effect (Jann, 1991). Tissue distribution studies of CLZ–SLN and suspension were carried out in Swiss albino mice after i.v. administration. Average size and zeta potential of SLN of different lipids with stearylamine ranged from 96.7 $\pm$3.8 to 163.3 $\pm$0.7 nm and from 21.3 $\pm$1.3 to 33.2 $\pm$0.6 mV, respectively. In the brain, the relative bioavailability increased significantly. The area under the curve (AUC) of CLZ–tripalmitin and CLZ–tripalmitin–stearylamine were 4.1- and 5.3-fold greater and mean residence time (MRT) increased 4.3 and 3.9 times, respectively, than CLZ suspension (Manjunath & Venkateswarlu, 2005).

In vitro drug release can be achieved for up to several weeks and enables the protection of drugs against chemical degradation. Camptothecin (CA) lactone opens rapidly to the carboxylate form with a $t_{1/2}$ value of 23.8 min in phosphate-buffered saline (PBS), pH 7.4, at 37°C (Burke & Mi, 1994).

The encapsulation efficiency of CA in SLN was 99.6% (Yang, Zhu, Lu, Liang, & Yang, 1999). In the brain, after i.v. administration in mice, the pharmacokinetic parameters of CA–SLN and CA–solution indicate that AUC of CA–SLN were 10.3-fold greater and MRT increased 5 times than CA–solution. It proved that CA-loaded SLN prevented it from hydrolysis.

Wang, Sun, and Zhang (2002) have reported the synthesis of 3′,5′-dioctanoyl-5-fluoro-2′-deoxyuridine to overcome the limited access of the drug 5-fluoro-2′-deoxyuridine (FUdR) and its incorporation into SLN (DO–FUdR). After i.v. administration, the brain area under the concentration/time curve of DO–FUdR–SLN and DO–FUdR were 11- and 5-fold higher than that of FUdR, respectively. These results indicated that DO–FUdR–SLN had a good (2.5 times of the free drug) brain-targeting efficiency *in vivo*.

Lipid nanocapsules

LNCs is another kind of lipid NPs (Heurtault, Saulnier, Pech, Proust, & Benoit, 2002). LNCs are composed of a liquid, oily core (medium-chain triglycerides) surrounded by hydrophilic (PEG_{660}-hydroxystearate) and lipophilic (phosphatidylethanolamine and phosphatidylcholine) surfactants. LNCs were prepared by a novel phase inversion-based technique. In addition, this technique also avoids the use of organic solvents and pharmaceutically acceptable excipients. The PEG_{660}-hydroxystearate located on the external surface constitutes a steric barrier against opsonization and inhibits the P-gp (Buckingham, Balasubramanian, Emanuele, Clodfelter, & Coon, 1995; Coon et al., 1991).

Various anticancer drugs such as etoposide, docetaxel, and paclitaxel (Khalid, Simard, Hoarau, Dragomir, & Leroux, 2006; Lamprecht & Benoit, 2006; Peltier, Oger, Lagarce, Couet, & Benoit, 2006) were encapsulated into these LNCs. Paclitaxel-loaded LNC led to a significant accumulation of anticancer molecules in the brain tissue (Koziara, Lockman, Allen, & Mumper, 2004). In a cancer cell culture model, etoposide or paclitaxel LNC showed a 4-fold or 40-fold higher efficiency, respectively, than the drug solution while blank LNCs were found to be less toxic than the pure drug at equivalent concentrations (Lacoeuille, Garcion, Benoit, & Lamprecht, 2007). Thus two novel types of immunonanocapsules decorated with OX26 MAb or OX26 MAb Fab' fragments have been developed (Beduneau et al., 2007). The specific association of immunonanocapsules to cells overexpressing Tf-R was demonstrated, suggesting their ability to deliver drugs to the brain.

SLN and LNC will represent an attractive carrier for an efficient delivery of various drugs into the brain. Indeed the capacity of entrapping the lipophilic or hydrophilic drugs and the choice of several administration routes make the SLN and LNC delivery system promising.

Polymeric nanoparticles

Solid biodegradable NPs were a step up from liposomes, targeting NPs opened a whole new field for drug delivery. Compared to liposomes, polymeric and lipid NPs are highly stable, both during storage and *in vivo* and their sustained release over a period of weeks is more easily achieved.

NPs for pharmaceutical and chemical use are defined as polymeric particles made of natural or synthetic polymers or supramolecular assemblies ranging in size between 10 and 100 nm depending on their intended use. NPs are highly advantageous over larger microparticles, because they are better suited for i.v. delivery. Different kinds of polymers such as poly(alkylcyanoacrylate) (PACA), chitosan, maltodextrins, polylactic acid (PLA), poly(glycolic acid) (PGA), and their copolymers (PLGA) have been used to prepare NPs (De Miguel et al., 2000; Olivier, Huertas, Lee, Calon, & Pardridge, 2002). NPs can be used to deliver hydrophilic drugs, hydrophobic drugs, and macromolecules, such as proteins and oligonucleotides. Drugs may be entrapped in the polymer, be adsorbed to the surface, or may be chemically attached.

A number of possible mechanisms for NP-mediated drug transport across the BBB have been described. These include endocytotic/transcytotic pathway with a possibility of inhibition of P-gp efflux mechanisms.

Poly(alkylcyanoakrylate) nanoparticles

Described for the first time in 1977 by Couvreur, Tulkens, Roland, Trouet, and Speiser (1977), the NPs made of PACA polymers are synthesized from monomers of (iso) butylcyanoacrylate or (iso) hexylcyanoacrylate. Various PACA NPs were designed including nanospheres and oil-containing and water-containing nanocapsules. Inclusion of the drug can be made during the polymerization process or by adsorption onto preformed NPs. The degradation and toxicity of NPs based on PACA are dependent on the alkyl chain (Olivier, 2005).

For the brain delivery with the poly(butylcyanoacrylate) (PBCA) NPs, the chemical nature of the overcoating surfactant is one of the determinant factor, because among the surfactants tested, only polysorbates led to a CNS pharmacological effect. Indeed, in earlier studies realized by Kreuter et al., neurotrophic molecules such as dalargin (Kreuter, Alyautdin, Kharkevich, & Ivanov, 1995; Schroeder, Sommerfeld, & Sabel, 1998) and loperamide (Alyautdin et al., 1997) have been loaded onto NPs with the aim of brain delivery. The results showed that only when this drug was adsorbed onto the surface of PBCA NPs further coated with the polysorbate-80 (PS-80), a pronounced analgesic effect was obtained, after peripheral administration. Furthermore, it has been reported that apolipoproteins (Apo-E and Apo-B) could be involved in the brain penetration of PBCA NPs overcoated with PS-80 (Kreuter et al., 2002). Then, only dalargin or loperamide-PBCA NPs coated with PS-80 and/or with Apo-B or Apo-E were able to achieve an antinociceptive effect. More recently, an attempt to target the anti-Alzheimer's drug rivastigmine in the brain by using PBCA NPs coated with 1% nonionic surfactant PS-80 was made. The studies demonstrate that the brain concentration of intravenously injected rivastigmine can be enhanced over 3.82-fold by binding to PBCA NPs (Wilson et al., 2008). To increase the plasma half-life and to improve the delivery of the drug PACA NPs, Peracchia et al. (1999) have covered the surface of NPs by hydrophilic polymer PEG. The brain biodistributions of PEGylated-PACA NPs, poloxamine 908-coated PACA NPs, and conventional nonlong-circulating PACA NPs were compared after i.v. administration. The results demonstrate that the concentration of PEGylated NPs in the CNS, especially in white matter, is greatly increased in comparison to conventional non-PEGylated NPs (Calvo et al., 2002). The i.v. administration of poly(methoxy-PEG_{2000}cyanoacrylate-co-hexadecylcyanoacrylate) 1:4 (PEG-PHDCA) nanospheres of size between 146 and 161 nm and with a zeta potential of −20 mV, in rats bearing a unilateral intracerebral gliosarcome, have shown that these NPs accumulate preferentially into the tumor tissue of the brain and to a lesser extent in peripheral tissue and the opposite hemisphere. In addition, this accumulation is also much higher compared to that observed for NPs uncovered with PEG (135–161 nm, −45 mV). Thus, two mechanisms have been proposed to explain the accumulation of PEG-PHDCA in the brain and the tumor sides. The first was attributed to the reduction of plasma clearance due to the presence of PEG, and the other was the capacities of the diffusion/convection of these NPs allowing their extravasation to the tumor sides. As for the presence of these NPs in the healthy tissue, these authors contribute to PEG-PHDCA an affinity to endothelial cells of the BBB (Brigger et al., 2002). Recently, it has been demonstrated that only when Apo-E or Apo-B-100 were preadsorbed onto PEG-PHDCA NPs, NPs were found to be more efficient than control NPs for penetrating into rat brain endothelial cells, suggesting the involvement of apolipoproteins and the LDL receptors (LDL-R) in the brain transport of PEG-PHDCA NPs (Kim et al., 2007).

However, while many *in vivo* and *in vitro* studies demonstrate the advantage applications of PACA NPs to cross the BBB and to deliver drugs to the brain, the mechanisms of transport of these NPs through the BBB have not yet been clearly elucidated. It was hypothesized that polysorbate-coated NPs were transported across the

BBB via endocytosis by the brain capillary endothelial cells. This endocytosis would be triggered by an Apo-E, reported to absorb on PS-20, 40, 60, or 80-coated NPs (Kreuter, 2001). In contrast, Olivier et al. (1999) have suggested that the PBCA NPs coated with PS-80 displayed some toxic effect toward the BBB, and it was also suggested that the Apo-E adsorption is not specific of PS-80-coated surfaces because it was shown to adsorb onto PEGylated PLA NPs (Olivier, 2005; Olivier et al., 1999).

To date, all published work underlines the importance of surface properties of these NPs and more specifically the benefits of the recovery of NPs surfaces with surfactants (PS-80, Pluronic® F-68, polaxamer), and the presence of the ligands (Apo-E, Apo-B). All these surface characteristics aim to improve the plasma half-life of complex NP-drug, directing its pharmacokinetics and biodistribution (Li & Huang, 2008; Roney et al. 2005). Gao and Jiang (2006) found out that NPs overcoated by PS-80 could significantly improve the drug level in both brain tissues and CSFs compared with uncoated ones and simple solution. In addition, the 70-nm NPs could slightly increase methotrexate delivery into the brain while no significant difference was observed among 170-, 220-, and 345-nm PS-80-coated PBCA NPs.

All these results demonstrate that these NPs are good candidates for further investigation.

PLGA nanoparticles

Poly(lactide-*co*-glycolide) (PLGA) is a biodegradable aliphatic polyester-based polymer able to form easily solid NPs, and it decomposes without induction of inflammation or immune reaction. Its use is approved by U.S. FDA. Moreover, this copolymer can be easily covalently linked to peptides. In particular polyester NPs, systems in which a drug can be entrapped or embedded in the NPs matrix or adsorbed on their surface, are being extensively investigated for different therapeutic applications such as for sustained drug, vaccine, and gene delivery (Moghimi, Hunter, & Murray, 2001; Panyam & Labhasetwar, 2003). Feng et al. have encapsulated the antineoplastic drug paclitaxel in PLGA NPs. In *in vitro* experiments with 29 different cancer cell lines (including both neural and nonneural cell lines), they showed targeted cytotoxicity 13 times greater than with the drug alone (Feng, Mu, Win, & Huang, 2004).

Even as PLGA are recognized as safe materials, encapsulation of drugs needs to design particulate carriers to reach the brain with engineered surface properties for targeting purpose. In the mid-1990s, long-circulating PEGylated-PLA or PLGA NPs have been made available that opened great opportunities for drug targeting (Gref et al., 1995). They will allow to control of the specific interactions with BBB as well as the nonspecific interactions with blood components and phagocytic cells (Gref, Rodrigues, & Couvreur, 2002). More recently, the synthesis of functionalized PEGylated-PLA/PLGA NPs opened new perspectives for targeted drug delivery in general, and for drug brain targeting in particular. PEGylated NPs are made of methoxypoly(ethylene glycol)-PLA/PLGA (mPEG-PLA/PLGA), that is, esters of PLA or PLGA with PEG of various molecular weights. These mPEG-PLA NPs by nasal administration to rats yielded 1.6- to 3.3-fold greater percentage in CSF, olfactory bulb, and other brain tissues, compared to nasal solution (Zhang et al., 2006). The availability of functionalized PEG-PLA permits to prepare target-specific NPs by conjugation of cell surface ligand. Using peptidomimetic antibodies to BBB transcytosis receptor, brain-targeted PEGylated immunonanoparticles can be synthesized that should make possible the delivery of entrapped actives into the brain parenchyma without inducing BBB permeability alteration. As cationic bovine serum albumin (CBSA) conjugated PEGylated-PLA NPs (CBSA-NPs), after injection in mice caudal vein, fluorescent microscopy showed a higher accumulation of CBSA-NPs in the lateral ventricle, third ventricle, and periventricular region than that of control NPs (BSA-NPs) (Lu et al., 2005).

For the functional peptides surface engineering, Luca et al. found an increased permeability of PLGA NPs when conjugated with five short peptides, which are part of sequences present on synthetic opioid peptides. These molecules were shown to be BBB permeable, and the permeability can be enhanced by the presence of glycosidic moieties (glucose, lactose, etc.). The presence of these peptides on the surface of NPs could allow the obtainment of a more specific mechanism of crossing BBB (receptor-mediated endocytosis) that could target CNS in a more specific way than the NPs developed before (Luca et al., 2005). The analysis of the images obtained in the *in vivo* experiments showed qualitatively that the fluorescent peptide- and β-D-glucose-covered PLGA NPs are able to reach the brain and to penetrate the cells of the cerebral tissue (Costantino et al., 2006). Tosi et al. (2007) used PLGA derivatized with the peptide H_2N-Gly-L-Phe-D-Thr-Gly-L-Phe-L-Leu-L-Ser (*O*-β-D glucose)-$CONH_2$ (g7) to prepare g7-NPs. Fluorescent studies showed that these NPs were able to cross the BBB. In this research, g7-NPs were loaded with loperamide obtaining sustained release of the embedded drug in the brain and labeled with Rhodamine-123; they found that these NPs are able to reach all the brain areas. *In vivo* experiments showed that streptavidin-biotin-peroxidase conjugated PLGA NPs could be easily detected in the brain parenchyma or in the liver cell population according to the infusion pathway (Tosi et al., 2005). Tetanus toxin C (TTC) fragment conjugated PEGylated-PLGA NPs were also shown to selectively target neuroblastoma cells *in vitro* (Townsend et al., 2007).

In our group, recent studies performed on an *in vitro* BBB model based on a coculture of endothelial cells and astrocytes showed that 90-nm Tf-NPs can enter in a specific manner the endothelial cells using a cholesterol-dependent pathway. Moreover the uptake of these NPs was 20-fold increased by Tf targeting. This targeting was inhibited with excess Tf suggesting that they enter the cells via the Tf-R. The studies suggest that the Tf-surface coated NPs can specifically interact with BBB via the Tf-R and are highly endocytosed by the caveolae pathway (manuscript under preparation).

Maltodextrin nanoparticles

Maltodextrin NPs are porous NPs with a polysaccharidic backbone. They are made from maltodextrins meshed with a reticulating agent and sized down using HPH (De Miguel et al., 2000; Fenart et al., 1999). They can be neutral, anionic, or cationic. These NPs can be used alone or associated with lipids (Loiseau, Imbertie, Bories, Betbeder, & De Miguel, 2002; Major, Prieur, Tocanne, Betbeder, & Sautereau, 1997). The system comprises a polysaccharide core that can be composed of neutral, positively, or negatively charged particles. It can be further surrounded by a phospholipid layer (De Miguel et al., 2000). Moreover lipids can be inserted in their core and can be used as drug reservoir of lipophilic drugs such as amphotericin B (Loiseau et al., 2002). The inclusion of drugs is generally made using the post-loading technique using premade NPs. Fenart et al. (1999) studied the ability of 60-nm maltodextrin NPs of different charges (neutral, anionic, and cationic) to cross the BBB *in vitro*. These particles were coated or not with a lipid bilayer made of dipalmitoyl phosphatidyl choline (DPPC) and cholesterol. They found that lipid coating of ionically charged NPs was able to increase BBB crossing three- or fourfold compared with uncoated particles. As these lipid-coated NPs were nontoxic toward BBB integrity, they were found to cross BBB by transcytosis by an unknown mechanism. Furthermore, a 27-fold increase in albumin transport was observed when albumin had previously been loaded in the cationic lipid-coated NPs. Jallouli, Paillard, Chang, Sevin, and Betbeder (2007) studied the mechanism of interaction of three porous maltodextrin NPs differing from their surface charge and inner composition on BBB. The data showed that at 4°C the three NPs bind in different areas on endothelial cells (Fig. 2).

Cationic NPs were found mainly around the paracellular area, neutral NPs mainly on the cell surface, and dipalmitoylphosphatidylglycerol (DPPG) NPs binding was found at both paracellular areas and on the surface of the cells. All these results obtained with three NPs of same size and backbone composition clearly reflect the complexity of interactions of glucidic NPs and brain endothelial cells. At 37°C neutral and cationic NPs were transcytosed. Filipin treatment contrary to cationic NPs inhibited neutral NP transcytosis suggesting that sterols are implied in their

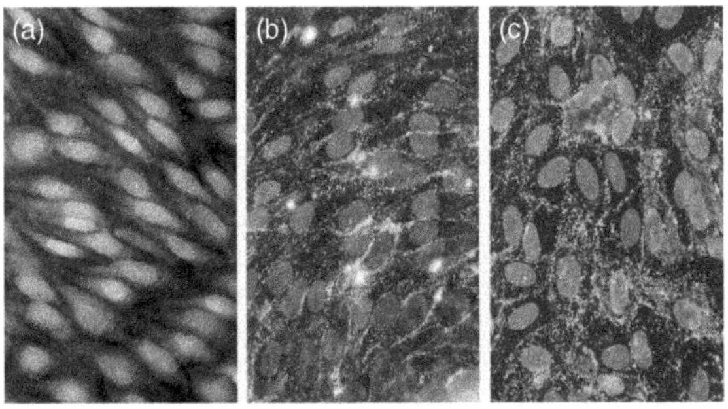

Fig. 2 Evaluation of NPs binding on BBB at 4°C using fluorescence microscopy. Neutral NPs were labeled with rhodamine (a) and cationic NPs (b) and DPPG-NPs (c) were labeled with FITC. The nuclei were labeled with Hoechst.

transcytosis pathway. This inhibition shows that neutral NPs, like LDL in this model, use the caveolae pathway (Dehouck et al., 1997). Moreover, only cationic NP transcytosis was inhibited by chlorpromazine, suggesting that they are endocytosed via the clathrin pathway. These results obtained with NPs of same size, but differing from their charge, clearly demonstrate the importance of the surface charge of NPs to interact with BBB. Other studies performed using the intranasal route showed that cationic maltodextrin NPs, at very low doses, can increase brain delivery of morphine. The mechanism implied was a direct nose–brain delivery as no increase of morphine was not observed in the blood. Strangely, it was demonstrated that the NPs were used as a permeation enhancer, which only works at very low doses, and not as a delivery system as there was no binding of the drug with the NP (Betbeder et al., 2000).

In summary, these results showed that 60-nm NPs can cross BBB by transcytosis using either the caveolae or the clathrin pathway. Both neutral and cationic NPs are promising candidates to deliver drugs to the brain. Further *in vivo* studies after i.v. administration should give a better understanding of their potential.

Conclusions

Great efforts have been made to develop strategies for delivering drugs to the CNS by enhancing their ability to cross the BBB. The current challenge is to develop drug delivery systems that ensure the safe and effective passage of drugs across the BBB. The delivery of drugs, peptides, proteins, and genes to the brain depends on brain-specific vectors. The development of such vectors requires the identification of new receptor–ligand and antigen–antibody interactions that are selective for the BBB. Knowledge of the pharmacogenomics of the BBB will undoubtedly lead to the discovery of new ways of promoting the delivery of pharmaceuticals to the CNS.

Despite the significant progress made so far in the formulation technology and the new methods of ligand conjugation, there are still numerous lacunas to be filled before colloidal carriers can be projected as the ultimate nanosystems for BBB-targeted drug delivery.

The molecular biology and genomics of the BBB continue to be elucidated, but remain incomplete. It is absolutely essential to complete our knowledge base of all carrier- and receptor-mediated transport systems active at the BBB. This information will be critical for the further development of brain drug targeted strategies for delivery of therapeutic agents across the BBB. New formulations of neuroactive drugs into colloid carriers are expected to improve their pharmacokinetic profile and allow for higher concentrations to be attained in the brain. While there is probably no single universal system for delivering drugs to the brain, the techniques described above promise to provide practical methods for the delivery of a range of therapeutic agents.

Acknowledgments

This work was funded by the PhD fellowship between France and China from the French research ministry.

References

Abbott, N. J., Rönnbäck, L., & Hansson, E. (2006). Astrocyte-endothelial interactions at the blood-brain barrier. *Nature Reviews Neuroscience*, 7, 41–53.

Afergan, E., Epstein, H., Dahan, R., Koroukhov, N., Rohekar, K., Danenberg, H. D., et al. (2008). Delivery of serotonin to the brain by monocytes following phagocytosis of liposomes. *Journal of Controlled Release.*

Allen, T. (1994). Long-circulating (sterically stabilized) liposomes for targeted drug delivery. *Trends in Pharmacological Sciences, 15*, 215–220.

Allen, T. M., & Chonn, A. (1987). Large unilamellar liposomes with low uptake into the reticuloendothelial system. *FEBS Letters, 223*, 42–46.

Alyautdin, R. N., Petrov, V. E., Langer, K., Berthold, A., Kharkevich, D. A., & Kreuter, J. (1997). Delivery of loperamide across the blood-brain barrier with polysorbate 80-coated polybutylcyanoacrylate nanoparticles. *Pharmaceutical Research, 14*, 325–328.

Banks, W. A., & Kastin, A. (1985). Peptides and the blood-brain barrier: Lipophilicity as a predictor of permeability. *Brain Research Bulletin, 15*, 287–292.

Bazile, D., Prud'homme, C., Bassoullet, M. T., Marlard, M., Spenlehauer, G., & Veillard, M. (1995). Stealth Me.PEG-PLA nanoparticles avoid uptake by the mononuclear phagocytes system. *Journal of Pharmaceutical Sciences, 84*, 493–498.

Beduneau, A., Saulnier, P., Hindre, F., Clavreul, A., Leroux, J.- C., & Benoit, J.-P. (2007). Design of targeted lipid nanocapsules by conjugation of whole antibodies and antibody Fab' fragments. *Biomaterials, 28*, 4978–4990.

Betbeder, D., Spérandio, S., Latapie, J. P., de Nadai, J., Etienne, A., Zajac, J. M., et al. (2000). Biovector nanoparticles improve antinociceptive efficacy of nasal morphine. *Pharmaceutical Research, 17*, 743–748.

Bickel, U., Kang, Y. S., Yoshikawa, T., & Pardridge, W. M. (1994). *In vivo* demonstration of subcellular localization of anti-transferrin receptor monoclonal antibody-colloidal gold conjugate in brain capillary endothelium. *Journal of Histochemistry and Cytochemistry, 42*, 1493–1497.

Bobo, R. H., Laske, D. W., Akbasak, A., Morrison, P. F., Dedrick, R. L., & Oldfield, E. H. (1994). Convection-enhanced delivery of macromolecules in the brain. *Proceedings of the National Academy of Sciences of the United States of America, 91*, 2076–2080.

Brigger, I., Morizet, J., Aubert, G., Chacun, H., Terrier-Lacombe, M. J., Couvreur, P., et al. (2002). Poly(ethylene glycol)-coated hexadecylcyanoacrylate nanospheres display a combined effect for brain tumor targeting. *Journal of Pharmacology and Experimental Therapeutics, 303*, 928–936.

Buckingham, L. E., Balasubramanian, M., Emanuele, R. M., Clodfelter, K. E., & Coon, J. (1995). Comparison of solutol HS 15, Cremophor EL and novel ethoxylated fatty acid surfactants as multidrug resistance modification agents. *International Journal of Cancer, 62*, 436–442.

Burke, T. G., & Mi, Z. (1994). The structural basis of camptothecin interactions with human serum albumin: Impact on drug stability. *Journal of Medicinal Chemistry, 37*, 40–46.

Calvo, P., Gouritin, B., Villarroya, H., Eclancher, F., Giannavola, C., Klein, C., et al. (2002). Quantification and localization of PEGylated polycyanoacrylate nanoparticles in brain and spinal cord during experimental allergic encephalomyelitis in the rat. *European Journal of Neuroscience, 15*, 1317–1326.

Chattopadhyay, N., Zastre, J., Wong, H. L., Wu, X. Y., & Bendayan, R. (2008). Solid lipid nanoparticles enhance the delivery of the HIV protease inhibitor, atazanavir, by a human brain endothelial cell line. *Pharmaceutical Research, 25*, 2262–2271.

Chekhonin, V. P., Zhirkov, Y. A., Gurina, O. I., Ryabukhin, I. A., Lebedev, S. V., Kashparov, I. A., et al. (2005). PEGylated immunoliposomes directed against brain astrocytes. *Drug Delivery, 12*, 1–6.

Coon, J. S., Knudson, W., Clodfelter, K., Lu, B., Weinstein, R. S., & Solutol, H. S. (1991). 15, nontoxic polyoxyethylene esters of 12-hydroxystearic acid, reverses multidrug resistance. *Cancer Research, 51*, 897–902.

Costantino, L., Gandolfi, F., Bossy-Nobs, L., Tosi, G., Gurny, R., Rivasi, F., et al. (2006). Nanoparticulate drug carriers based on hybrid poly(D,L-lactide-co-glycolide)-dendron structures. *Biomaterials, 27*, 4635–4645.

Couvreur, P., Tulkens, P., Roland, M., Trouet, A., & Speiser, P. (1977). Nanocapsules: A new type of lysosomotropic carrier. *FEBS Letters, 84*, 323–326.

Dehouck, B., Fenart, L., Dehouck, M. P., Pierce, A., Torpier, G., & Cecchelli, R. (1997). A new function for the LDL receptor: Transcytosis of LDL across the blood-brain barrier. *Journal of Cell Biology, 138*, 877–889.

De Miguel, I., Imbertie, L., Rieumajou, V., Major, M., Kravtzoff, R., & Betbeder, D. (2000). Proofs of the structure of lipid coated nanoparticles (SMBV) used as drug carriers. *Pharmaceutical Research, 17*, 817–824.

Emerich, D. F., Snodgrass, P., Pink, M., Bloom, F., & Bartus, R. T. (1998). Central analgesic actions of loperamide following transient permeation of the blood-brain barrier with Cereport (RMP-7). *Brain Research, 801*, 259–266.

Fabel, K., Dietrich, J., Hau, P., Wismeth, C., Winner, B., Przywara, S., et al. (2001). Long-term stabilization in patients with malignant glioma after treatment with liposomal doxorubicin. *Cancer, 92*, 1936–1942.

Fenart, L., Casanova, A., Dehouck, B., Duhem, C., Slupek, S., Cecchelli, R., et al. (1999). Evaluation of effect of charge and lipid coating on ability of 60-nm nanoparticles to cross an *in vitro* model of the blood-brain barrier. *Journal of Pharmacology and Experimental Therapeutics, 291*, 1017–1022.

Feng, S. S., Mu, L., Win, K. Y., & Huang, G. (2004). Nanoparticles of biodegradable polymers for clinical administration of paclitaxel. *Current Medicinal Chemistry, 11*, 413–424.

Fillebeen, C., Descamps, L., Dehouck, M. P., Fenart, L., Benaïssa, M., Spik, G., et al. (1999). Receptor-mediated transcytosis of lactoferrin through the blood-brain barrier. *Journal of Biological Chemistry, 274*, 7011–7017.

Fundaro, R., Cavalli, A., Bagoni, D. Vighetto, Zara, G. P., & Gasco, M. R. (2000). Non-stealth and stealth solid lipid nanoparticles(SLN) carrying doxorubicin: Pharmacokinetic and tissue distribution after i.v. administration to rats. *Pharmaceutical Research, 42*, 337–343.

Gao, K., & Jiang, X. (2006). Influence of particle size on transport of methotrexate across blood brain barrier by polysorbate 80-coated polybutylcyanoacrylate nanoparticles. *International Journal of Phamaceutics*, *310*, 213–219.

Gref, R., Domb, A., Quellec, P., Blunk, T., Müller, R. H., Verbavatz, J. M., et al. (1995). The controlled intravenous delivery of drugs using PEG-coated sterically stabilized nanospheres. *Advanced Drug Delivery Reviews*, *16*, 215–233.

Gref, R., Rodrigues, J., & Couvreur, P. (2002). Polysaccharides grafted with polyesters: Novel amphiphilic copolymers for biomedical applications. *Macromolecules*, *35*, 9861–9867.

Heurtault, B., Saulnier, P., Pech, B., Proust, J.-E., & Benoit, J.-P. (2002). A novel phase inversion-based process for the preparation of lipid nanocarriers. *Pharmaceutical Research*, *19*, 875–880.

Hoekstra, D., & Martin, O. C. (1982). Transbilayer redistribution of phosphatidylethanolamine during fusion of phospholipid vesicles. Dependence on fusion rate, lipid phase separation, and formation of nonbilayer structures. *Biochemistry*, *23*, 6097–6103.

Hussain, A. A. (1998). Intranasal drug delivery. *Advanced Drug Delivery Reviews*, *29*, 39–49.

Huwyler, J., Yang, J., & Pardridge, W. M. (1997). Receptor mediated delivery of daunomycin using immunoliposomes: Pharmacokinetics and tissue distribution in the rat. *Journal of Pharmacology and Experimental Therapeutics*, *282*, 1541–1546.

Illum, L. (2000). Transport of drugs from the nasal cavity to the central nervous system. *European Journal of Pharmaceutical Science*, *11*, 1–18.

Jallouli, Y., Paillard, A., Chang, J., Sevin, E., & Betbeder, D. (2007). Influence of surface charge and inner composition of porous nanoparticles to cross blood-brain barrier *in vitro*. *International Journal of Phamaceutics*, *344*, 103–109.

Jann, M. W. (1991). Clozapine. *Pharmacotherapy*, *11*, 179–195.

Khalid, M. N., Simard, P., Hoarau, D., Dragomir, A., & Leroux, J.-C. (2006). Long circulating poly(ethylene glycol)-decorated lipid nanocapsules deliver docetaxel to solid tumors. *Pharmaceutical Research*, *23*, 752–758.

Kim, H. R., Andrieux, K., Gil, S., Taverna, M., Chacun, H., Desmaële, D., et al. (2007). Translocation of poly (ethylene glycol-co-hexadecyl)cyanoacrylate nanoparticles into rat brain endothelial cells: Role of apolipoproteins in receptor-mediated endocytosis. *Biomacromolecules*, *8*, 793–799.

Koziara, J. M., Lockman, P. R., Allen, D. D., & Mumper, R. J. (2004). Paclitaxel nanoparticles for the potential treatment of brain tumors. *Journal of Controlled Release*, *99*, 259–269.

Kratzer, I., Wernig, K., Panzenboeck, U., Bernhart, E., Reicher, H., Wronski, R., et al. (2007). Apolipoprotein A-I coating of protamine-oligonucleotide nanoparticles increases particle uptake and transcytosis in an *in vitro* model of the blood-brain barrier. *Journal of Controlled Release*, *117*, 301– 311.

Kreuter, J. (2001). Nanoparticulate systems for brain delivery of drugs. *Advanced Drug Delivery Reviews*, *47*, 65–81.

Kreuter, J., Alyautdin, R. N., Kharkevich, D. A., & Ivanov, A. A. (1995). Passage of peptides through the blood-brain barrier with colloidal polymer particles (nanoparticles). *Brain Research*, *674*, 171–174.

Kreuter, J., Shamenkov, D., Petrov, V., Ramge, P., Cychutek, K., Koch-Brandt, C., et al. (2002). Polipoprotein-mediated transport of nanoparticle-bound drugs across the blood-brain barrier. *Journal of Drug Targeting*, *10*, 317–325.

Kroll, R. A., Pagel, M. A., Muldoon, L. L., Roman-Goldstein, S., & Neuwelt, E. A. (1996). Increasing volume of distribution to the brain with interstitial infusion: Dose, rather than convection, might be the most important factor. *Neurosurgery*, *38*, 752–754.

Lacoeuille, F., Garcion, E., Benoit, J. P., & Lamprecht, A. (2007). Lipid nanocapsules for intracellular drug delivery of anticancer drugs. *Journal of Nanoscience and Nanotechnology*, *7*, 4612–4617.

Lamprecht, A., & Benoit, J. P. (2006). Etoposide nanocarriers suppress glioma cell growth by intracellular drug delivery and simultaneous P-glycoprotein inhibition. *Journal of Controlled Release*, *112*, 208–213.

Li, S. D., & Huang, L. (2008). Pharmacokinetics and biodistribution of nanoparticles. *Molecular Pharmaceutics*, *5*, 496– 504.

Loiseau, P. M., Imbertie, L., Bories, C., Betbeder, D., & De Miguel, I. (2002). Design and antileishmanial activity of amphotericin B-loaded stable ionic amphiphile biovector formulations. *Antimicrobial Agents and Chemotherapy*, *46*, 1597–1601.

Lu, W., Zhang, Y., Tan, Y. Z., Hu, K. L., Jiang, X. G., & Fu, S. K. (2005). Cationic albumin-conjugated pegylated nanoparticles as novel drug carrier for brain delivery. *Journal of Controlled Release*, *107*, 428–448.

Luca, C., Francesca, G., Giovanni, T., Francesco, R., Vandelli, M. A., & Flavio, F. (2005). Peptide-derivatized biodegradable nanoparticles able to cross the blood-brain barrier. *Journal of Controlled Release*, *108*, 84–96.

Major, M., Prieur, E., Tocanne, J. F., Betbeder, D., & Sautereau, A. M. (1997). Characterization and phase behaviour of phospholipid bilayers adsorbed on spherical polysaccharidic nanoparticles. *Biochimica et Biophysica ACTA*, *1327*, 32–40.

Manjunath, K., & Venkateswarlu, V. (2005). Pharmacokinetics, tissue distribution and bioavailability of clozapine solid lipid nanoparticles after intravenous and intraduodenal administration. *Journal of Controlled Release*, *107*, 215–228.

Maruyama, K., Takizawa, T., Yuda, T., Kennel, S. J., Huang, L., & Iwatsuru, M. (1995). Targetability of novel immunoliposomes modified with amphipathic poly(ethylene glycol)s conjugated at their distal terminals to monoclonal antibodies. *Biochimica et Biophysica ACTA*, *1234*, 74–80.

Mehnert, W., & Mäder, K. (2001). Solid lipid nanoparticles: Production, characterization and applications. *Advanced Drug Delivery Reviews*, *47*, 165–196.

Miller, T. P., Grogan, T. M., Dalton, W. S., Spier, C. M., Scheper, R. J., & Salmon, S. E. (1991). P-Glycoprotein expression in malignant lymphoma and reversal of clinical drug resistance with chemotherapy plus high-dose verapamil. *Journal of Clinical Oncology*, *9*, 17–24.

Moghimi, S. M., Hunter, A. C., & Murray, J. C. (2001). Long-circulating and target specific nanoparticles: Theory to practice. *Pharmacological Reviews, 53*, 283–318.

Müller, R.H., & Lucks, J.S. (1996). Arzneistoffträger aus festen Lipidteilchen, Feste Lipidnanosphären (SLN). European Patent No. 0605497.

Neuwelt, E. A., & Rapoport, S. I. (1984). Modification of the blood-brain-barrier in the chemotherapy of malignant brain tumors. *Federation Proceedings, 43*, 214–219.

Oldendorf, W. H. (1971). Brain uptake of radiolabeled amino acids, amines, and hexoses after arterial injection. *American Journal of Physiology, 221*, 1629–1639.

Olivier, J. C. (2005). Drug transport to brain with targeted nanoparticles. *NeuroRx, 2*, 108–119.

Olivier, J. C., Fenart, L., Chauvet, R., Pariat, C., Cecchelli, R., & Couet, W. (1999). Indirect evidence that drug brain targeting using polysorbate 80-coated polybutylcyanoacrylate nanoparticles is related to toxicity. *Pharmaceutical Research, 16*, 1836–1842.

Olivier, J. C., Huertas, R., Lee, H. J., Calon, F., & Pardridge, W. M. (2002). Synthesis of pegylated immunonanoparticles. *Pharmaceutical Research, 19*, 1137–1143.

Omori, N., Maruyama, K., Jin, G., Li, F., Wang, S. J., Hamakawa, Y., et al. (2003). Targeting of post-ischemic cerebral endothelium in rat by liposomes bearing polyethylene glycol-coupled transferrin. *Neurological Research, 25*, 275–279.

Panyam, J., & Labhasetwar, V. (2003). Biodegradable nanoparticles for drug and gene delivery to cells and tissue. *Advanced Drug Delivery Reviews, 55*, 329–347.

Pardridge, W. M. (1999). Non-invasive drug delivery to the human brain using endogenous blood-brain barrier transport system. *Pharmaceutical Science & Technology Today, 2*, 49–59.

Pardridge, W. M. (2001). Neuroprotection in stroke: Is it time to consider large molecule drugs? *Drug Discovery Today, 6*, 751–753.

Pardridge, W. M. (2003). Blood-brain barrier drug targeting: The future of brain drug development. *Molecular Interventions, 3*, 90–105.

Pardridge, W. M., Oldendorf, W. H., Cancilla, P., & Frank, H. J. (1986). Blood-brain barrier: Interface between internal medicine and the brain. *Annals of Internal Medicine, 105*, 82–95.

Peira, E., Marzola, P., Podio, V., Aime, S., Sbarbati, A., & Gasco, M. R. (2003). *In vitro* and *in vivo* study of solid lipid nanoparticles loaded with superparamagnetic iron oxide. *Journal of Drug Targeting, 11*, 19–24.

Peltier, S., Oger, J. M., Lagarce, F. Couet, W., & Benoit, J. P. (2006). Enhanced oral paclitaxel bioavailability after administration of paclitaxel-loaded lipid nanocapsules. *Pharmaceutical Research, 23*, 1243–1250.

Peracchia, M. T., Harnisch, S., Pinto-Alphandary, H., Gulik, A., Dedieu, J. C., Desmaële, D., et al. (1999). Visualization of *in vitro* protein-rejecting properties of PEGylated stealth polycyanoacrylate nanoparticles. *Biomaterials, 20*, 1269–1275.

Potmesil, M. (1994). Camptothecins: From bench research to hospital wards. *Cancer Research, 54*, 1431–1439.

Ramge, P., Unger, R. E., Oltrogge, J. B., Zenker, D., Begley, D., Kreuter, J., et al. (2000). Polysorbate-80 coating enhances uptake of polybutylcyanoacrylate (PBCA)-nanoparticles by human and bovine primary brain capillary endothelial cells. *European Journal of Neuroscience, 12*, 1931–1940.

Rice, A., Michaelis, M. L., Georg, G., Liu, Y., Turunen, B., & Audus, K. L. (2003). Overcoming the blood-brain barrier to taxane delivery for neurodegenerative diseases and brain tumors. *Journal of Molecular Neuroscience, 20*, 339–343.

Roney, C., Kulkarni, P., Arora, V., Antich, P., Bonte, F., Wu, A., et al. (2005). Targeted nanoparticles for drug delivery through the blood-brain barrier for Alzheimer's disease. *Journal of Controlled Release, 108*, 193–214.

Schroeder, U., Sommerfeld, P., & Sabel, B. A. (1998). Efficacy of oral dalargin-loaded nanoparticle delivery across the blood-brain barrier. *Peptides, 19*, 777–780.

Schmidt, J., Metselaar, J. M., Wauben, M. H., Toyka, K. V., Storm, G., & Gold, R. (2003). Drug targeting by long-circulating liposomal glucocorticosteroids increases therapeutic efficacy in a model of multiple sclerosis. *Brain, 126*, 1895–1904.

Shi, N., Zhang, Y., Zhu, C., Boado, R. J., & Pardridge, W. M. (2001). Brain-specific expression of an exogenous gene after i.v. administration. *Proceedings of the National Academy of Sciences of the United States of America, 98*, 12754–12759.

Stockhorst, U., Fries, D. D., Steingrueber, H. J., & Scherbaum, W. A. (2004). Insulin and the CNS: Effects on food intake, memory, and endocrine parameters and the role of intranasal insulin administration in humans. *Physiology and Behavior, 83*, 47–54.

Tosi, G., Costantino, L., Rivasi, F., Ruozi, B., Leo, E., Vergoni, A. V., et al. (2007). Targeting the central nervous system: *In vivo* experiments with peptide-derivatized nanoparticles loaded with Loperamide and Rhodamine-123. *Journal of Controlled Release, 122*, 1–9.

Tosi, G., Rivasi, F., Gandolfi, F., Costantino, L., Vandelli, M. A., & Forni, F. (2005). Conjugated poly (D,L-lactide-co-glycolide) for the preparation of *in vivo* detectable nanoparticles. *Biomaterials, 26*, 4189–4195.

Townsend, S. A., Evrony, G. D., Gu, F. X., Schulz, M. P., Brown, R. H., Jr., & Langer, R. (2007). Tetanus toxin C fragment-conjugated nanoparticles for targeted drug delivery to neurons. *Biomaterials, 28*, 5176–5184.

Vorbrodt, A. W., & Dobrogowska, D. H. (1999). Interaction of glycated albumin-gold complexes with mouse brain microvascular endothelium. *Folia Histochemica et Cytobiologica, 37*, 3–10.

Wang, J. X., Sun, X., & Zhang, Z. R. (2002). Enhanced brain targeting by synthesis of 3′,5′-dioctanoyl-5-fluoro-2′-deoxyuridine and incorporation into solid lipid nanoparticles, *European Journal of Pharmaceutics and Biopharmaceutics, 54*, 285–290.

Wilson, B., Samanta, M. K., Santhi, K., Kumar, K. P., Paramakrishnan, N., & Suresh, B. (2008). Poly (n-butylcyanoacrylate) nanoparticles coated with polysorbate 80 for the targeted delivery of rivastigmine into the brain to treat Alzheimer's disease. *Brain Research*, *1200*, 159–168.

Yang, S. C., Zhu, J. B., Lu, Y., Liang, B. W., & Yang, C. Z. (1999). Body distribution of camptothecin solid lipid nanoparticles after oral administration. *Pharmaceutical Research*, *16*, 751–757.

Zhang, Q. Z., Zha, L. S., Zhang, Y., Jiang, W. M., Lu, W., Shi, Z. Q., et al. (2006). The brain targeting efficiency following nasally applied MPEG-PLA nanoparticles in rats. *Journal of Drug Targeting*, *14*, 281–290.

H.S. Sharma (Ed.)
Progress in Brain Research, Vol. 180
ISSN: 0079-6123

CHAPTER 2

Quantum dot nanotechnologies for neuroimaging

Gabriel A. Silva*

Departments of Bioengineering and Ophthalmology and Neurosciences Program, University of California, San Diego, CA, USA

Abstract: Functionalized quantum dot nanocrystals provide an opportunity for high signal-to-noise ratio specific labeling of cells with micron-scale spatial resolution, and extend the cellular imaging toolbox available to the cellular neurobiologist. In this review we discuss previous work from our group aimed at optimizing quantum dot labeling protocols specific to neurons and neural glial cells, labeling and imaging of intact neural retinal tissue sections in a rat model of retinal degeneration focused on the formation of the glial scar following focal reactive gliosis, and on the characterization and estimation of the number of functionally available antibodies for biological binding conjugated to quantum dots following two popular conjugation schemes.

Keywords: quantum dots; neurons; glial cells; imaging; nanotechnology

Introduction

Nanotechnology and nanoengineering have the potential to produce significant scientific and technological advances in diverse fields including biology and medicine. In a broad sense, they can be defined as the science and engineering involved in the design, syntheses, characterization, and application of materials and devices whose smallest functional organization in at least one dimension is on the nanometer scale, ranging from a few to several hundred nanometers. A nanometer is one billionth of a meter, or three orders of magnitude smaller then a micron, roughly the size scale of a molecule itself (e.g., a DNA molecule is about 2.5 nm long, while a sodium atom is about 0.2 nm). The potential impact of nanotechnology stems directly from the spatial and temporal scales being considered: materials and devices engineered at the nanometer scale imply controlled manipulation of individual constituent molecules and atoms in how they are arranged to form the bulk macroscopic substrate. This in turn, results in nanoengineered substrates and devices that can be designed to exhibit specific and controlled bulk chemical and physical properties as a result of the control over their molecular synthesis and assembly.

In this chapter we collect and adapt a number of primary studies and published reviews by the author and colleagues into a single manuscript focusing on functionalized quantum dot imaging and specific labeling of neural cells and tissues, including both neurons and various types of glial cells (Pathak, Cao, Davidson, Jin, & Silva, 2006; Pathak, Davidson, & Silva, 2007; Pathak et al., 2009; Silva, 2004; Silva, 2006). Specifically, this chapter discuses work done by the author's lab to optimize protocols for

* Corresponding author.
Tel.: +1.858.822.4591;
E-mail: gsilva@ucsd.edu

DOI: 10.1016/S0079-6123(08)80002-7

specific labeling and imaging of fixed dissociated neural cells and cytoarchitecturally intact neural retinal tissue sections (Pathak et al., 2006, 2009), illustrates the use of such protocols and methods, and discusses the direct quantitative estimation of the number of functionally available antibodies conjugated to quantum dot nanocrystals for applications to biological labeling (Pathak et al., 2007). The interested reader is referred to other review articles by the author and colleagues that discuss applications of nanotechnologies to neuroscience (Blumling & Silva, in press; Provenzale & Silva, in press; Silva, 2004, 2005, 2006, 2007, 2008a, 2008b; Yu & Silva, 2008).

Semiconductor quantum dots and their utility for neurobiological imaging

Semiconductor fluorescent quantum dots are nanometer-sized functionalized particles that display unique physical properties that make them particularly well suited for visualizing and tracking molecular processes in cells using standard fluorescence (Biju, Itoh, Anas, Sujith, & Ishikawa, 2008; Gao, 2003; Jaiswal, Goldman, Mattoussi, & Simon, 2004; Wu et al., 2003). They are readily excitable and have broad absorption spectra with very narrow emission spectra, allowing multiplexing of many different colored quantum dots; they display minimal photobleaching, thereby allowing molecular tracking over prolonged periods; they also display a blinking property that allows the identification of individual quantum dots. As a result, single-molecule binding events can be identified and tracked using optical fluorescence microscopy, allowing the pursuit of experiments that are difficult or not possible given other experimental approaches.

Quantum dots are nanometer-sized particles composed of a heavy metal core, such as cadmium selenium or cadmium telluride with an intermediate unreactive zinc sulfide shell and a customized outer coating of different bioactive molecules tailored to a specific application (Fig. 1). The composition and very small size of quantum dots (5–8 nm) gives them unique and very stable fluorescent optical properties that are readily tunable by changing their physical composition or size. The

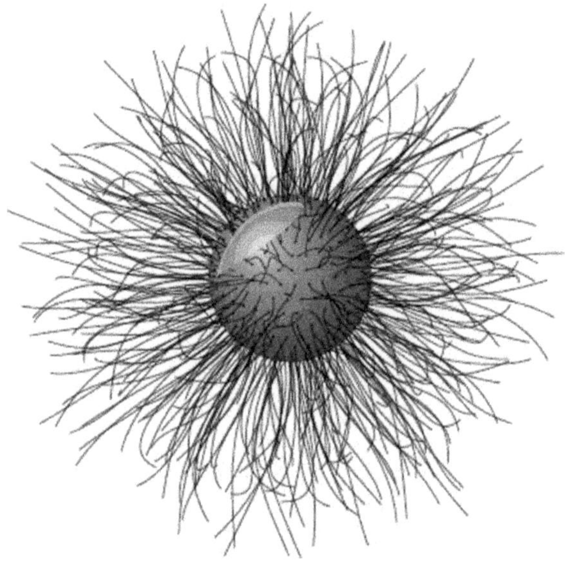

Fig. 1. Cartoon structure of a typical functionalized quantum dot. The heavy metal core is shielded from the biological environment by an outer shell. The outer shell is in turn chemically functionalized with biologically relevant molecules such as antibodies and other peptides (black) for specific binding to target epitopes, for example, on cells. Reproduced from Blumling and Silva (in press).

photochemical properties of quantum dots allow selective fluorescent tagging of proteins similar to classical immunocytochemistry (ICC). However, the use of quantum dots is associated with minimal photobleaching and a much higher signal-to-noise ratio. Their broad absorption spectra but very narrow emission spectra allows multiplexing of many quantum dots of different colors in the same sample, something which cannot be achieved with traditional fluorophores. The physics responsible for these effects are beyond the scope of this brief introduction, but the small size of quantum dot particles results in large but specific energy jumps between the energy band gaps of excited electron–hole pairs in the semiconductor core. This effect results in scaled changes of absorption and emission wavelengths as a function of particle size, so that small changes in the radius of quantum dots translate into very distinct changes in color. This physical property represents another major advantage over traditional organic fluorophores that, in general, require distinct chemistries to produce different colors. For biological applications quantum dots can be chemically functionalized to target proteins at high ligand–receptor densities. Recent work has shown that, at least in some cellular systems, quantum dots conjugated with natural ligands are readily internalized into cells, do not interfere with intracellular signaling, and are non-toxic.

For neuroscience quantum dots represent a tool of significant potential. Besides offering an alternative to traditional ICC, they are particularly valuable for studies of neurons and glia. Quantum dots can be used to visualize, measure, and track individual molecular events using fluorescence microscopy; and they provide the ability to visualize and track dynamic molecular processes over extended periods (e.g., from seconds to many minutes). These properties are difficult to achieve using other techniques or approaches. For example, quantum dots are useful for experiments that are limited by the restricted anatomy of neuronal and glial interactions, such as the small size of the synaptic cleft; or between an astrocyte process and a neuron. Because of their extremely small size and optical resolution, they are also well suited for tracking the molecular dynamics of intracellular and/or intercellular molecular processes over long timescales. However, it should be appreciated that the hydrodynamic radius of functionalized quantum dots is larger (15–20 nm) than their actual size of 5–8 nm. Recent studies using quantum dots in neuroscience illustrate the potential of this technology. Triller and colleagues used antibody functionalized quantum dots to track the lateral diffusion of glycine receptors in cultures of primary spinal cord neurons (Dahan et al., 2003). They were able to track the trajectory of individual glycine receptors for tens of minutes at spatial resolutions of 5–10 nm, demonstrating that the diffusion dynamics varied depending on whether the receptors were synaptic, perisynaptic, or extrasynaptic. Vu, Desai and colleagues tagged nerve growth factor (βNGF) to quantum dots and used them to promote neuronal-like differentiation in cultured PC12 cells (Vu et al., 2005). Ultimately, these approaches could be used to visualize and track functional responses in neurons. However, as with any new technology there are caveats. For example, Vu et al. reported that βNGF conjugated to quantum dots had reduced activity compared to free βNGF. Other groups are pushing the technology forward and providing new quantum dot based tools. Brinker and colleagues developed a technique to produce biocompatible water-soluble quantum dot micelles that retain the optical properties of individual quantum dots. These micelles showed uptake and intracellular dispersion in cultured hippocampal neurons (Fan et al., 2005). Ting and colleagues are developing a modified quantum dot labeling approach that addresses the relatively large size of antibody–quantum dot conjugates and the instability of some quantum dot–ligand interactions. Their technique tags cell surface proteins with a specific peptide (a 15 amino acid polypeptide called acceptor protein) that can be directly biotinylated as a target for streptavidin-conjugated quantum dots (Howarth, Takao, Hayashi, & Ting, 2005). Using this approach they were able to specifically label and track α-amino-3-hydroxyl-5-methyl-4-isoxazole-propionate (AMPA) receptors on cultured hippocampal neurons.

Specific labeling and imaging of dissociated neurons and glial cells

We have previously discussed in detail our quantum dot labeling protocols for labeling neurons and glia (Pathak et al., 2006). We conjugated anti-β-tubulin III and antiglial fibrillary acidic protein (GFAP) antibodies to 605-nm quantum dots and labeled primary cortical neurons, PC12 cells, primary cortical astrocytes, and r-MC1 retinal Muller glial cells. β-tubulin III and GFAP are ubiquitous cytoskeletal proteins specific to neurons and macroglia, respectively, but the protocols should label any protein of interest. Table 1 summarizes the detailed methods.

Using our protocols we were able to get excellent specific labeling of β-tubulin in neurons and PC12 cells and GFAP in astrocytes and Muller cells, with negligible nonspecific binding or background (see Fig. 2). Labeling with unconjugated or primary antibody omitted streptavidin-conjugated quantum dots showed no labeling at all (data not shown). β-tubulin and GFAP labeling using functionalized quantum dots displayed

Table 1. Summary of quantum dot labeling protocol for neurons and glia

Preprocessing and fixing
Remove media from wells by gently aspirating.
Wash cells with warmed PBS.
Fix cells with 4% paraformaldehyde (Electron Microscopy Sciences, catalog #157 15-S) in PBS for 10 min at room temperature.
Wash cells 3× with PBS.
Permeabilize cells with 0.2% Triton X-100 (Fisher Scientific, catalog #BP151-100) in PBS for 5 min.
Wash cells 3× for 5 min with PBS.
Incubate with 10% horse serum in PBS for 30 min at room temperature.
Rinse with PBS.
Apply Streptavidin/Biotin Blocking Kit (Vector Labs, catalog #SP-2002).
Primary incubation
Rinse with PBS.
Add biotinylated molecule of interest (e.g., antibodies; use ProtOn Biotin Labeling Kit or similar for biotinylation; Vector Labs, catalog #PLK-1202).
Incubate 2 h at room temperature. (Biotinylated secondary antibody for 1 h — alternative three-step labeling protocol.)
Remove antibodies by gently aspiration and rinse 3× with PBS.
Quantum dot incubation
Add streptavidin-conjugated quantum dots (We used Quantum Dot Corporation's 605-nm quantum dots here, catalog #1010-1) in 10% horse serum.
Incubate 1 h at room temperature.
Rinse 3× with PBS.
Mount with 90% glycerol (Sigma, catalog #G-6279) in PBS.

Reproduced from Pathak et al. (2006).

similar labeling patterns to those expected using standard ICC controls visualized with fluorophore-tagged secondary antibodies (Fig. 2g and h). For comparable imaging conditions, quantum dot-labeled cells were brighter and displayed more detailed and sharper microstructural anatomy. The pattern of quantum dot labeling was typical for that observed in other cell types, displaying a dense punctuate pattern and fine details of both intracellular intermediate filaments and cellular processes, unlike traditional fluorophores which tend to have a diffused appearance due to the broad point spread function of their fluorescence signal. Nonspecific artifact labeling using some quantum dot protocols may label neural cells incorrectly due to nonspecific putative electrostatic interactions. We observed this when conjugating antibodies directly to quantum dots, which resulted in unconjugated quantum dots nonspecifically staining the nucleus of Muller cells (see Fig. 2i). Nonspecific binding was also observed when using other published protocols for nonneural cells (Wu et al., 2003). Blocking conditions also need to be carefully optimized since most standard blocking approaches did not work satisfactorily in our hands, including 1–5% bovine serum albumin, 10% horse serum, and 10% fetal bovine serum among others, which resulted in a high level of nonspecific quantum dot binding to the cells (data not shown). Another advantage to labeling with quantum dots is that each individually visualized dot in a fluorescence micrograph represents one to three individual quantum dots, based on our own calculations and those of others (Chan & Nie, 1998). This means that qualitative and potentially quantitative information can be measured for individual binding events between quantum dot-conjugated molecules and their cellular molecular targets, a direct result of the underlying physics (Michalet et al., 2005; West & Halas, 2003) that cannot be done with standard ICC (see Dahan et al., 2003 for an example).

Efficacy of different antibody conjugation methods to quantum dots

One critical issue that has not been addressed is experimentally determining the number of antibodies bound to quantum dots which are functionally

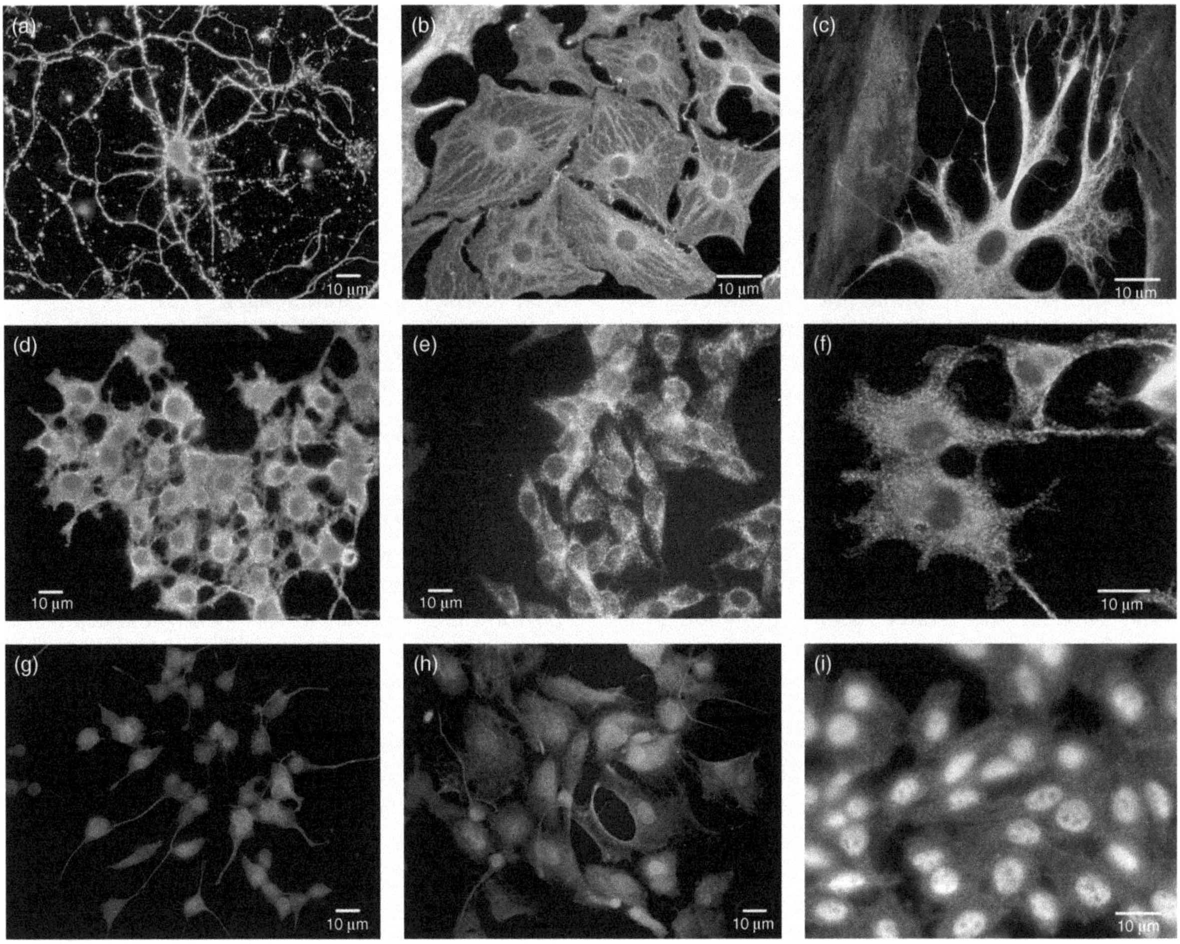

Fig. 2. Fluorescent labeling of neurons and glia with antibody-conjugated 605-nm quantum dots. (a) Primary cortical neurons specifically labeled for β-tubulin. (b, c) Primary cortical astrocytes specifically labeled for GFAP. (d, f) PC12 cells labeled for β-tubulin. (e) r-MC1 neural retinal Muller glial cells specifically labeled for GFAP. (g) PC12 cells labeled for β-tubulin using standard ICC. (h) Primary spinal cord astrocytes labeled for GFAP using standard ICC. (i) An example of artifactual nonspecific labeling in r-MC1 Muller cells with anti-GFAP-conjugated 605-nm quantum dots. In this case, putative nonspecific electrostatic interactions between quantum dots and cellular proteins led to intense nuclear staining and mild cytoplasmic staining using other quantum dot conjugation protocols described for mammalian cells. All imaging parameters were constant for the different experimental conditions, with an acquisition/exposure time of 30 ms for all panels except (i), which was taken with an acquisition time of 100 ms. Reproduced from Pathak et al. (2006).

available for target protein binding (Pathak et al., 2007). This is critical for the analysis and proper interpretation of biological data labeled using these kinds of methods. While other groups have qualitatively characterized antibody functionalized quantum dots using TEM, AFM, UV spectroscopy, and gel electrophoresis, and in some cases have suggested estimates of the putative number of total antibodies bound to quantum, no calculations of the number of functional antibodies bound to quantum dots based on quantitative experimental results have been reported. We previously reported derived numbers of functional IgG antibodies conjugated to quantum dots based on calculations of quantitative electrophoresis experiments using two different conjugation schemes: A common direct covalent conjugation using a reduced disulfide maleimide reaction and biotinylated antibodies bound to streptavidin-functionalized quantum dots. Antibody–quantum dot complexes were run in a sodium dodecyl sulfate polyacrylamide gel electrophoresis (SDS-PAGE) to separate the functional component of conjugated antibodies from the quantum dots. We then blotted the antibody fragments onto a membrane to determine the identity and amount of the antibodies, and quantitatively compared the degree of functional binding to known protein standards to derive the number of bound antibody.

The number of functional antibodies covalently bound to commercially available quantum dots was on average much less than one functional IgG molecule per quantum dot (0.076 ± 0.014) and therefore of limited utility for biological experiments. In contrast, antibodies bound to quantum dots via the streptavidin–biotin system resulted in higher numbers of functional antibodies, with 0.60 ± 0.14 IgG molecules per quantum dot for a 1:1 IgG : quantum dot molar ratio and 1.3 ± 0.35 IgG molecules per quantum dot for a 2:1 ratio. In addition to these specific results, our methods may be of broader interest because our approach is easily extendable for experimentally deriving the number of functional antibodies or peptides bound to other classes of nanoparticles (e.g., magnetic nanoparticles).

We begin by considering the covalent conjugation of antibodies to quantum dots. Prior to their conjugation, antibodies were reduced using dithiothreitol (DTT), which generates three distinct fragments identifiable by their molecular weights: A 25-kDa light chain, which importantly includes the functional specific epitope binding site for a particular IgG molecule, a 50-kDa heavy chain, and a 75-kDa partially cleaved chain consisting of a heavy chain and a light chain held together by an unreduced disulfide bond (Fig. 3a). Following this, individual fragments were covalently bound to quantum dots via an N-Succinimidyl 4-(maleimidomethyl)cyclohexanecarboxylate (SMCC) linkage bond which cannot be broken by DTT treatment, an important consideration for the interpretation of the experimental results that follow. This gives rise to three possible binding scenarios to quantum dots (Fig. 3a): Covalently bound light chains, covalently bound heavy chains, and covalently bound heavy–light chain partial fragments, of which only the latter can undergo further DTT reduction to remove the light chain fragment from heavy chains that remain bound to quantum dots, or heavy chains removed from light chains bound to quantum dots.

We first confirmed that antibodies were indeed covalently bound to the quantum dots by running IgG–quantum dot complexes though SDS-PAGE with and without DTT. For DTT reduced conditions we observed light chains cleaved from covalently bound partial fragments (Fig. 4a, lanes 4–6). As expected, this separation occurred minimally in lanes without DTT (Fig. 4a, lanes 2 and 3). The presence of a weak band at the 25-kDa position in nonreduced lanes (Fig. 4a, lanes 2 and 3) was due to low concentrations of reducing agents in the gel and running buffers. Interestingly, we saw no heavy chains being dissociated from light chain-bound partial fragments. It is unclear why this was the case, although we hypothesize that the probability of the heavy chain portion of a partial fragment binding to a quantum dot is considerably higher than the light chain portion because there is twice the surface area for heavy chain binding and it is a condition that is sterically favored (since the bend in the partial chain may tend to hide the light chain from the quantum dot). Another potential explanation for the lack of heavy chain is methodological. Given the intensity of other bands in the membranes, small amounts of free heavy chain may have gone undetected given the exposure time we used to develop the membrane, which if had been longer may have shown the presence of heavy chains but would have overexposed the other darker bands resulting in uninterpretable smearing. Additional evidence that heavy chains covalently bound to quantum dots originating from partial fragments remained bound to the quantum dots is inferred by a nonspecific colloidal blue protein stain which labels any protein in the gel that did not transfer to the membrane (Fig. 4b). Since blue bands appeared at the

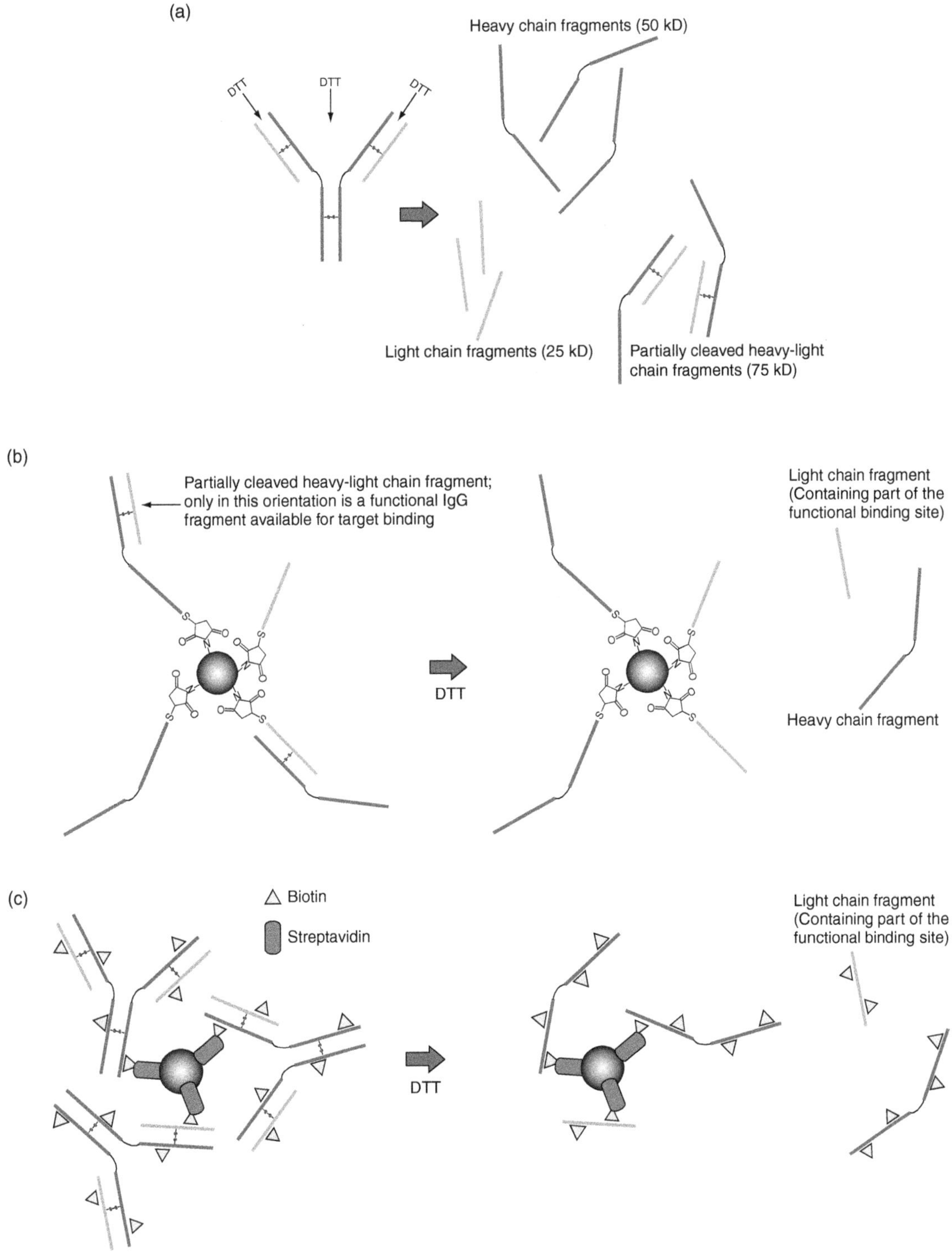

Fig. 3. Antibody reduction and conjugation to quantum dots. (a) Schematic of antibody cleavage sites by DTT at disulfide linkages. The fragments that can result from DTT reduction include the light chain, heavy chain, and partially cleaved fragments due to incomplete reduction. (b) Schematic of direct SMCC covalent conjugation of antibodies to quantum dots. Further reduction with DTT following the primary reduction associated with the conjugation reaction yields the light chains which are counted in the derivation of the average number of functional IgG molecules originally on quantum dots. (c) Similar schematic for biotinylated antibodies conjugated to streptavidin-coated quantum dots. Reproduced from Pathak et al. (2007).

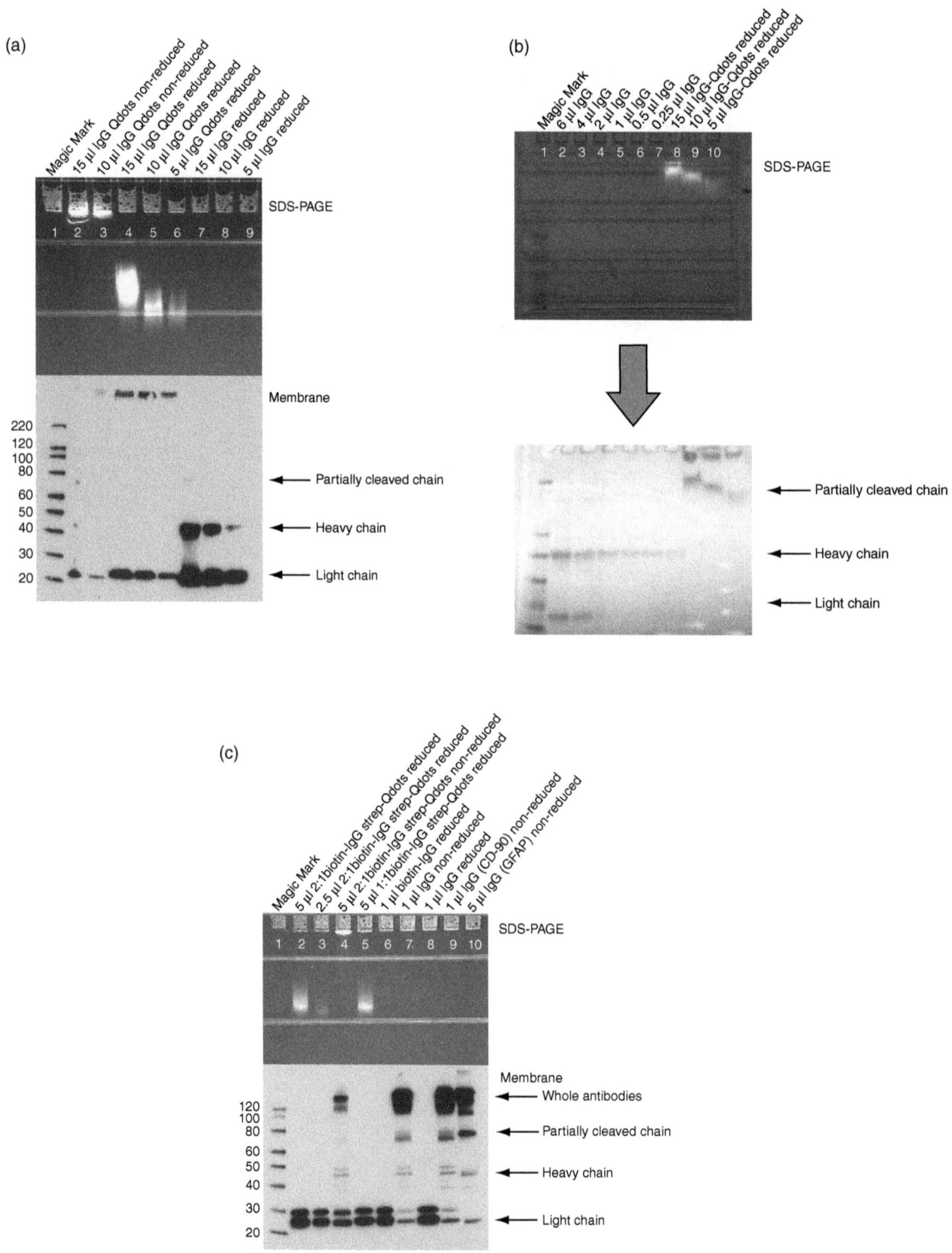

Fig. 4. Separation of IgG antibodies into fragments using SDS-PAGE and membrane transfer under different experimental and control conditions. (a) Covalently conjugated IgG to quantum dots via an SMCC linker and controls. (b) Colloidal nonspecific stain for proteins in gels for the direct conjugation method. (c) Biotinylated IgG bound to streptavidin-coated quantum dots and controls. Reproduced from Pathak et al. (2007).

position in the gels that corresponded to the quantum dots, some amount of residual protein did remain on their surface. Given that most of the light chains were cleaved, since they transferred strongly to the membrane, this residual protein is most likely heavy chain. Regardless, for the purposes of calculating the amount of functional antibody on quantum dot surfaces this is of minimal importance, since it is the amount of available light chain that we are interested in since it is the light chain that contains the ligand-binding epitope. Anther important consideration to note is that the amount of partial fragments initially available for binding to quantum dots following the initial DTT reduction was very low, as evident in the reduced unconjugated IgG controls (Fig. 4a, lanes 7–9). This point is an important consideration for why the number of available functional antibody in the covalently conjugated condition was calculated to be so low. (Note that no partial fragments were visible for the quantum dot lanes because the entire partial chain cannot be cleaved intact from the quantum dot since the SMCC linkage cannot be broken by DTT.) If antibodies had been electrostatically attached to quantum dots, several bands would have shown up in nonreduced lanes (Fig. 4, lanes 2 and 3) because the gel would have electrostatically separated the antibodies from the quantum dots according to their molecular size and weight. Further indirect evidence that antibodies were covalently bound is implied by the fact that quantum dots in nonreduced lanes did not travel through the gels but remained in the loading wells due to the large size of the unreduced complex (visible as hyperintense signals in the loading wells for lanes 2 and 3 of the SDS-PAGE).

We ran the same experiments with biotinylated antibodies and streptavidin-coated quantum dots at 2:1 and 1:1 antibody to quantum dot molar ratios. Biotinylated antibodies have biotin molecules throughout the entire antibody, which results in the IgG molecules being conjugated to quantum dots presumably in all possible spatial arrangements (Fig. 4c). Importantly and very differently from the direct covalent conjugation reaction, using the biotin–streptavidin system the entire antibody molecule is conjugated to the quantum dot; it is not reduced into its light chain and heavy chain fragments prior to binding. Similar to the covalent antibody conjugation method, nonreduced conditions resulted in quantum dots remaining in the loading wells (Fig. 4c, lane 4) while reduced conditions allowed quantum dots to run through the gels (Fig. 4c, lanes 2, 3, and 5). Some amount of antibody did transfer in nonreduced conditions for biotin–streptavidin IgG–quantum dot complexes because the reducing agents in the running buffers and the gel caused the light chain to dissociate in the same manner as for the covalent conjugation. However, since all bands were much stronger in the biotin–streptavidin method in general, bands for the nonreduced condition were correspondingly stronger. Bands in non-DTT-treated antibody lanes (i.e., Fig. 4c, lanes 7, 9, and 10) show the reduction process in greater detail since reduction agents in the running buffers reduced the antibodies less efficiently than DTT-treated conditions (Fig. 4a, lanes 2, 3, 5, 6, and 8).

Based on these data and the qualitative models introduced above that describe the different putative binding scenarios for antibodies directly covalently conjugated to quantum dots and for antibodies bound to quantum dots via biotin and streptavidin (Fig. 3b), we derived the average number of functional IgG conjugated to quantum dots. We use the term "functional antibody" to describe the amount of Fc light chain, which includes a part of the target protein-binding epitope that is physically oriented outward from a quantum dot and presumably able to interact with its ligand. Molecularly, roughly the first 110 amino acids at the amino terminal end of both heavy and light chains form the variable V regions which contain highly variable segments called complementary-determining regions. The pairwise association of V regions from both heavy and light chains is what actually forms the antigen-binding site. As such, only a partial fragment bound to the quantum dot would be functional. Furthermore, because of the structure of the antigen-binding site, a partial fragment covalently bound to the quantum dot oriented with the light chain facing the nanoparticle would almost surely prevent ligand binding. Since it is the Fc light chain portion of the antibody that actively binds to proteins, quantifying the amount of light chain fragments not directly bound to quantum dots and oriented outward gives a good approximation of the functional activity of antibody–quantum dot complexes.

To determine the number of functional IgG bound to quantum dots, we measured the density of the 25-kDa light chain bands, and compared them to controls of known antibody concentrations. Using image analysis software that measures the band density of electrophoresis gels (ImageQuant TL, GE Healthcare), we fitted curves to known concentrations of unconjugated IgG to obtain standard curves of IgG band densities (Fig. 4a and b; all $r^2 \geq 0.89$). Using these curves, we then determined the concentration of IgG bands associated with covalently bound IgG and 2:1 and 1:1 IgG : quantum dot molar ratio streptavidin–biotin conjugation conditions (Fig. 5c and d). Finally, we calculated the number of functional antibodies bound to the quantum dots for each condition (Fig. 5e).

For covalently conjugated IgG we calculated that on average there is much less than one antibody molecule (0.076 ± 0.014) per quantum dot. In other words, adding 10 μL of antibodies directly conjugated to quantum dots is equivalent to adding 0.455 μL from a 0.5 mg/mL stock. This suggests that covalently conjugated antibodies have low amounts of functionally available antibodies and are of inadequate sensitivity for reliable specific labeling of target proteins. In contrast, the number of antibodies bound to quantum dots via the strepavidin–biotin system resulted in a more biologically reasonable 0.60 ± 0.14 IgG molecules per quantum dot for a 1:1 IgG : quantum dot molar ratio and, as would be expected, 1.3 ± 0.35 IgG molecules per quantum dot for a 2:1 ratio. This is equivalent to a functional volume of 0.943 μL of antibody for a 2:1 molar ratio or 0.53 μL for 1:1 molar ratio starting from 4 μL of biotinylated antibody conjugated to streptavidin quantum dots from a 0.5 mg/mL stock concentration. We acknowledge that these numbers are an approximation, since light chains near the quantum dot surface attached to a heavy chain bound to the quantum dot as part of a partial fragment would be sterically unavailable for antigen binding but could still dissociate following DTT reduction. However, this may represent a small source of error because it may be sterically difficult for bound heavy–light chain domains to bind to the quantum dot, therefore thermodynamically favoring the functional partial fragment orientation (see Fig. 3b). In any case, this error would contribute to an overestimation of the number functional antibodies conjugated to a quantum dot, and therefore represent an upper bound on the number of putative functional antibodies, further emphasizing the significance of the results we present here.

These results are significantly less than suggested estimates of about 2–10 antibodies conjugated per quantum dot. To the best of our knowledge, no conjugation reaction can control the binding orientation of IgG molecules. Consequently, due to Brownian motion the number of bound functional antibodies is almost certainly less than the number of total bound IgG. This is not considered by TEM imaging approaches that measure the size (i.e., diameter) of antibody–quantum dot complexes in order to estimate the number of bound antibodies. An important question is: Why did covalent conjugations result in lower numbers of functional antibodies compared to streptavidin–biotin conjugations? One possible explanation is that DTT-reduced antibody fragments attaching to the surface of quantum dots leave few opportunities for light chain fragments to be properly oriented outward and available for protein binding, since of the three reduced fragment types only partial fragments result in functional antibodies and even then the orientation of the partial fragment binding to the quantum dot surface must be correct to allow the light chain fragment to point outward in order to interact with its ligand. In biotin–streptavidin conjugations the antibody is never cleaved, leaving the whole molecule bound to the quantum dot surface and structurally offering more opportunities for light chain fragments to bind their targets. It is plausible that other covalent conjugation chemistries result in higher yields of functional antibodies, comparable to those we report for streptavidin–biotin

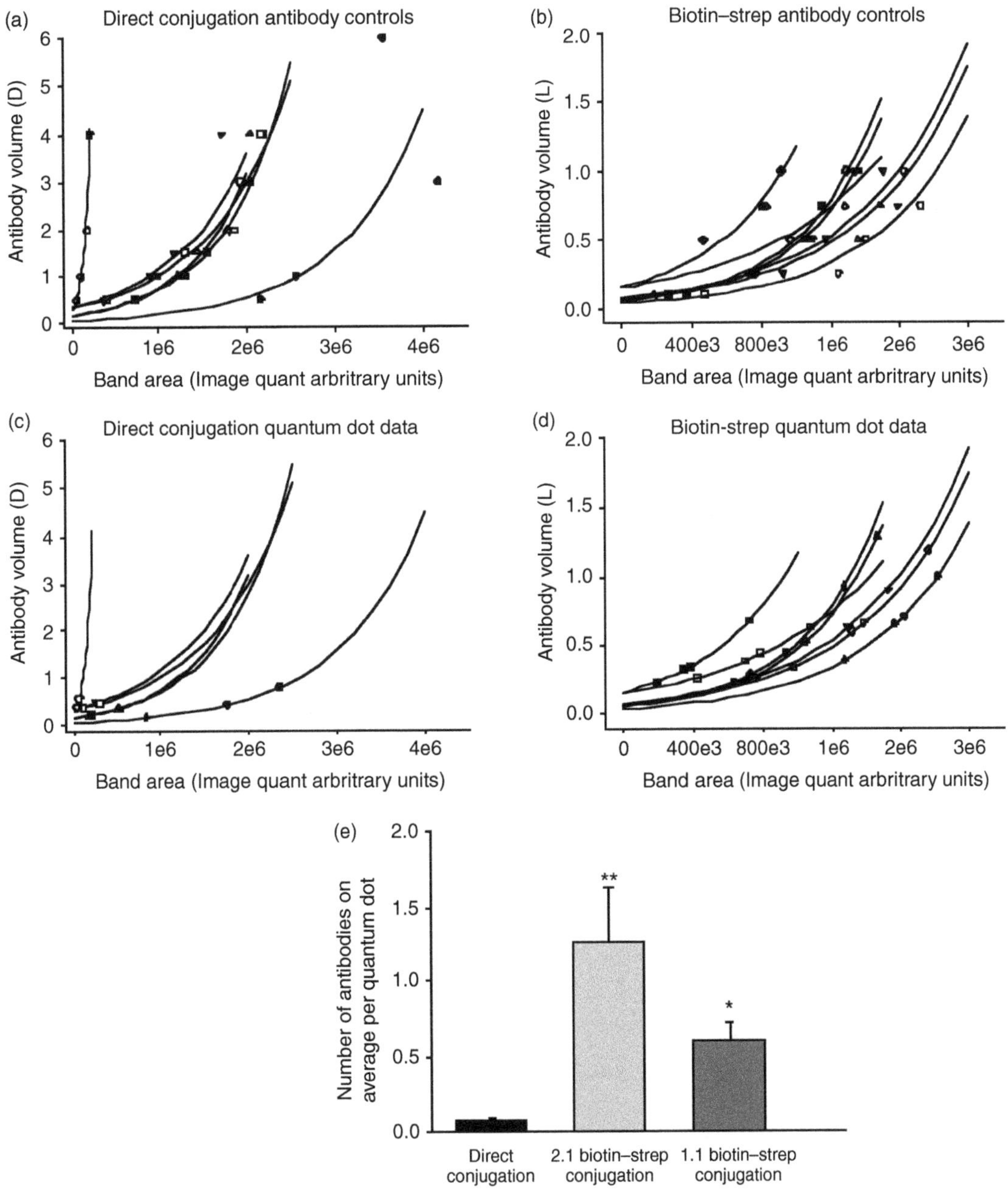

Fig. 5. Derivation of the average number of functional antibodies on both covalently conjugated and streptavidin–biotin-conjugated quantum dots based on measurements of the bound density for different concentrations. (a, b) Fitted linear log control curves ($\ln y = ax - b$) for known volumes of unconjugated IgG antibody band densities in SDS-PAGE gels. Note that the data for each gel was fitted with its own curve in order to control for intergel variability. Each symbol represents a different gel ($n = 6$ gels for covalently conjugated IgG conditions containing a total of 32 unconjugated IgG controls and 13 IgG–quantum dot complexes, and 7 gels for streptavidin–biotin IgG–quantum dot complexes containing a total of 35 unconjugated IgG controls and 28 IgG–quantum dot complexes). (c, d) Corresponding derived volumes from SDS-PAGE band densities for conjugated and streptavidin–biotin-conjugated antibody–quantum dot complexes using the curves plotted in panels (a) and (b), respectively. (e) Calculated values for the average number of antibodies conjugated to quantum dots for both conditions based on the derived measurements of functional antibody volumes (* and ** $p < 0.01$). Reproduced from Pathak et al. (2007).

conjugates or even higher, but it cannot simply be assumed so since, as we show here, at least one well established and commonly used covalent conjugation reaction results in very low numbers of functional antibody on quantum dots. We propose that functionalized quantum dot labeling of biological preparations need to be preceded by the experimental determination of the number of functionalized antibodies per quantum dot, especially given the variability in conjugation methods between different labs. These considerations have a direct impact on the quality, interpretation, and relevance of biological or physiological results obtained using quantum dot labeling nanotechnologies.

Labeling reactive gliosis in retinal tissue sections

In the normal neural retina GFAP expression is associated with the astrocyte layer in the inner nuclear layer and the endfeet of Müller glial cells near the retinal capillaries. Quantum dot labeling of GFAP in control sections of rat retina showed only Müller cell endfeet and astrocytes were GFAP positive, with no GFAP upregulation and no nonspecific binding (Fig. 6a and b). The high specificity and signal to noise ratio of our quantum dot labeling protocol is particularly emphasized in Fig. 6 because the micrographs were taken in wide-field nonconfocal mode which collects light from the entire thickness of the tissue slice, unlike in confocal mode

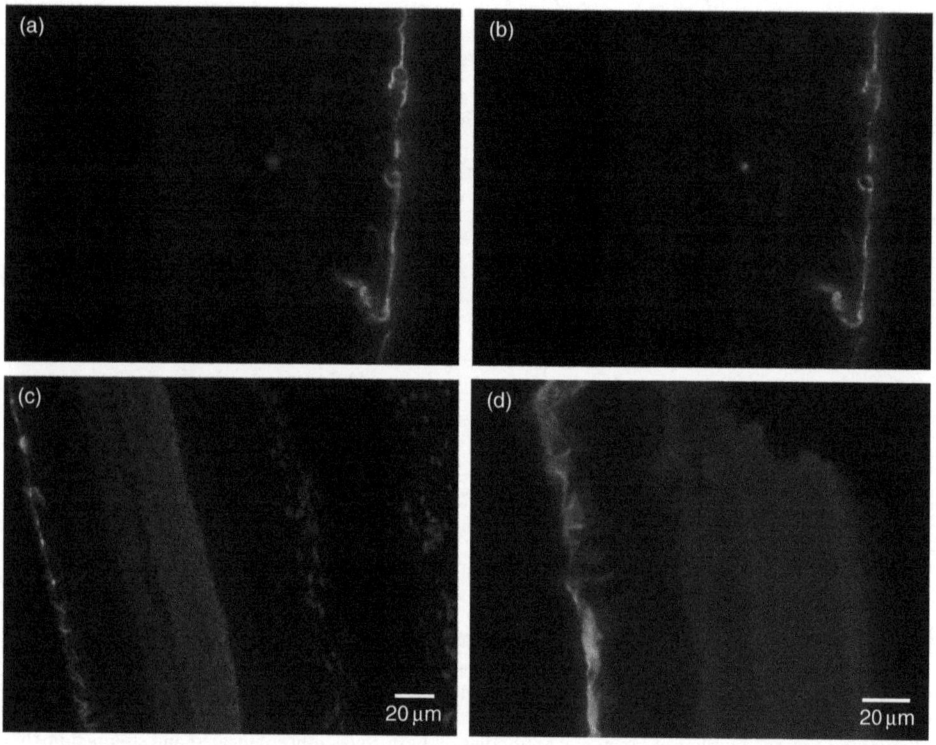

Fig. 6. Control labeling of the noninjured rat neural sensory retina for GFAP. (a, b) Anti-GFAP antibody functionalized quantum dot conjugates specifically label only Müller cell endfeet-associated retinal capillaries and astrocytes associated in the inner nuclear layer associated with retinal ganglion cells. Two slices from a wide-field nonconfocal image stack are shown, and display no observable nonspecific labeling despite the use of nonconfocal mode. (c, d) Widefield nonconfocal standard ICC using an anti-4′,6-diamidino-2-phenylindole(GFAP)-conjugated primary antibody and FITC fluorophore-tagged secondary antibody. A nonspecific nuclear DAPI stain was used to visualize the other retinal layers. Note the more diffuse labeling using FITC compared to the quantum dots and the presence of some nonspecific labeling in the distal layers of the retina. Panels (a), (b), and (d) were taken at 40× and 50-ms exposure times, while panel (c) was taken at 20× at a 50-ms exposure. All micrographs are 10-μm slices. Reproduced from Pathak et al. (2009).

where stray light is physically excluded from the plane of focus. This is critical because, as discussed above, one of the biggest difficulties associated with immunospecific quantum dot labeling of neural cells is nonspecific interactions and clumping between quantum dot particles that can produce false positive results (Pathak et al., 2006, 2007). Methodologically, nonspecific binding proved to be more of an issue with the methanol fixed samples; all samples shown in the results were fixed using paraformaldehyde followed by Triton X-100 to remove excessive cross-linking of proteins induced by fixation. It is essential for the bulkier quantum dot conjugates, compared to the smaller size of traditional fluorophores, to experience a low amount of cross-linking in the tissue in order to avoid clumping. It is a common mistake to assume that quantum dot nanoparticles are smaller than fluorescent dyes, when in fact they are 10–20 times larger (depending on the color) than Fluorescein isothiocyanate (FITC). Serially sectioned 10-μm slices throughout the thickness of the retina, of which two consecutive slices are shown in Fig. 6a and b, showed no observable nonspecific labeling. These results are similar to control retinal sections labeled using traditional ICC with a primary antibody specific to the target antigen and a FITC-tagged secondary antibody that binds to the primary antibody and acts as a fluorescent reporter (Fig. 7c and d), although the FITC labeling was somewhat more qualitatively diffuse and did show some degree of nonspecific labeling despite our best attempts.

Upregulation of GFAP in Müller cells and astrocytes occurs only under pathological conditions and is considered the hallmark of the

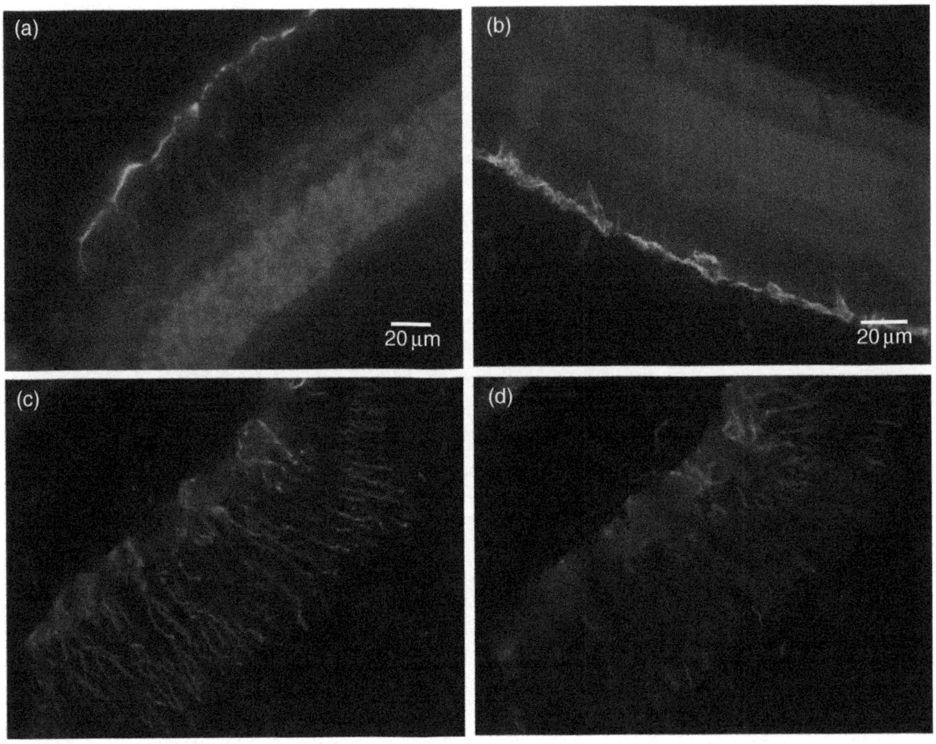

Fig. 7. Specific labeling of GFAP upregulation in the rat neural retina at a laser-induced lesion site imaged using standard FITC ICC. (a, b) Wide-field nonconfocal standard ICC using an anti-GFAP-conjugated primary antibody and FITC fluorophore-tagged secondary antibody. A nonspecific nuclear DAPI stain was used to visualize the other retinal layers. (c, d) Confocal imaging of two different lesions with a 1.6-μm optical slice and an acquisition time of 2 s, comparable with the data shown in Fig. 2. Panel (c) shows a slice near the center of a lesion, while panel (d) shows a slice closer to the boundary of a lesion. Note the particularly high background and diffuse labeling in (d). Reproduced from Pathak et al. (2009).

reactive glial response. GFAP upregulation in Müller cells is particularly apparent because it spans the length of their cell bodies throughout most of the thickness of the retina up to the inner limiting membrane. In the rat laser-induced choroidal neovascularization (CNV) model we used, gliosis and glial scarring occur as secondary processes and result in a strong upregulation of GFAP. Figure 8 shows a confocal z-stack of a 10-μm tissue section with an imaged slice thickness of 1 μm centered at a laser-induced lesion site. To the best of our knowledge, these results represent the first successful specific labeling *in situ* of an intact neural tissue preparation. The intense upregulation of GFAP in both Müller cells and astrocytes indicated a strong reactive response to the induced trauma. Our quantum dot labeling protocol was optimized to ensure even tissue penetration, minimal nonspecific antigen labeling, and maximal specific antigen retrieval. Given this, the fact that the upregulation in GFAP for all lesions we looked at extended over a cross-sectional thickness of the retina of about 10 μm, as demonstrated in Fig. 8 by the drop in fluorescence signal in the confocal stack by slice 10, suggests that the reactive volume of the

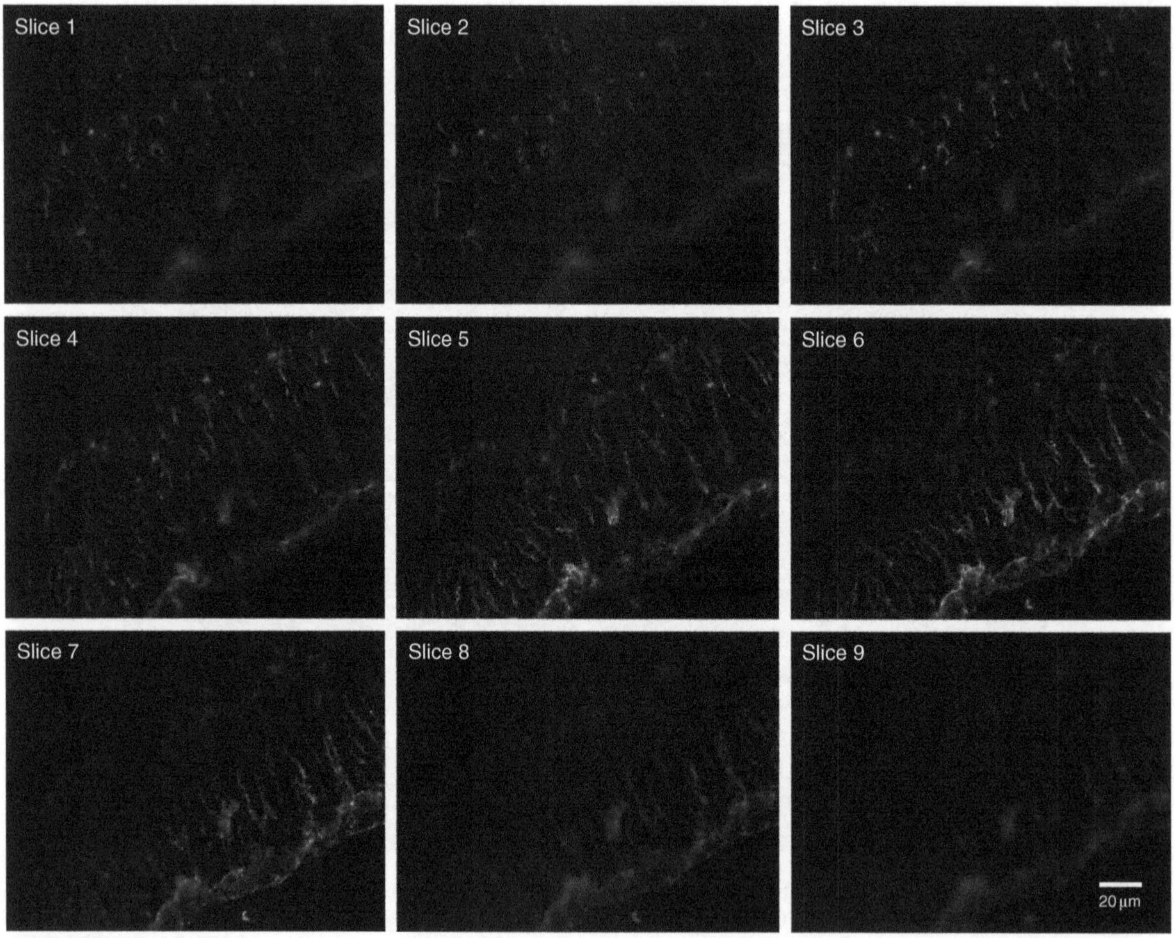

Fig. 8. Specific labeling of GFAP upregulation in the rat neural retina at a laser-induced lesion site imaged using anti-GFAP quantum dot conjugates. A serial cross section of the retina 10 μm thick that encompassed one of the induced lesions was imaged at 1-μm-thick optical slices using confocal microscopy using a 1.8-s exposure time. Reproduced from Pathak et al. (2009).

neural retina in response to the laser-induced injury averaged between 9 and 10 μm in cross-sectional width. The high signal-to-noise ratio of the quantum dot-labeling procedure also putatively provides greater observable and therefore measurable cellular detail throughout the volume of the glial response. In the representative stack in Fig. 8, the upregulation progresses from proximal Müller cell processes near the boundary of the lesion site (progressively from slice 1 to 3) to the entire length of the Müller cells and astrocyte layer near the center of the lesion (slices 4–6), followed by a progressive visible decrease in reactivity near the other side of the lesion boundary (progressively from slice 7 to 9). Therefore, this labeling method should be amenable to quantitatively measuring the extent and thickness of glial scars and presumably other neuronal- and glial-specific markers in neural tissue preparations at high spatial resolutions due to the cellular specificity and low background of the procedure. Such an approach would conceivably allow better quantitative measurements and statistics of both physiologically normal and, as illustrated here, pathological cellular processes. This quantum dot-labeling procedure is considerably superior to nonconfocal wide-field epifluorescence microscopy of retinal sections for specific labeling and imaging of GFAP upregulation in gliosis due to diffuse labeling and higher nonspecific background in the latter (Fig. 7a and b), as introduced above. In our hands the quantum dot-labeling procedure was subjectively less diffuse, more intense, had a noticeably lower nonspecific background, and showed more cellular detail then the best optimized standard FITC immunocytochemistry labeling we could achieve (Fig. 7c and d). This was especially true near the border of imaged lesions where GFAP upregulation gradually decreased and there was less signal intensity (Fig. 7d). In these cases, FITC labeling appeared to display considerably higher nonspecific background, which makes this approach difficult for identifying the edges of GFAP upregulation and gliosis lesion boundaries, resulting in less confidence in any derived measurements of lesion volumes and the spread of the reactive glial response.

Acknowledgments

This work was supported by funds from NIH grants NINDS NS054736.

References

Biju, V., Itoh, T., Anas, A., Sujith, A., & Ishikawa, M. (2008). Semiconductor quantum dots and metal nanoparticles: Syntheses, optical properties, and biological applications. *Analytical and Bioanalytical Chemistry*, *391*, 2469–2495.

Blumling, J., & Silva, G. A. (in press). Diagnostic and therapeutic nanotechnologies with applications to the retina and optic nerve. *Journal of Neuro-Ophthalmology*.

Chan, W. C., Nie, S. (1998). Quantum dot bioconjugates for ultrasensitive nonisotopic detection. *Science*, *281*, 2016–2018.

Dahan, M., Levi, S., Luccardini, C., Rostaing, P., Riveau, B., & Triller, A. (2003). Diffusion dynamics of glycine receptors revealed by single-quantum dot tracking. *Science*, *302*, 442– 445.

Fan, H., Leve, E. W., Scullin, C., Gabaldon, J., Tallant, D., Bunge, S., et al. (2005). Surfactant-assisted synthesis of water-soluble and biocompatible semiconductor quantum dot micelles. *Nano Letters*, *5*, 645–648.

Gao, X. (2003). Molecular profiling of single cells and tissue specimens with quantum dots. *Trends in Biotechnology*, *21*, 371–373.

Howarth, M., Takao, K., Hayashi, Y., & Ting, A. Y. (2005). Targeting quantum dots to surface proteins in living cells with biotin ligase. *Proceedings of the National Academy of Sciences, US A*, *102*, 7583–7588.

Jaiswal, J. K., Goldman, E. R., Mattoussi, H., & Simon, S. M. (2004). Use of quantum dots for live cell imaging. *Nature Methods*, *1*, 73–78.

Michalet, X., Pinaud, F. F., Bentolila, L. A., Tsay, J. M., Doose, S., Li, J. J., Sundaresan, G., Wu, A. M., Gambhir, S. S., Weiss, S. (2005). Quantum dots for live cells, in vivo imaging, and diagnostics. *Science*, *307*, 538–544.

Pathak, S., Cao, E., Davidson, M. C., Jin, S., & Silva, G. A. (2006). Quantum dot applications to neuroscience: New tools for probing neurons and glia. *Journal of Neuroscience*, *26*, 1893–1895.

Pathak, S., Davidson, M. C., & Silva, G. A. (2007). Characterization of the functional binding properties of antibody conjugated quantum dots. *Nano Letters*, *7*, 1839–1845.

Pathak, S., Tolentino, R., Kim, N., D'Amico, L., Barron, E., Cheng, L., et al. (2009). Quantum dot labeling and imaging of glial fribrillary acidic protein intermediate filaments and gliosis in the rat neural retina and dissociated astrocytes. *Journal of Nanoscience and Nanotechnology*, *9*, 5047–5054.

Provenzale, J. M., & Silva, G. A. (2009). Applications of nanoparticles to central nervous system imaging and therapy. *American Journal of Neuroradiology*, 10.3174/ajnr. A1590:1-9.

Silva, G. A. (2004). Introduction to nanotechnology and its applications to medicine. *Surgical Neurology*, *61*, 216–220.
Silva, G. A. (2005). Nanotechnology approaches for the regeneration and neuroprotection of the central nervous system. *Surgical Neurology*, *63*, 301–306.
Silva, G. A. (2006). Neuroscience nanotechnology: Progress, opportunities and challenges. *Nature Reviews. Neuroscience*, *7*, 65–74.
Silva, G. A. (2007). Nanotechnology approaches for drug and small molecule delivery across the blood brain barrier. *Surgical Neurology*, *67*, 113–116.
Silva, G. A. (2008a). The central nervous system. *Drug Discovery Today: Disease Models*, *5*, 1–3.
Silva, G. A. (2008b). Nanotechnology approaches to crossing the blood-brain barrier and drug delivery to the CNS. *BMC Neuroscience*, *9*(Suppl. 3), S4.
Vu, T. Q., Maddipati, R., Blute, T. A., Nehilla, B. J., Nusblat, L., & Desai, T. A. (2005). Peptide-conjugated quantum dots activate neuronal receptors and initiate downstream signaling of neurite growth. *Nano Letters*, *5*, 603–607.
West, J. L., Halas, N. J. (2003). Engineered nanomaterials for biophotonics applications: improving sensing, imaging, and therapeutics. *Annu. Rev. Biomed. Eng. 5*, 285–292.
Wu, X., Liu, H., Liu, J., Haley, K., Treadway, J., Larson, J., et al. (2003). Immunofluorescent labeling of cancer marker Her2 and other cellular targets with semiconductor quantum dots. *Nature Biotechnology*, *21*, 41–46.
Yu, D., & Silva, G. A. (2008). Stem cell sources and therapeutic approaches for central nervous system and neural retinal disorders. *Neurosurgical Focus*, *24*, E11.

H.S. Sharma (Ed.)
Progress in Brain Research, Vol. 180
ISSN: 0079-6123

CHAPTER 3

Colloidal systems for CNS drug delivery

Luca Costantino*, Giovanni Tosi, Barbara Ruozi, Lucia Bondioli, Maria Angela Vandelli and Flavio Forni

Department of Pharmaceutical Sciences, University of Modena and Reggio Emilia, Modena, Italy

Abstract: The pharmaceutical treatment of central nervous system (CNS) disorders is the second largest area of therapy, following cardiovascular diseases. Nowadays, noninvasive drug delivery systems for CNS are actively studied. The development of these new delivery systems started with the discovery that properly surface-engineered colloidal vectors, and in particular liposomes and polymeric nanoparticles, with a diameter ~200 nm, were shown to be able to cross the blood–brain barrier (BBB) without apparent damage, and to deliver drugs or genetic materials into the brain. However, even if this ability was confirmed by confocal microscopy and measured by biodistribution experiments or by means of the pharmacological effect exerted by the embedded drugs, a clear understanding of the main characteristics of the colloidal systems that are important for BBB crossing is still lacking. It is also shown that the presence of the drug is able to modify the surface of these systems, with unpredictable results on the colloidal systems biodistribution; thus, the results obtained in the absence of the loaded drug have to be taken cautiously. Moreover, since the loaded drug is only a fraction of the colloidal system that is administered, the presence of the carrier in the body and into CNS, especially in the case of long-term therapies, might cause adverse effects not yet fully understood. Thus, even if promising results have been obtained, and some colloidal systems loaded with a drug are the US Food and Drug Administration (FDA) approved for human use (but not for brain targeting), a long way of research has to be done in order to use these drug delivery systems for the treatment of CNS pathologies.

Keywords: nanoparticles; liposomes; blood–brain barrier; drug delivery; central nervous system

Introduction

The pharmaceutical treatment of central nervous system (CNS) disorders is the second largest area of therapy, following cardiovascular disease. Moreover, the development of new drug molecule is expensive and time-consuming (Bawa, 2008). Hence, there is a need for evolving an existing drug molecule from a conventional form to a novel delivery system that can significantly improve its performance in terms of activity, efficacy by targeting to the specific site, safety, and patients' compliance.

In order to penetrate to the brain tissue, CNS drugs must pass through the blood–brain barrier (BBB). Many of the compounds that otherwise

* Corresponding author.
Tel.: +39-059-2055125; Fax: +39-059-2055131;
E-mail: luca.costantino@unimore.it

DOI: 10.1016/S0079-6123(08)80003-9

would be effective in treating CNS diseases are excluded from reaching a sufficient concentration in the brain tissue and producing the desired therapeutic effect. Thus, drug delivery to the brain is a challenge, because this tissue benefits from a very efficient protective barrier. Even if several disorders and diseases that affect the brain can lead to some loss of BBB integrity, this barrier represents an insurmountable obstacle for many drugs able to treat pathologies in which the BBB integrity is preserved (Kerns & Di, 2008).

The delivery of the drug unable to cross the BBB to the brain has been traditionally approached with either medicinal chemistry or neurosurgical-based invasive brain drug delivery. Nowadays, noninvasive delivery systems are actively studied. The development of these new delivery systems started with the discovery that colloidal systems, and in particular liposomes and polymeric nanoparticles, with a diameter ~200 nm, thus much greater than drug molecules (it is generally recognized that in order for the passive transcellular BBB permeation of compounds to be possible, the molecular weight should be less than 500 Da) were shown to be able to cross the BBB without apparent damage. These colloidal systems can be useful for the targeting of large molecule therapeutics (peptides, antisense drugs, etc.). Due to their poor stability in biological fluids, besides their poor pharmacokinetic characteristics, they might be formulated in these protective nanocontainers.

However, even if the research showed that it is possible to target CNS with nanoparticulate systems, there are some problems connected to the development of these carriers. The main one is represented by the carrier's interaction with the reticuloendothelial system (RES) that exerts a rapid clearance from blood circulation. This defensive system is activated by the recognition of these nonself circulating elements by the opsonins, which make effective a setting in motion of macrophages. This activation is linked to the size and to the surface properties of the nanosystems. Another problem, the selective delivery to CNS among other districts of the body, is at present under active study.

Classification of colloidal carriers for CNS drug delivery

Several colloidal carriers have been developed as drug delivery agents for CNS (Fig. 1):

A) Nanoparticles (Np): polymeric and solid lipid nanoparticles (SLNs): nanoparticles are solid colloid matrix-like particles ranging in size from 1 to 1000 nm, made of polymers (Soppimath, Aminabhavi, Kulkarni, & Rudzinski, 2001) (polymeric Np), lipids (Kaur, Bhandari, Bhandari, & Kakkar, 2008) (lipid Np), or serum albumin (Hawkins, Soon-Shiong, & Desai, 2008) (serum albumin Np). The loaded drugs can be released in a controlled manner through surface or bulk erosion, diffusion through the polymer matrix, swelling followed by diffusion, or in response to local environment (the release kinetic depends on the drug physicochemical characteristics).
B) Polymeric micelles: nanostructures formed by amphiphilic copolymers having an A–B diblock structure with (A) the hydrophilic (shell) and (B) the hydrophobic polymers (core) (Liu et al., 2008).
C) Liposomes: small vesicles composed of unilamellar or multilamellar phospholipid bilayers surrounding aqueous compartments (Bangham, Standish, & Watkins, 1965). Liposomes are surely the most studied colloidal systems; they are applied in different fields of medical research and represent the first innovative approach for the treatment of challenging pathologies, like cancer, HIV, strokes, and many other diseases which require selective therapies. However, the good *in vitro* results are not related to the *in vivo* results, leading to a very low applicability of the liposome approach in the treatment protocols, owing to some drawbacks, connected with the *in vivo* delivery. These drawbacks are mainly represented by the very fast elimination and degradation when injected into the bloodstream and the metabolism of lipids which are constituents of the liposomes. Moreover, differently to the Np, they are unable to maintain therapeutic drug

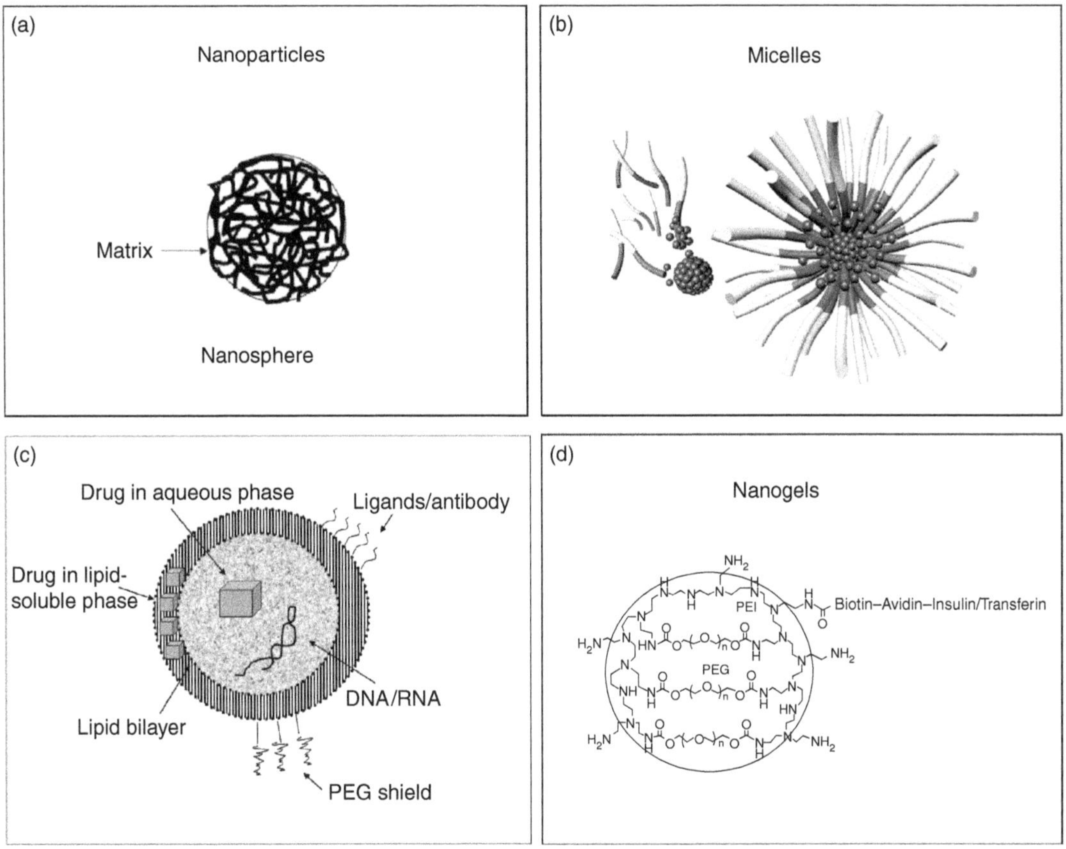

Fig. 1. Drug delivery agents for CNS.

concentrations for a prolonged time (they are unable to exert a sustained release of drug) (Senior, 1987).

Polymersomes have an architecture similar to that of liposomes, but they are composed of synthetic polymer amphiphiles, including polyesters poly(D,L-lactic acid) (PLA)-based copolymers (Discher & Eisenberg, 2002; Discher et al., 1999).

D) Nanogels: nanoscale size polymer network of cross-linked ionic polyethyleneimine and nonionic poly(ethylene glycol) (PEG) chains (Kabanov & Batrakova, 2004) used for the delivery of anionic oligodeoxynucleotides (ODNs) to the brain.

To date, several liposomal formulations are present in therapy (see Table 1); among other colloidal systems, an albumin-based Np formulation (albumin-bound paclitaxel Np, Abraxane, 130 nm diameter) has been approved by FDA in 2005 for the treatment of breast cancer in patients who fail combination chemotherapy for metastatic disease or relapse within 6 months of adjuvant chemotherapy. Albumin is a natural carrier of hydrophobic molecules, and it is thought to facilitate endothelial transcytosis of unbound and albumin-bound plasma constituents (Hawkins et al., 2008). None of these systems target the brain.

Assays for the assessment of the BBB crossing by colloidal systems

Several methods have been used in order to determine the ability of the colloidal carriers to cross the

Table 1. Liposomal formulation approved by FDA

Active substance	Product name	Indication	FDA approval date
Doxorubicin	Doxil[a]	Refractory Kaposi's sarcoma; ovarian cancer; recurrent breast cancer	1995
Doxorubicin	Caelyx[a]		1995
Daunorubicin	Daunoxome	Kaposi's sarcoma	1996
Doxorubicin	Myocet[b]	Combinational therapy of recurrent breast cancer	Approved in Europe and Canada
Amphotericin B	Ambisome	Several fungal infections	1997
Propofol	Deprivan	Anesthetic	1989
Cytarabine	Depocyt	Lymphomatous meningitis	1999
Verteporfin	Visudyne	Treatment of the wet form of age-related macular degeneration (AMD)	2000
Econazole	Pevaryl Lipogel	Cutaneous candidiasis	Not available
Influenza surface antigen inactivated vaccine	Inflexal V[c]	Influenza vaccine	1997
Hepatitis A antigen inactivated vaccine	Epaxal[b,c]	Hepatitis A vaccine	Available in Canada and elsewhere
Vincristin	Onco TCS	Non-Hodgkin's lymphoma	2005

[a] PEGylated liposome.
[b] Not yet approved by the FDA.
[c] Virosome.

BBB; among these, the *in vivo* assays are widely used. Following this method, a colloidal system carrying a labeled compound (fluorescent or radioactive) is administered i.v.; then, in the case of a fluorescent marker, confocal and fluorescent microscopies of brain tissue (Figs. 2 and 3) reveal the presence of the colloidal systems in brain parenchyma (qualitative assay); in the case of a radioactive

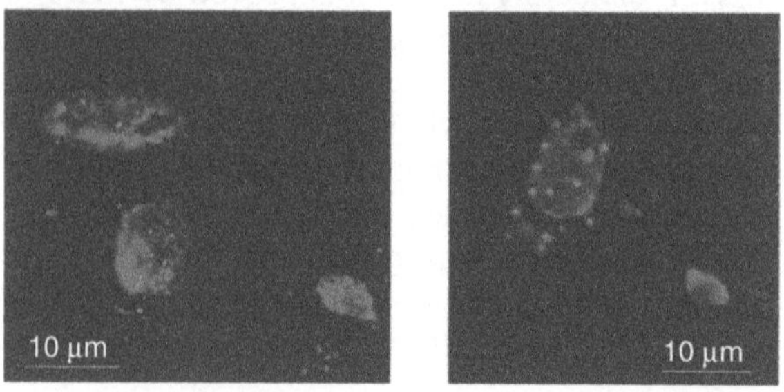

Fig. 2. Colocalization confocal image of peptide-decorated Np (Costantino et al., 2005) in rat cerebral parenchyma after intravenous (i.v.) administration; Np (red-spots) have been found in CNS, close to nuclei (blue structures due to DAPI-labeling). The colocalization study is based on the images obtained by analysis of the deepness of the sample (30 slides in a 5-μm sample) and evaluated in the colocalized points (white spots).

marker, the radioactivity is determined in the organ homogenates and the assay can be quantitative.

Another *in vivo* assay that avoids the problems linked to the biodistribution of the colloidal systems in the whole animal is the "*in situ* rat brain perfusion technique" (Takasato, Rapaport, & Smith, 1984). This method provides high-quality *in vivo* BBB permeability data; following this method, a catheter is placed in the common carotid artery of an anesthetized animal, and the external carotid artery is ligated, sometimes along with the pterygopalatine, while internal carotid remains open. Then the perfusate containing the colloidal systems replaces the blood, and at the end of the experiment the brain is removed and the analyte in the half of the brain that was perfused is measured (Koziara, Lockman, Allen, & Mumper, 2003; Lockman, Koziara, Mumper, & Allen, 2004; Lockman, Koziara, et al., 2003; Lockman, Oyewumi, et al., 2003). Moreover, a carotid arterial infusion technique coupled with a capillary depletion method (Triguero, Buciak, & Pardridge, 1990) can be used to determine the effective passage of the BBB, avoiding the presence of the carriers into the microvasculature.

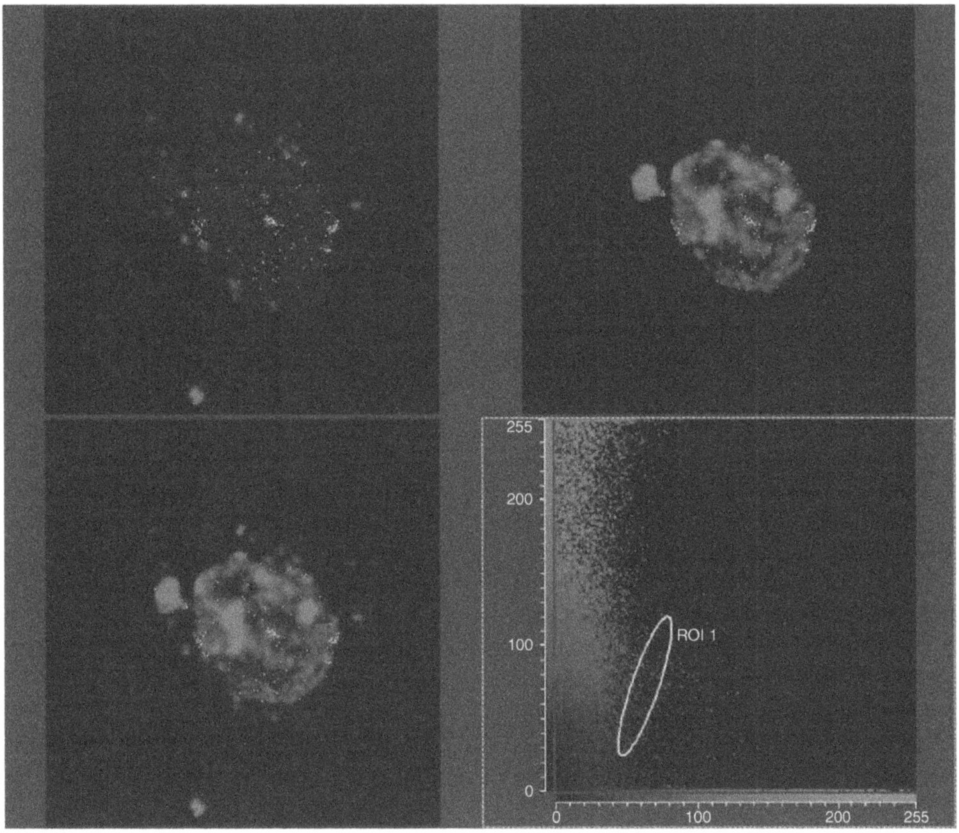

Fig. 3. Fluorescent microscopy image of peptide-decorated Np (Costantino et al., 2005) into rat brain parenchyma after i.v. administration: red spots represent Np, while blue structures represent nuclei (DAPI-labeling) of the cerebral parenchyma.

Cells-based *in vitro* methods comprise the bovine microvessel endothelial cell (BMEC) method, in which the endothelial cells of brain microvessels are cultured, forming monolayers with tight junctions, or the coculture methods (all these two methods are based on primary cultures). Following these methods, the expression levels of transporters can vary from preparation to preparation. Recently, a rat BBB model that shows a good *in vitro/in vivo* correlation has been developed that is based on a coculture of rat brain endothelial cells (RBECs) and rat astrocytes, a model that will help in understanding the mechanism of colloidal carriers translocation and in designing new types of colloidal carriers as brain delivery systems (Garcia-Garcia, Gil, Andrieux, Desmaele, & Nicolas, 2005).

Parameters of colloidal systems involved in the interaction with biological systems

To date, even if colloidal systems were shown to be able to act as CNS drug delivery agents, an exhaustive knowledge of these systems is still lacking. The parameters involved in the interactions of the colloidal systems with body components (besides the presence of specific ligands for body receptors) are

- *Shape*

  While the interaction of spherical particles with cells and within animals has been studied extensively (the preparations of colloidal systems, especially for CNS studied to date, led always to spherical objects), the effect of shape received little attention (Champion,

Katare, & Mitragotri, 2007). Recently, filomicelles were shown to possess different pharmacokinetic characteristics with respect to their spherical counterpart. They were taken up by cells and were shown to be able to act as drug delivery agents (Geng et al., 2007). Thus, it has to be expected that the "shape" issue will be extensively studied in the future.

- *Size (and the polydispersion)*

  Since the biodistribution and the pharmacokinetics depend on the physicochemical properties of particles, and on their size (see for example, Gaumet, Gurny, & Delie, 2007), it is difficult to validate the data from biodistribution studies without accurate particle size determination (a combination of methods, one of which should be a microscopic method, is highly recommended (Gaumet, Vargas, Gurny, & Delie, 2008); moreover, a proper determination of size distribution (and, obviously, samples with a narrow Polydispersity indices) is needed in order to correlate effect and size. Size is an important parameter to be considered for BBB crossing, since even unfunctionalized gold Np are able to cross the BBB; a systematic study about the biodistribution of gold particle size showed that those with a diameter of 15 and 50 nm, the smallest ones among those studied, are able to cross BBB (Sonavane, Tomoda, & Makino, 2008).

- *Surface charge (zeta potential, z-p)*

  The surface charge of the colloidal systems could vary in function of the ions present in the medium and in the different buffers, distilled water, or plasma in which the Np charge is evaluated, either for albumin particles (Roser, Fischer, & Kissel, 1998) or for poly(D,L-lactide-co-glycolide) (PLGA) Np (Panyam, Zhou, Prabha, Sahoo, & Labhasetwar, 2002). The effect of this parameter on BBB permeability is not clear, and it is difficult to correlate this parameter with the results obtained *in vivo*, because, in the case of polymeric Np, it is determined very often in distilled water or buffers, or in other solutions; thus the value that has been obtained might not reflect the *in vivo* situation, in which the Np adsorb proteins from plasma at pH 7. In the case of liposomes that target CNS, this parameter has never been determined. Moreover, it has been reported that the binding, uptake, and transcytosis of 60-nm porous Np is dependent not only on their charge but also on their inner composition; it has been suggested that the Np charge modifies the mechanism of BBB crossing (studies performed using poysaccharidic, 60 nm diameter, Np; z-p measured in 15 mM NaCl; experiments performed in coculture of BCECs and glial cells) (Jallouli, Paillard, Chang, Sevin, & Betbeder, 2007). Some systematic studies on the effect of charge on BBB permeability were conducted: the biodistribution of SLNs loaded with labeled etoposide intravenously administered in mice (Reddy, Sharma, Chuttani, Mishra, & Murthy, 2004) showed that positively charged SLNs (z-p +5 mV, in the presence of the drug, pH 7.4 phosphate-buffered saline (PBS); diameter 391 nm) possess a higher brain accumulation than the negatively charged ones (z-p −47 mV, in the presence of the drug, pH 7.4 PBS; diameter 362 nm). The etoposide brain concentration using positively charged SLN was 10- and 14-fold higher than the negatively charged ones, at 1 and 4 h after administration, respectively (Reddy et al., 2004).

  The favorable effect of positive charges on Np was also supported by the pharmacokinetic studies after i.v. administration of Clozapine-loaded tripalmitin stearylamine SLN in mice, with (z-p +23.2 mV, pH 6.8 to 7.0; 163 nm) and without stearylamine (z-p +0.2 mV in the presence of the drug, pH 6.8 to 7.0; 233 nm) (Manjunath & Venkateswarlu, 2005). The positively charged SLN preparations demonstrated the highest relative bioavailability in the brain in comparison with other organs (liver, spleen, heart, kidney).

  Another work on the evaluation of the effect of charge or lipid coating on the BBB-crossing ability by Np was conducted *in vitro* on brain capillary coculture (Fenart et al., 1999). In this work, neutral, anionic, and cationic nanoparticles made of cross-linked maltodextrins derivatized or not (neutral) with phosphates (anionic), quaternary ammonium (cationic) ligands, or coated with lipids shown to possess different behaviors: lipid coating of ionically charged Np was able to increase BBB crossing three- to fourfold in comparison with uncoated Np, while coated neutral Np did not cross the BBB efficaciously.

Moreover, cationic bovine serum albumin (CBSA), which was investigated in isolated brain capillaries and evaluated with internal carotid perfusion/capillary depletion *in vivo*, indicated a good accumulation profile in the brain (Kumagai, Eisenberg, & Pardridge, 1987; Triguero et al., 1990). CBSA appeared to have favorable pharmacokinetic properties with a longer serum half-life and greater degree of selectivity to brain tissue as compared to other organs (liver, heart, lungs) (Bickel, Yoshikawa, & Pardridge, 2001). The mechanism of BBB crossing seems to be absorptive-mediated transcytosis (AMT), an endocytosis initiated by the binding of polycationic substances to negative charges on the plasma membranes (Gonatas, Stieber, Hickey, Herbert, & Gonatas, 1984).

In a complementary work (Wei, Tanb, Hua, & Jiang, 2005), CBSA conjugated with PEG-poly (lactide) (PEG-PLA) nanoparticle (CBSA-NP, z-p −8.92 mV, NaCl solution (1 mM); 84.4 nm diameter), was designed for brain drug delivery. The results demonstrated little toxicity of high amounts (200 μg/mL) of CBSA-Np against BCECs (200 μg/mL of CBSA-Np) and a permeability of CBSA-Np about 7.76 times higher than that of BSA-Np.

Some researches demonstrated that Np surface charge could strongly alter BBB integrity and permeability (Lockman et al., 2004); in fact it was found that after *in situ* brain perfusion, neutral (−14 mV, filtered water; diameter 75 ± 53 nm) SLN showed no effects on the BBB integrity along with low amount (10 μg/mL) of negatively (−60 mV, filtered water; diameter 127 ± 71 nm) charged SLN; on the contrary, higher amount of negatively charged SLN (20 mg/mL) and positively (+45 mV, filtered water; diameter 97 ± 69 nm) charged SLN were found to produce loss of BBB integrity. In this case the standard deviation in the diameter of the Np has been included in order to underline the weakness of the conclusions that can be drawn on the basis of this kind of Np.

This parameter is very important for the interaction with RES that removes the colloidal systems from the blood. In fact, less phagocytosis by macrophages was observed with nanovectors provided with z-potential close to zero (Roser, Fisher & Kissel, 1998; Tabata & Ikada, 1988).

- *Surface hydrophilicity*

This parameter (Gaumet et al., 2007), important for the *in vivo* distribution of particulate drug carriers, able to influence the adsorption of plasma proteins (Gessner et al., 2000) has never been investigated for the Np able to target CNS.

- *Role and amount of surfactants present on polymeric Np surface*

Another aspect that has been scarcely studied so far is the determination of the quantity of surface-active agents that is present on the Np surface and the kinetic of its release after the i.v. administration of the Np. The surface-active agents, commonly employed in the preparation of the polymeric Np (such as polysorbate 80 (PS-80), Pluronic F-68, and poly(vinylalcohol) (PVA)) are able to exert biological effects, besides the adsorption of apolipoproteins, key factor for the ability of BBB crossing. In fact Pluronic inhibits P-glycoprotein-mediated drug efflux system (polymeric micelles of Pluronic improve brain distribution of drugs) (Batrakova & Kabanov, 2008).

Residual PVA associated with PLGA Np affects their physical properties and cellular uptake (Sahoo, Panyam, Prabha, & Labhasetwar, 2002). Moreover, it was noted that residual PVA is able to influence different pharmaceutical properties of the polyester PLGA Np, such as particle size, z-p, polidispersity index, surface hydrophobicity, and protein loading, and also slightly influences the *in vitro* release of the encapsulated protein (Sahoo et al., 2002). Even if PVA has never been used for CNS Np, these findings strongly suggest the need of a careful study about the effects of the surfactants present on Np that target CNS and their amount.

- *Conformation, thickness, and density of PEG chains present on the colloidal surface*

  In order to obtain a coating that prevents opsonization, the subsequent recognition by the macrophages of the RES, and thus their clearance from the blood, PEG was covalently attached to the surface of the colloidal systems; PEG can assume mushroom or brush conformations, and these influences in a different way complement activation and phagocyte recognition (Owens & Peppas, 2006). Moreover, most research indicates that a surface PEG chain molecular weight of 2000Da or greater is required to achieve increased mononuclear phagocytic system (MPS)-avoidance characteristics (Owens & Peppas, 2006). It is known that the presence of PEG is able to confer to colloidal systems the ability to cross the BBB, but it is not known which characteristics have to possess this coverage (density, conformations, length of the PEG chains).
- *Surface heterogeneity (in the case of polymeric Np)*

  The Np can be heterogeneous with respect to their surface properties, a phenomenon that can influence the interactions with biological systems (Moghimi & Szebeni, 2003). Thus this parameter should be considered for the Np able to target CNS.

These parameters have been deeply investigated in the case of the interaction of the colloidal systems with macrophages (Chellat, Merhi, Moreau, & Yahia, 2005; Moghimi & Szebeni, 2003; Vonarbourg, Passirani, Saulnier, & Benoit, 2006), but for delivery agents that target CNS, they have not always been determined (see Tables 2 and 3), and a systematic study is lacking; thus, it is impossible to draw appropriate conclusions about their importance on BBB crossing. Moreover, studies of these characteristics in the presence of the drug that has to be delivered are very few. This is a very important situation since a part of the drug remains adsorbed on the surface of the colloidal systems, thus modifying their characteristics.

For example, a comparison between unloaded and doxorubicin-loaded PEG- poly(hexadecylcyanoacrylate) (PHDCA) Np showed the marked effect exerted by the drug (and also by the binding of serum proteins) on diameter and charge. Unloaded Np: diameter 139.5 ± 32.3 nm (after contact with serum 158.4 ± 54.2 nm); z-p: –24.4 mV (after contact with serum z-p -29.5 ± 0.2 mV); doxorubicin-loaded Np: diameter: 175.9 ± 32.5 nm (after contact with serum: plurimodal, indicating that aggregation occurred); z-p: $+15.5 \pm 0.5$ mV (after contact with serum z-p -29.1 ± 0.3 mV). In comparison with the unloaded Np, the doxorubicin-loaded Np featured a completely different biodistribution profile in the MPS, as well as their extravasation in the 9L gliosarcoma interstitium and plasma concentration. In this case a negative preclinical result was observed in an orthotopic murine brain tumor model; the Np accumulated to a 2.5-fold lesser extent into 9L tumor as compared to the unloaded Np. This effect has been attributed to the presence of the drug, which modifies Np surface, as evidenced by the z-p measured in water (Brigger et al., 2004); the effect on z-p and the modification of Np biodistribution caused by the presence of the drug (doxorubicin) were noted also for poly(butylcyanoacrylate) (PBCA)-PS-80 Np (Ambruosi, Gelperina, et al., 2006).

It was also shown that this drug, depending on the surfactant type present on the Np (Poloxamer 188 (Pluronic F-68) or PS-80 (Tween 80)), modifies the plasma protein adsorption on Np (loaded during emulsion–polymerization procedure), which in turn could modify Np pharmacokinetic (Petri et al., 2007).

Thus, even if promising results for CNS drug delivery have been obtained with colloidal systems, more in-depth studies are needed in order to fully characterize these systems.

A section dealing with the preparations of the most studied colloidal systems for CNS drug targeting (mainly polymeric Np and liposomes) has been included in order to allow the understanding of some of the unresolved problems that affect these colloidal carriers. Moreover, it is emerging that data obtained in the absence of loaded drugs have to be considered cautiously since the drug

Table 2. Successful Np drug delivery to CNS by the Np. The release kinetic of the embedded drug has never been determined

Drug	Polymer	Loading	Np diameter (nm)	z-potential (mV)	Assay (i.v. administration)	Drug content resp. to polymer (%)	References
Doxorubicin. HCl	PBCA + Polysorbate 80*	Polymerization of BCA/Dextran 1% in presence of drug (pH acidic)	270	ND	Drug levels in brain in healthy rats	ND	Gulyaev et al. (1999)
Doxorubicin	PBCA + Polysorbate 80	Polymerization of BCA/Dextran 1% in presence of drug (pH acidic)	270		Glioblastoma	ND	Steiniger et al. (2004)
Doxorubicin	PBCA + Polysorbate 80	Polymerization of BCA/Dextran 1% in presence of drug (pH acidic)	213 and 215	ND	Rat glioma	23	Ambruosi, Gelperina, et al. (2006)
Doxorubicin	PBCA + Polysorbate 80	Adsorption of drug onto preformed Np	210 (unloaded); 225 (drug loaded)	+ 8mV	Biodistribution	2.9	Ambruosi, Khalansky, et al. (2006)
MRZ 2/576	PBCA + Polysorbate 80*	Polymerization of BCA/Dextran 1% in presence of drug (pH acidic)	228	ND	MES test after i.v. administration	ND	Friese, Seiller, Quack, Lorenz, and Kreuter (2000)
Dalargin ((Tyr-D-Ala-Gly-Phe-Leu-Arg)	PBCA + Polysorbate 80	Adsorption on preformed Np (PBCA/Dextran)	230	ND	Antinociceptive assay	1.35	Kreuter, Alyautdin, Kharkevich, and Ivanov (1995)
Loperamide	Peptide-derivatized PLGA/F68	Nanoprecipitation in the presence of drug	155	–15.2 (Simil plasma fluid pH 7.4)	Antinociceptive assay	15.1	Tosi et al. (2007)

(*Continued*)

Table 2. (*Continued*)

Drug	Polymer	Loading	Np diameter (nm)	z-potential (mV)	Assay (i.v. administration)	Drug content resp. to polymer (%)	References
Loperamide	PBCA + Polysorbate 80*	Polymerization of BCA/Poloxamer188 1% in presence of drug (pH acidic)	290	ND	Antinociceptive assay	1.8	Alyautdin et al. (1997)
Loperamide (not specified the salification)	Albumin ApoA-I/ B100Np	Adsorption on the preformed Np	220–240	–35/–42 (water)	Antinociceptive assay	8.6	Kreuter et al. (2007)
Loperamide (not specified the salification)	Albumin-ApoE Np	Adsorption on the preformed Np	340	ND	Antinociceptive assay	8.4	Michaelis et al. (2006)

*In this case it has been clearly specified that Polysorbate 80 was added to a preformed and drug loaded Np suspension, and the mixture was stirred for a given time before the administration.

Table 3. Successful drug/genetic material delivery to CNS by liposomes

Drug/gene	Lipids	Ligand	Loading of drug	Liposome diameter	Drug content resp. to total lipid (%)	Assay	Release kinetic	References
Doxorubicin	DSPC, CHOL, DSPE, DSPE-PEG, DSPE-PEG-COOH	Tf	pH gradient (Mayer, Bally, & Cullis, 1986) in liposomes obtained by thin layer evaporation method + freeze-thaw extrusion	122 nm	ND	*In vitro* studies (cellular uptake) glioma cells	ND	Eavarone, Yu, and Bellamkonda, (2000)
5-FU	Soya PC, ST, CHOL	Tf	Cast film method followed by sonication (Bangham, 1968)	1.12–2.10 μm	ND	*In vitro* drug release, stability in serum, brain uptake of radiolabeled formulations	66.7 ± 2.91% in 24 h	Soni, Kohli, and Jain (2005, 2008)
Horseradish peroxidise (HRP)	EPC, CHOL, DSPE-PEG, PEG-maleimide-DSPE	Tf	Postinsertion technique: micelles of PEG and PEG-Tf were inserted into preformed liposomes obtained by TLE methods and extrusion	100 nm	5–13 μg of drug/μmol of phospholipid	*In vitro* studies; association of liposomes by brain capillary endothelial cells (BCEC)	ND	Visser et al. (2005)
Doxorubicin	DSPC, DPPE-PEG-amine, CHOL	Folic acid	Ammonium sulfate gradient (Bolotin et al., 1994)	130 nm	ND	*In vitro* experiments; drug uptake and cell death in coculture	ND	Saul, Annapragada, Natarajan, and Bellamkonda (2003)
Daunomycin	DSPC, CHOL, DSPE-PEG-mal	OX26 mAb	pH gradient (Mayer et al., 1986) in liposomes obtained by thin layer evaporation methods + freeze-thaw extrusion	65–115 nm	ND	Pharmacokinetic experiments in rats	ND	Huwyler, Wu, and Pardridge, (1996) and Huwyler, Yan, and Pardridge (1997)

(*Continued*)

Table 3. (*Continued*)

Drug/gene	Lipids	Ligand	Loading of drug	Liposome diameter	Drug content resp. to total lipid (%)	Assay	Release kinetic	References
Luciferase DNA and β-galactosidase DNA	POPC, DDAB, DSPE-PEG, DSPE-PEG-maleimide	OX26 mAb	Plasmid DNA was added to the lipids used to prepare liposomes via TLE methods followed by freeze-thaw extrusion	45–114 nm	ND	Pharmacokinetic and organ uptake after i.v. administration into rats.	ND	Shi and Pardridge (2000) and Shi, Boado, and Pardridge (2001)
SV-40-β-galactosidase; SV-40 luciferase; hEGFR antisense mRNA (clone 882)	POPC, DDAB, DSPE-PEG, DSPE-PEG-maleimide	83-14 mAb	Plasmid DNA was added to the lipids used to prepare liposomes via repetitive freeze/thaw extrusion	85 nm	ND	*In vitro* cellular uptake and intracellular delivery of genes (U87 human glioma cells)	ND	Zhang, Lee, Boado and Pardridge (2002)
SV-40-β-galactosidase; SV-40 luciferase	POPC, DDAB, DSPE-PEG, DSPE-PEG-maleimide	83-14 mAb	Plasmid DNA was added to the lipids used to prepare liposomes via TLE methods followed by freeze/thaw extrusion	85 nm	ND	Brain expression of genes in the monkey brain after i.v. administration	ND	Zhang, Schlachetzki, and Pardridge (2003)
β-galactosidase and luciferase plasmid DNA	POPC, DDAB, DSPE-PEG, DSPE-PEG-maleimide	8D-3 mAb	Plasmid DNA was added to the lipids used to prepare liposomes via TLE methods followed by freeze/thaw extrusion	ND	ND	Distribution and organs uptake after iv administration into mice.	ND	Shi, Zhang, Zhu, Boado, and Pardridge (2001)
EGFR (human epidermal growth factor receptor gene) antisense mRNA	POPC, DDAB, DSPE-PEG, DSPE-PEG-maleimide	8D-3 mAb 83-14 mAb	Plasmid DNA was encapsulated in liposomes with repetitive freeze/thaw cycles	85 nm	ND	*In vivo* experiments; evaluation of EGFR expression within the experimental brain tumors by	ND	Zhang, Zhu, and Pardridge (2002)

GFAP-TH or SV-40-TH plasmid	POPC, DDAB, DSPE-PEG, DSPE-PEG-maleimide	OX26 mAb or 83-14 mAb	Plasmid DNA was encapsulated in liposomes with repetitive freeze/thaw cycles	85–100 nm	ND	*In vitro* and *in vivo* experiments; evaluation of TH gene expression and analysis of brain tumors using immunocytochemistry and confocal microscopy	ND	Zhang, Calon, Zhu, Boado, and Pardridge (2003) and Zhang, Schlachetzki, Zhang, Boado, and Pardridge (2004)
EGFR shRNA (plasmid encodes for a short hairpin RNA)	POPC, DDAB, DSPE-PEG, DSPE-PEG-maleimide	8D-3 mAb 83-14 mAb	Plasmid DNA was encapsulated in liposomes with repetitive freeze/thaw cycles	85–100 nm	ND	*In vivo* experiments: analysis of brain tumors using immunocytochemistry	ND	Zhang et al. (2004)

Abbreviations: HSPC, fully hydrogenated soy phosphatidyl choline; MPEG-DSPE, α-(2-(1,2-distearoyl-sn-glycero(3)phosphooxy)ethylcarbamoyl)-ω-methoxypoly(oxyethylen)-40, sodium salt; DPPE-PEG-amine, 1,2-dipalmitoyl-sn-glycerophosphoethanolamine-poly(ethylene glycol)amine; POPC, 1-palmitoyl-2-oleoyl-sn-glycerol-3-phosphocholine; DDAB, didodecyldimethylammonium bromide; DSPE, distearoylphosphatidylethanolamine; DSPC, distearoylphosphatidylcholine; Soya PC, L-α-Soya phosphatidylcholine; PS, phosphatidylserine; DM, dextran-magnetite; HSPC, hydrogenated soybean phospholipids; CHOL, cholesterol; DSPE-PEG, 1,2-dioleoyl-sn-glycerol-3-phosphor-ethanolamine-n-(poly(ethyleneglycol)); DSPE-PEG-NHS, 1,2-dioleoyl-sn-glycerol-3-phosphor-ethanolamine-n-(poly(ethyleneglycol)-hydroxy succinamide); DSPE-PEG-mal, 1,2-dioleoyl-sn-glycerol-3-phosphor-ethanolamine-n-(poly(ethyleneglycol)-maleimide); ST, sterylamine; EPC, egg phosphatidylcholine.

characteristics modify the Np surface, thus their pharmacokinetics (doxorubicin possesses a positive charge at physiological pH owing to its amino group, and its presence has a marked effect on Np, see below).

Several data regarding BBB crossing by colloidal systems have been obtained in the presence of brain tumor experimental models. It is well known that many CNS diseases affect the BBB status (Garcia-Garcia, Andrieux, Gil, & Couvreur, 2005), and in particular, gliomas are characterized by both neovascularization and vascular hyperpermeability; thus colloidal carriers can target the tumor by passive extravasation across the impaired endothelium (enhanced permeability and retention effect, EPR effect); the results obtained regarding the ability of the colloidal systems to cross the BBB can be overestimated with respect to intact BBB; however, the comparison between the amount of colloidal systems (or drugs embedded into the colloidal systems) present in tumors and the amount of colloidal systems in normal brain regions (the other hemisphere) can give information regarding the ability of these systems to cross the intact BBB (Brigger et al., 2002).

Surface modification of colloidal carriers for CNS drug delivery

In order to be able to cross the BBB, the colloidal systems have to possess appropriate targeting moieties on their surface. The surface modifications of the colloidal systems that have been shown to be able to cross the BBB can be classified into four classes (the strategies marked with an asterisk have been tested in the presence of a loaded drug or genetic material) (Scheme 1).

Surface-active agents

This has been the first successful approach for polymeric Np drug delivery to CNS. The presence of a surface active agent (PS, Pluronic, PVA, etc.) is necessary for the preparation of polymeric Np and stabilization of their suspensions; however, it was recognized very soon that only PS are able to confer to acrylate-based polymeric Np the ability to cross the BBB, while a coating with poloxamers (184, 188, 388, 407), poloxamine 908, cremophors (EZ or RH40), or polyoxyethylene(23)-laurylether was not effective (Alyautdin et al., 1997); this effect has been attributed to PS ability to adsorb ApoE from the plasma (Kreuter et al., 2002); the particles then seem to mimic Low Density Lipoproteins (LDL) particles and could interact with LDL receptors present at the BBB.

Proofs that ApoE is involved in PBCA-PS-80 Np endocytosis are based on the fact that the adsorption of ApoE on the Np surface is specific for the surfactant PS, and also for the CNS activity; moreover, ApoE-deficient mouse showed a reduced pharmacological response to the drug embedded into the Np with respect to wild-type mouse (Kreuter, 2001).

Further proofs about the role of ApoE in BBB crossing derived from the studies of a peptide derived from the LDL receptor-binding domain of ApoE covalently coupled onto PEG-derivatized liposomes. This kind of surface modification is able to mediate the uptake of liposomes into brain capillary endothelial cells (BCECs) (Sauer, Dunay, Weisgraber, Bienert, & Dathe, 2005).

Then, it was hypothesized that the ability of Np to cross the BBB would be due also to the adsorption of apolipoproteins A1 and B100 from the plasma followed by recognition with the scavenger receptor class B type I (SR-BI) or with lipoprotein receptors located at the BBB, respectively (Kreuter, 2001; Kreuter et al., 2007), since the covalent linkage of these proteins to BSA Np allowed the obtainment of albumin Np able to cross the BBB (Kreuter et al., 2007; Michaelis et al., 2006). The results correlated with the previously observed involvement of apolipoprotein A-I in the brain delivery of doxorubicin loaded in the surfactant-coated PBCA Np (Petri et al., 2007) as well as the interaction of lipid drug conjugate (LDC) Np with the brain vessels endothelial cells (Gessner, Olbrich, Schroeder, Kaiser, & Muller, 2001). Moreover, it was shown that the adsorption of apolipoprotein A-1 on protamine oligonucleotide Np increases particle uptake and transcytosis in an *in vitro* model of BBB (Kratzer, Wernig, Panzenboeck, Bernhart, & Reicher, 2007).

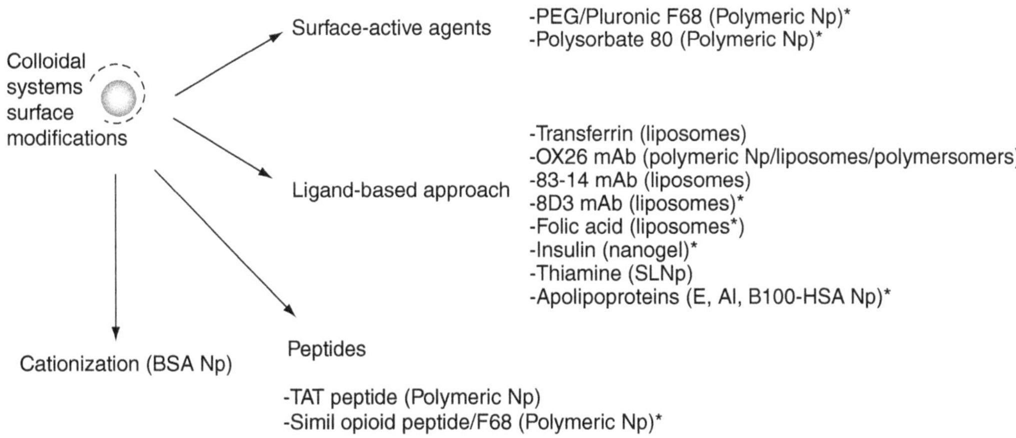

Scheme 1. Modification of colloidal carrier surface in order to target CNS.

It was then discovered that also PEG moieties, that have been covalently linked to the surface of the colloidal carriers because they confer stealth properties (PEG prolongs the blood circulation time of proteins, peptide drugs, liposomes, and nanoparticles) (Harris & Chess, 2003), are able to provide the polymeric Np with the ability to cross the BBB. Even in this case, it was noted that ApoE and ApoB100 adsorb on the surface of poly(methoxypoly (ethylene glycol) cyanoacrylate-co-hexadecyl cyanoacrylate) (PEG-PHDCA 1:4) polymeric Np (Kim et al., 2007). The ability of polymeric Np to bind apolipoproteins is not limited to PEG or PS-80-covered surfaces, since this property is shared also by PEG-PCL and PCL-dextran Np (Lemarchand et al., 2006); however, these Np have not been tested for brain drug delivery.

Moreover, PBCA Np coated with another surface-active agent, Pluronic F-68, considerably enhances the antitumor effect of doxorubicin against an intracranial glioblastoma in rats (Petri et al., 2007). Both surfactants (PS-80 and Pluronic F-68) induced a considerable adsorption of apolipoprotein A-I. However, when used as a coating agent for PBCA Np loaded with dalargin, it failed to enhance delivery of dalargin across the BBB, whereas with the dalargin-loaded Np, PS-80 coating was very effective (Kreuter, Petrov, Kharkevich, & Alyautdin, 1997).

At present, none of these Np has been tested in humans; it should be remembered that there is a species-dependent protein adsorption pattern on Np (Gessner et al., 2001); thus, it is difficult to forecast the success of these Np in humans. Moreover, it was suggested that PS-80-PBCA Np may compromise the integrity of the BBB (Olivier et al., 1999), but this toxic effect has been observed a high PS-80 doses (Gelperina et al., 2002; Kreuter et al., 2003); however, recently, PS-80, used as a solvent for the administration of the antitumor drug docetaxel, has been linked to the appearance of adverse effects (Hawkins et al., 2008).

The examination of the structures of surface-active agents (and polymers covalently linked to them) that showed to be able to confer to the colloidal systems the ability to cross the BBB (Fig. 4) showed that they are very similar, and it could be hypothesized a similar mode of action (adsorption of apolipoproteins from plasma).

However, it is possible that also the polymer type plays a role in polymeric Np BBB crossing ability, because PS-80 coated poly(methylmethacrylate) nanoparticles are not distributed into the brain after i.v. administration (Lode et al., 2001); replacing PS-80-coated PBCA Np with PS-80-coated polystyrene Np completely abolished dalargin brain delivery (Olivier et al., 1999).

PEGylated liposomes have proven their ability to deliver drugs to CNS, but they were studied

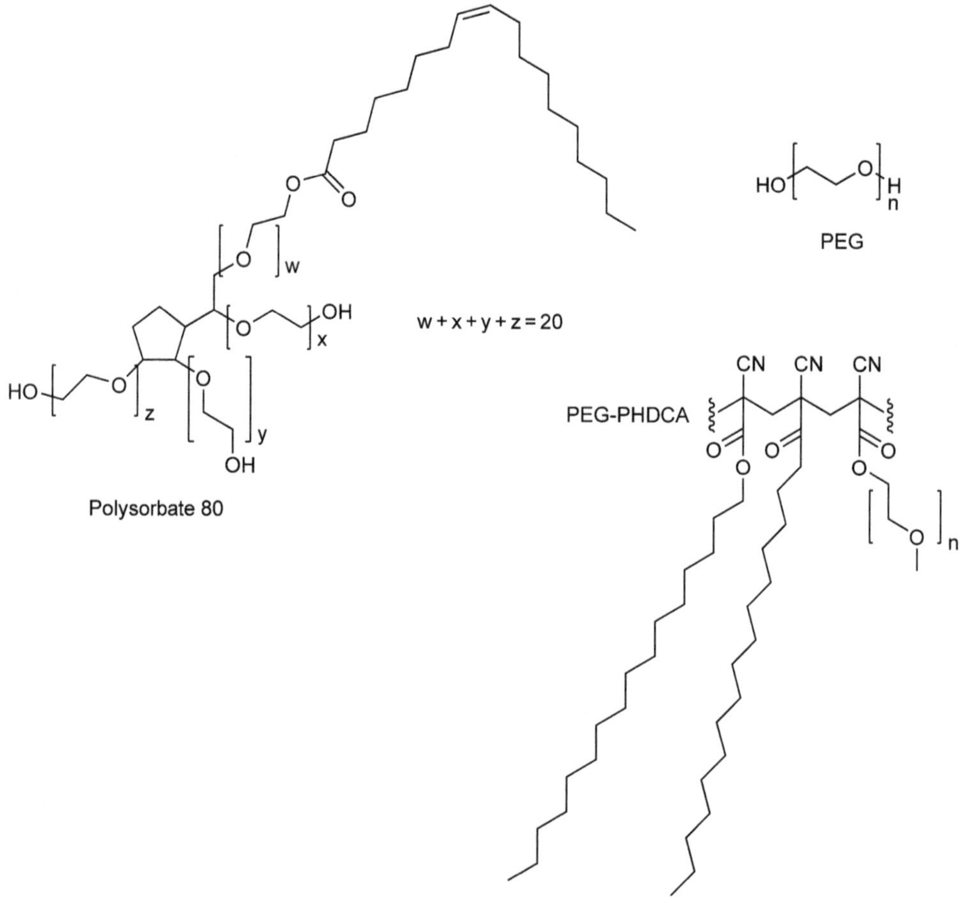

Fig. 4. Structure of surface-active agents (and polymers covalently linked to them) starting material for the preparation of colloidal carriers able to cross the BBB.

only in pathologies in which there is a disruption of the BBB; a liposomal formulation of PEGylated liposomes encapsulating doxorubicin (Caelyx) is used in clinical practice, showing effectiveness in glioblastomas and metastatic tumors; other PEGylated liposomes were found to be useful drug carriers for the treatment of experimental autoimmune encephalitis. No attempts were made in order to determine if these systems are able to bind circulating apolipoproteins (owing to the presence of PEG), or if these liposomes are able to cross the intact BBB. It has been hypothesized that their long circulating properties, owing to reduced clearance by RES system, allow them to selectively extravasate in pathological sites like tumors or inflamed regions with a leaky endothelium (Garcia-Garcia, Andrieux, et al., 2005).

Ligand-based approach

This approach has been followed in order to increase the selectivity for brain targeting by the colloidal systems. It is based on the presence of specific receptors on brain capillaries, such as receptors for insulin, insulin-like growth factors, angiotensin II, atrial natriuretic peptide, brain natriuretic peptide, interleukin 1 and transferrin (Tf); thus these molecules were conjugated with the cargo (colloidal particles) in order to access the brain parenchyma via receptor-mediated transcytosis (Bickel et al., 2001). Following the interaction with the receptor, the ligands are internalized by means of receptor-mediated endocytosis.

Transferrin

Tf is a Fe-binding protein; the transferrin receptor (TfR) is heterogeneously distributed into the brain (de Boer, van der Sandt, & Gaillard, 2003). The unusually high expression of Tf and insulin receptors on the surface of the normal BBB provides a potential advantage for the delivery of drugs into the brain. Various therapeutic agents have been chemically linked to Tf in order to deliver it to brain (Qian, Li, Sun, & Ho, 2002). Tf itself is, however, limited as a brain drug transport vector since the TfRs are almost saturated under physiological conditions due to high endogenous plasma concentration of Tf, while the antibodies that bind to the TfR have been shown to selectively target BBB endothelium due to the high levels of TfR expressed by these cells (Qian et al., 2002). In the field of colloidal systems, this strategy has been applied mainly to liposomes.

Tf-coupled liposomes were used to enhance the delivery of 5-fluorouracil to the brain. The brain uptake mediated by these carriers was found higher when compared with the free drug and noncoupled liposomal formulations (Soni et al., 2005, 2008). Significantly increased gliomal doxorubicin uptake was achieved by drug encapsulation within Tf-coupled liposomes compared to non-coupled liposomes (Eavarone et al., 2000).

Liposomes decorated with Tf loaded with horseradish peroxidase (HRP) are able to translocate across the BBB in *in vitro* experiments (Visser et al., 2005) (BCEC culture).

OX26 Ab

One of the best studied and applied ligand for the CNS drug delivery is the Ab OX26, a monoclonal Ab to the rat TfR (Friden et al., 1991). It was found that there was no *in vivo* competition of the colloidal carriers with endogenous Tf (Huwyler et al., 1996), demonstrating that OX26 recognized a binding site different from the natural ligand Tf (Huwyler et al., 1997; Pardridge, 1995).

The conjugation of OX26 to nanocarriers has been applied in a number of researches due to the significant *in vitro/in vivo* results that showed an increase in brain accumulation. Particularly, Huwyler et al. (1996) demonstrated that OX26 Ab was able to mediate the delivery of daunomycin to the rat brain by using an immunoliposome-based drug delivery system. Moreover, the expression of exogenous gene in rat brain can be achieved after i.v. administration using PEGylated liposomes conjugated with the OX26 mAb (Shi and Partridge, 2000). To explain these results, it is possible to consider the BBB TfR as a bidirectional transcytosis system, mediating the movement of the PEGylated immunoliposomes through the endothelial barrier into the brain interstitial space; PEGylated immunoliposomes use a receptor-mediated transcytosis across the BBB and a receptor-mediated endocytosis into neurons within the brain (owing to the expression of TfR on the neuronal plasma membrane).

This Ab was also linked to PEG-PLA polymeric Np (Olivier, 2005), but this new Np has not been tested *in vivo*. It was successfully used as a vector in the delivery of other large molecules across the BBB (Qian et al., 2002), and to PEG-polycaprolactone polymersomes (self-assembled vesicles of amphiphilic block copolymers with thicker and tougher membranes than lipids) (Pang et al., 2008) (in this case a model peptide was incorporated into these particles (100 nm diameter) and its pharmacological activity assessed in rats: the peptide NC-1900 was able to improve the scopolamine-induced learning and memory impairments via i.v. administration).

8D3 mAb

8D3 mAb is a monoclonal Ab to the mouse TfR active as BBB transport vector in mice. Using this ligand, Shi and colleagues (Shi et al., 2001 Shi, Zhang, et al., 2001) demonstrated the brain-specific expression of different exogenous genes after i.v. administration.

83-14 mAb

83-14 mAb is the murine monoclonal Ab to the human insulin receptor (HIR-mAb). The HIR is expressed at both the BBB interface with human brain gliomas and the plasma membrane of human glioma cells. For gene targeting in humans, the HIR-mAb is preferred over the TfR mAb. Firstly, OX26 mAb (targeting rat TfR) and 8D3 Ab (targeting mouse TfR) are generally active in mice and rats, but they are not active in humans and primates; then, it was proved that the mAb directed at the human insulin receptor is transported across the primate BBB nearly 10-fold faster than the mAb directed at the human TfR (Pardridge, 2001). Thus PEGylated immunoliposomes targeting of an Epidermal growth factor receptor (EGFR) antisense gene to U87 human glioma cells derivatized with the 83-14 murine monoclonal antibody resulted in a 70–80% inhibition in cancer cell growth (Zhang et al., 2002). Moreover, the HIR-mAb PEGylated immunoliposomes that showed a net anionic superficial charge gave levels of gene expression comparable to that observed with a cationic and toxic lipid (lipofectamine).

In a second work, a specific promoter (SV-40) was also added in order to perform brain expression of β-galactosidase gene in primates brain (Zhang et al., 2003). The results on monkeys after an i.v. administration demonstrated a high expression of β-galactosidase gene in the brain; there were also evidences of peripheral gene expression, in particular in those tissues with permeable vasculature (liver and spleen) expressing insulin receptors. Comparing the efficiency of gene expression, the use of HIR-mAb exerted a 50-fold higher expression than TfR-MAb PEGylated immunoliposomes.

An important improvement step in the immunoliposome strategy was realized by the decoration of liposomes with two different peptido-mimetic Ab: one Ab is responsible for the brain tissue targeting and the second induces cellular uptake or increases cellular transport. Zhang et al. (2002), with the aim of delivery mRNA specifically to the brain, applied coupled conjugation technology using 8D3 mAb to TfR (transport across the BBB), and 83-14 mAb to human insulin receptor (improving the transport across the plasma membranes and nuclear membrane of human brain cancer cells). The tests were carried out on rats bearing human glioma cells. The results clearly showed that without the dual targeting it is not possible for the gene to reach a deep localization because the cancer cells lied distal from the microvasculature barrier, while, by using this coupled technology, the *in vivo* experiments showed a 100% survival after weekly administrations. The same approach was recently used in gene therapy using RNA interference (RNAi) toward oncogenic gene such as EGFR (Zhang et al., 2004). The RNAi-encapsulated immunoliposomes approach resulted in a 95% suppression of EGFR function in cultured glioma cells and in an 88% increase in survival time of mice with advanced intracranial brain cancer after i.v. administration.

Moreover, Zhang and colleagues (Zhang et al., 2003; Zhang et al., 2004) applied the PEGylated-immunoliposomal technology to the treatment of Parkinson's disease (PD). The contemporaneous use of brain-specific promoter (GFAP), compared with generic promoter (SV-40), and of mAb for TfR (OX26) and for HIR (83-14 mAb) was evaluated after an i.v. injection of PEGylated immunoliposomes into rats. The results showed a high reduction of the effect of PD, reversing the motor impairment; moreover, the couple OX26 mAb and SV-40 led to an expression not only in the brain, but also in the liver, while the couple OX26 mAb and GFAP avoided the expression in the liver.

Insulin

Insulin is not produced in the brain, but insulin is present in the brain, which arises from the blood via transport across the BBB in the endothelial insulin receptor. This ligand has been used in order to successfully target nanogel loaded with ODN to the brain (Vinogradov, Batrakova, & Kabanov, 2004).

Mannose

Mannose-decorated liposomes, based on the affinity of this carbohydrate for the transporter GLUT-1 that promote the passage of glucose from the blood to the brain, are able to cross the BBB only using a mouse (Umezawa & Eto, 1988); *in vivo* rat brain accumulation was not enhanced in comparison with nondecorated liposomes (Mora et al., 2002).

Thiamine

Thiamine is a water-soluble micronutrient that is essential for normal cell function, growth, and development; it has been considered as a ligand for drug delivery since all eukaryotic cells have a specific transport mechanism for it. On the basis of the BBB thiamine transport capacity and kinetics (Koziara et al., 2003; Lockman et al., 2004), this molecule has been suggested as a brain drug delivery vector (Smith, 1993). Solid lipid Np covered with thiamine (Lockman, Oyewumi, et al., 2003) showed that there is an association between the BBB thiamine transporter (rat brain perfusion technique) and Np; even if these Np are able to enter into the brain, *in vivo* biodistribution experiments (whole animals) did not show any difference with the control Np (without thiamine).

Apolipoproteins

In this case the colloidal carriers functionalized with apolipoproteins mimic lipoprotein particles; they were recognized by receptors on the brain endothelial cells of BBB and act as Trojan horses.

HSA Np covalently linked to ApoE are able to deliver to the brain the drug loperamide (Michaelis et al., 2006); the same effect was obtained with HSA Np covalently linked to ApoA-I and B100 (Kreuter et al., 2007); in this case the delivery to brain can be mediated by the interaction with the SR-BI expressed by the brain vessels' endothelial cells. The same effect is also shared by functionalized liposomes (Sauer et al., 2005, 2006).

Folic acid

Folate receptor is overexpressed in a large number of tumors such as carcinoma and brain tumors (Weitman et al., 1992). The targeting of the folate receptor has been shown considerably promising in mediating uptake of a variety of drugs, and several experiments showed that there is a significant uptake difference by folate-targeted liposomes between cells with overexpression and those with no expression of folate receptors (Anderson, Eliot, Stevenson, & Rogers, 2001; Lee & Low, 1995; Wang, Lee, Cauchon, Gorenstein, & Low, 1995). The last case is represented by brain tumors characterized by low levels of the folate receptors; however, Saul et al. (2003) demonstrated that it is possible to target doxorubicin to the C6 rat glioma cells (that show limited expression of the folate receptor) via folate receptors minimizing nonspecific uptake in normal cells by means of an optimization of the number of targeting ligand on liposomal surface.

Peptides

Few peptides have been studied so far in order to prepare peptide-decorated Np able to cross the BBB.

TAT peptide (YGRKKRRQRRR)

TAT peptide is the protein transduction domain from the transcriptional activator TAT protein of the human immunodeficiency virus type-1 (HIV-1). It was conjugated with polymeric Np (PLGA), and the Np thus obtained were shown to be able to enter into cells in culture (Nam, Park, Han, & Chang, 2002); it was also conjugated to proteins with molecular weight ranging from 36 to 119 KDa (Schwarze, Ho, Vocero-Akbani, & Dowdy, 1999), to quantum dots (Santra et al., 2005), and to PEG-cholesterol (to obtain micelles) (Liu et al., 2008), and the conjugates were shown to be able to cross *in vivo* the BBB; this peptide has also been used to enable transport of fluorescent-labeled silica Np *in vivo* across the BBB (Santra et al., 2004). No data regarding brain selectivity of these decorated colloidal carriers are available.

Simil-opioid peptide (H_2N–Gly-L-Phe-D-Thr-Gly-L-Phe-L-Leu–L-Ser–O-β-D-glucose-$CONH_2$)

This peptide has been discovered on the basis of the ability of the synthetic opioid peptide MMP-2200 (H_2N–L-Tyr-D-Thr-Gly-L-Phe-L-Leu-L-Ser–*O*–β-D-lactose–$CONH_2$) (Polt & Palian, 2001) to cross the BBB; in order to remove the opioid activity, linked to the N-terminal Tyr (Casy & Parfitt, 1986), this amino acid was substituted with Phe to avoid a potential opioid effect. This glycopeptide was then conjugated with PLGA to prepare modified polymers able to form Np. These Np were shown to be able to deliver the model drug loperamide into CNS (it was evaluated the antinociceptive effect) in a much greater quantity with respect to other loperamide-loaded Np previously studied (PS-80-PBCA Np, albumin-ApoE Np, albumin-ApoA-I/ApoB100) (Tosi et al., 2007). The mechanism of BBB crossing of these Np (receptor-mediated endocytosis, adsorptive endocytosis, or others) is at present unknown.

Cationization

The binding of polycationic substances to the negative charges on the plasma membranes or by the binding of extracellular lectins is able to promote transcytosis (AMT) (Gonatas et al., 1984). Studies with CBSA showed a good accumulation profile in the brain (*in situ* perfusion-capillary depletion technique) (Kumagai et al., 1987); the same effect was also shown by CBSA-PLA Np (Lu, Wan, She, & Jiang, 2007) and cationic CBSA-PEG Np (Lu et al., 2005). CBSA appeared to have favorable pharmacokinetic properties with a longer serum half-life and a greater degree of selectivity to brain tissues as compared to other organs (liver, heart, lung) (Bickel et al., 2001), in analogy to CBSA-coupled PEGylated liposomes that were taken up into brain endothelium (Thole, Nobmann, Huwyler, Bartmann, & Fricker, 2002).

Moreover, in an *in vivo* test (Lu, Sun, Wan, She, & Jiang, 2006), a plasmid was loaded on cationic BSA PEGylated Np and injected into rat tail vein; after 1–2 days an important gene expression in the brain glioma parenchyma was observed; repeated i.v. injections of Np carrying the gene promoter-induced apoptosis and significant delayed tumor growth.

Another peptide-based cationized polymer, polylysine, has been used to coat iron oxide Np for their transport across the BBB (Xiang, Zhu, & Li, 2001; Xiang et al., 2003).

Some of these colloidal systems (those marked with an asterisk (*) in Scheme 1) have been tested in the presence of a loaded drug. This is important because the surface properties of the drug-loaded Np are influenced not only by the coating surfactants but also by the drug present on the surface, which in turn can alter the surface properties. The alteration of the nanoparticle surface properties by drugs, as for instance indicated by a significant change in the surface charge (Ambruosi, Yamamoto, & Kreuter, 2005; Brigger et al., 2004), was shown to have a strong influence on the biodistribution (Ambruosi et al., 2005; Brigger et al., 2004; Panzenboeck et al., 2002).

Table 2 reports the pharmacological results of the delivery of drugs to CNS by means of Np. Among the drugs that have been used for the determination of the drug-targeting ability of the colloidal systems by means of the evaluation of the pharmacological effect, loperamide (an opioid unable to cross the BBB) and doxorubicin (an antitumoral drug) have been studied very often.

In the case of peptide-derivatized PLGA Np loaded with loperamide, the animals showed a strong analgesic effect, reaching the maximum of 60% at 240 min after administration and assuring an antinociception effect for 300 min, thus demonstrating a sustained release of the drug into the CNS. This long-lasting effect has never been shown before; in fact, previous studies (Alyautdin et al., 1997; Michaelis et al., 2006) showed that using Np either made of PBCA coated with PS-80 or of albumin (at approximately the same dose), the maximum antinociceptive effect at the same doses of loperamide is recorded early on, respectively, at 15 min and between 15 and 60 min. Then, the antinociceptive effect rapidly decreased over the time of the experiments.

The pharmacological activity that has been evaluated in the presence of doxorubicin was the reduction in growth of brain tumors, but in this case the tumor growth can compromise the BBB function. Similar to other solid tumors, gliomas are characterized by both neovascularization and vascular hyperpermeability (Schlageter, Molnar, Lapin, & Groothuis, 1999; Vajkoczy & Menger, 2004) and the carrier can target the tumor by passive extravasation across the impaired endothelium (EPR effect); thus a direct comparison of the colloidal systems performed on the basis of a pharmacological (antitumoral) activity of a drug could lead to invalid conclusions in the presence of an intact BBB.

Table 3 shows the results obtained in the delivery of drugs to CNS by liposomes.

Ability of the colloidal carriers to cross the BBB

This chapter was included in order to allow a comparison among the biodistributions of the colloidal carriers, focused on CNS localization, as drug delivery agents. This comparison is not based on the pharmacological effect exerted by the embedded drug, because, especially for polymeric Np, drugs can be released with different kinetics. Thus the pharmacological effect exerted by the loaded drugs on different colloidal systems can be different. For example, doxorubicin loaded into PEG-PHDCA Np is released at different rates if the drug is loaded in an acidic solution (drug in the protonated form) or in a basic solution (drug in a more lipophilic neutral form) (Brigger et al., 2004). In general, depending on formulations *in vitro*, drug release from polyesters lasts from a few hours (Allemann, Leroux, Gurny, & Doelker, 1993; Peracchia et al., 1997; Ubrich, Bouillot, Pellerin, Hoffmann, & Maincent, 2004) or a few days (Dong & Feng, 2004) to several weeks (Allemann et al., 1993; Feng, Mu, Win, & Huang, 2004; Fishbein, Chorny, Rabinovich, Banai, & Gati, 2000; Horisawa et al., 2002). Thus, a different pharmacological effect can be obtained.

It is difficult to summarize a direct comparison between the ability of colloidal carriers to reach the brain. In the field of polymeric Np, several pharmacokinetic and biodistribution studies were conducted, using unloaded or drug-loaded Np, and the label for the determination of the amount of Np that reaches the brain was placed on the polymer or it was represented by the embedded drug.

Thus, biodistribution experiments were conducted using labeled polymer (without the embedded drug) or by means of the evaluation of the embedded drug.

Experiments conducted with labeled polymer ([^{14}C] PEG-PHDCA Pluronic F-68 Np (Brigger et al., 2002; Calvo et al., 2001), [^{14}C]-PBCA PS-80 Np (Ambruosi, Khalansky et al., 2006)) showed that the maximum level of Np that reaches the brain is less than 1% of injected dose; in this last case evidence is given that the presence of the embedded drug (doxorubicin) modifies Np biodistribution; the same low level of Np that reaches the brain was obtained in experiments that evaluated the biodistribution of doxorubicin (quantified by HPLC) loaded into PBCA PS-80 Np (Gulyaev et al., 1999). Experiments conducted with ^{3}H-dalargin-loaded PBCA-PS-80 Np showed a little increase in dalargin brain concentration that is significant only after 5 min from their administration (Schroeder, Schroeder, & Sabel, 2000); on the contrary, peptide-decorated Np labeled with rhodamine-123 (a fluorescent dye) are able to deliver it up to 20% of the injected dose into the brain (levels obtained after 1.5 h after i.v. administration into rats) (Vergoni et al., in press). Obviously, in these cases a loss of the embedded substances before the BBB crossing is to be expected; thus the results should be underestimated.

In the case of liposomes, biodistribution studies are fewer than those concerning polymeric Np. Huwyler et al. (1996, 1997) studied the plasma pharmacokinetics and brain delivery of free drug (daunomycin), conventional liposomes, sterically stabilized liposomes, and immunoliposomes after single i.v. injection in the rat. The results suggested that the brain uptake of PEGylated liposomes is zero even if the AUC of drug is greatly increased; on the contrary, the use of PEG-conjugated OX26 immunoliposomes results in a measurable concentration of drug in the brain tissue at 60 min after injection (~0.03% ID/g of brain tissue).

The *in vitro* studies demonstrated the ability of immunoliposomes to associate to the BBB model; studies carried out by Visser et al. (2005) described the ability of some liposomes-Tf encapsulated protein drug (HRP) to associate and bind BCECs four- to eightfold higher than uncoupled liposomes. The comparative uptake between free and encapsulated doxorubicin to C6 glioma cells showed that the drug encapsulated within Tf-PEG liposomes exhibited 70% uptake as compared to 14% of PEG-liposomes, 34% of albumin-PEG-liposomes and 54% conventional liposomes (Eavarone et al., 2000). The permeability of NGF encapsulated into targeted sterically stabilized liposome (NGF-SSL-T) on the *in vitro* model of BBB was twofold higher than that of Nerve Growth Factor-Sterically Stabilized Liposomes (NGF-SSL) (Xie et al., 2005).

In *in vivo* studies, the ability to deliver the drugs was frequently expressed as brain uptake versus controls: the brain uptake of 5-FU encapsulated into Tf-coupled liposomes was 17 times higher as compared to free drug and 13 times higher as compared to noncoupled liposomal formulation (Soni et al., 2005, 2008); the brain concentration of encapsulated NGF by targeted sterically stabilized liposomes (NGF-SSL-T) was 10 times more than that of free NGF and 4 times more than that of sterically stabilized liposomes (NGF-SSL) (Xie et al., 2005); Amphotericin B concentration in the brain after amphotericin-loaded targeted liposomes administration (AmB-L-PEG-RMP 7) showed to be higher than that obtained after administration of Amphotericin loaded PEGylated liposomes (AmB-L-PEG) or of a physical mixture (AmB-L-PEG + RMP 7) (Zhang, Xie, Li, Wang & Hou, 2003).

Brain uptake of other liposomes was determined only in the presence of drug-loaded liposomes, and not in the case of genetic materials. In this last case, the ability of the liposome to carry genetic material was evaluated on the basis of the effect of the loaded genetic material (Shi et al., 2000; Zhang, Lee et al., 2002; Zhang, Schlachetzki, et al., 2004; Zhang, Zhang, et al., 2004; Zhang, Zhu, et al., 2002; Shi, Zhang et al., 2001).

Nanogel surface-modified with insulin or Tf are able, in a mouse model for biodistribution studies *in vivo*, to cause an accumulation of ODN in brain increased by over 15-fold while in liver and spleen decreased by twofold compared to free ^{3}H-ODN 1 h after their i.v. administration (Vinogradov et al., 2004).

Polymeric brain-targeted Np

Type of polymers: polyesters, polyalkylcyanoacrilates, albumin

Since, especially in the case of polymeric Np, a great quantity of polymer is administered with respect to the embedded drug (this value depends on the physicochemical characteristics of the drug; it is always of the order of few parts % with respect to the polymer), a chapter about the polymer characteristics has been included.

The polymer to be used in a nanoparticulate technology should have a high stability in blood, nontoxic qualities, nonthrombogenic and nonimmunogenic activities, absence of activation of neutrophils, a high biodegradability, the possibility of avoiding RES systems, and finally the possibility of a broad spectrum of drugs to be encapsulated, along with a scalable and cheap scale-up process.

Polyesters

PLGA or poly(D,L-lactide) (PLA) polymers are FDA approved and therefore are two of the most promising polymers for the preparation of Np, but clinical trials about the use of these Np as drug carriers are still lacking. The degradation of PLA or PLGA occurs by an autocatalytic cleavage of the ester bonds through spontaneous hydrolysis into oligomers and monomers of lactic and glycolic acids, which are substrates of the Krebs cycle (Li, 1999). Depending on their Molecular Weight (MW) and their conjugation with other polymers (such as PEG) these biodegradable polymers show different times of elimination from the body (half-lives in the order of 3–10 weeks; the experiments were done in phosphate buffer, pH 7.4, at 37°C, depending on the ratio of lactic acid to glycolic acid present in the copolymer, using large specimens of polymeric material (Bazile et al., 1992; Li, 1999). Studies conducted with microspheres (10–40 μm diameter) of PLGA injected into rat brain assessed their biocompatibility to the brain tissue, even if they remain in the brain for about 2 months (Menei et al., 1993; Nicholas et al., 2002). These microspheres are much greater than most of the Np that target the brain (diameter of the order of 0.2 μm); to the best of our knowledge, nothing is known about the degradation time of these small Np.

Researches are in progress in order to determine the effect of peptide modification on biodegradability and toxicity with respect to the starting polymer PLGA.

Alkylcyanoacrylates

One of the most studied polymers used in the nanotechnology applied to CNS delivery of drugs is the alkyl derivates of cyanoacrylate polymer (PACA). In particular, PBCA and PHDCA have been used to prepare nanoparticles for BBB crossing with different technological modification.

PACA are (Vauthier, Dubernet, Fattal, Pinto-Alphandary, & Couvreur, 2003; Vauthier, Labarre, & Ponchel, 2007) at present not approved by the FDA for an i.v. administration, although some of these polymers have been described to be devoid of toxicity (Kante et al., 1982; Kattan et al., 1992; Lherm, Müller, Puisieux, & Couvreur, 1992; Lukowski, Müller, Müller, & Dittgen, 1992; Müller, Lherm, Herbort, Blunk, & Couvreur, 1992; Müller, Lherm, Herbert, & Couvreur, 1990). Moreover PBCA Np were shown to be cleared rapidly from the body after 24 h (Grislain et al., 1983).

Three mechanisms of degradation of PACA Np have been described: the first, which is

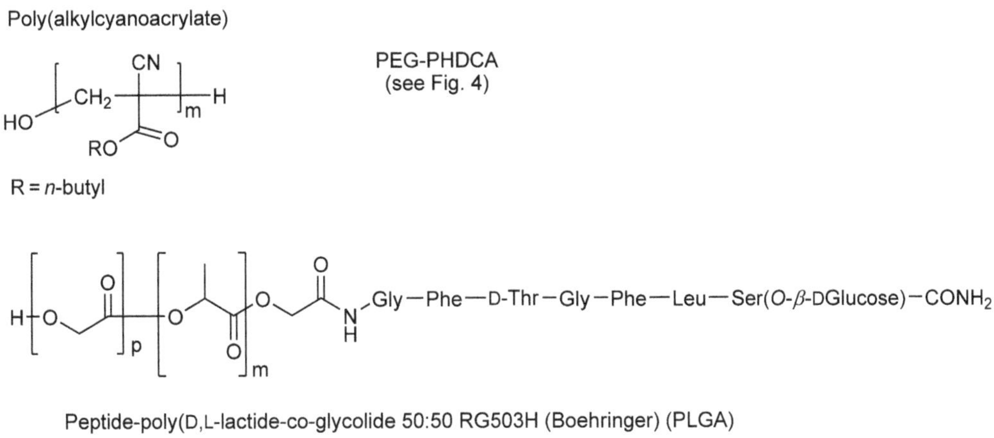

Fig. 5 Polymers used in the preparation of brain-targeted Np.

believed to be the most important, is based on the enzymatic degradation by esterases of the serum, lysosomes, and pancreatic juice, giving rise to alkylalcohol and poly(cyanoacrylic acid) by-products.

The second is based on the unzipping polymerization followed by an instantaneous repolymerization into a lower molecular weight polymer, while the third, which at the early stages of PACA Np development was believed to be the most important, is the Knoevenagel reaction. The degradation products are formaldehyde and cyanoacetic ester and the reaction yields only 5% of degradation after 24 h. These by-products are all potentially harmful to cells as they are small- or medium-sized nonself molecules. *In vitro* toxicity studies have been performed on several cell lines, namely macrophages (Gaspar, Préat, Opperdoes, & Roland, 1992; Kante et al., 1982), L929 fibroblasts (Huang & Lee, 2006; Lherm et al., 1992), hepatocytes (Fernandez-Urrsuno, Fattal, Feger, Couvreur, & Thérond, 1995; Fernandez-Urrsuno, Fattal, Porquet, Feger, & Couvreur, 1997), cancer cells DC3F (Kubiak, Couvreur, Manil, & Clausse, 1989), mesenchymal cells (Gipps, Groscurth, Kreuter, & Speiser, 1987), Swiss 3T3 cells (Maassen, Fattal, Müller, & Couvreur, 1993; Tseng, Tabata, Hyon, & Ikada, 1990), and murine cerebral microvessels (Vinters & Ho, 1988). Since macrophages are characterized by a high phagocytic propensity, they represent good models to study the toxicity of PACA Np intended for intracellular targeting (Gaspar et al., 1992; Kante et al., 1982). The results revealed that polyisohexylcyanoacrylate induced a respiratory burst in macrophages, and thus macrophage activation (Gaspar et al., 1992). Another study showed that polymethylcyanoacrylate NPs provoked perforation of the macrophage membrane. In the same conditions, perforation was not observed with polyisobutylcyanoacrylate NPs. The contrasting results can be explained by the different degradation rate that is correlated to the alkyl chain length (Kante et al., 1982). The influence of the alkyl chain length on the PACA Np cytotoxicity was investigated more accurately in a L929 fibroblast cell model analyzing cell growth inhibition. Polyethyl-, polyisobutyl-, polymethyl-, and polyisohexylcyanoacrylates were investigated for cell growth. The first two polymers showed a considerably higher toxicity compared to polyisohexylcyanoacrylate, while polymethylcyanoacrylate is only slightly less toxic.

The observed toxic effects could be due to the combination of two mechanisms: PACA degradation and adhesion. In the presence of longer alkyl chains, Np undergo a slower degradation favoring cell adhesion leading to high local degradation product concentrations (Lherm et al., 1992). Besides the length of the alkyl chain, the particle size also influences the toxicity. In fact, it was found that smaller particle sizes provoke higher cytotoxicity. This has been explained by the large surface area that leads to a faster degradation and a higher degradation product concentration (Lherm et al., 1992).

Another very useful model to study PACA Np toxicity is the liver (Fernandez-Urrusuno et al., 1997, 1995) since it is responsible for a rapid uptake upon Np i.v. administration. PACA Np i.v. administrations initiate an inflammation process accompanied by metabolic disorders in the glucidic metabolism and by massive glycoprotein undersialylation. This process, reversible 14 days after the end of the treatment, can be correlated to the Np phagocytosis by Kupffer cells. In fact, *in vivo*, PACA Np are supposed to principally distribute in the Kupffer cells initiating their activation, while the direct contact with hepatocytes is not likely to occur. For this reason, PACA degradation products are not mainly responsible for the low level of reduced glutathione. Indeed, the decrease of the reduced glutathione level could be due to a phagocytosis-induced burst respiratory effect.

It can be concluded that the PACA Np cytotoxicity is likely due to the burst release of degradation by-products, directly related to the length of the polymer alkyl chains. Even if Np made of longer alkyl chain polymers are less toxic, they could adhere to cells, leading to a high local degradation product concentration. For brain targeting, less toxic, slow degrading PACA Np should be more suitable, although their chronic administration would lead to "waste disposal" in brain endothelial cells and/or other brain structures.

Brain-targeted liposomes

Type of lipids

With the aim of both optimizing the properties of liposomal bilayers and stabilizing the encapsulation efficacy of the drugs, liposomes for CNS delivery were frequently prepared mixing phospholipids to the family of phosphatidylcholine (PC), phosphatidylethanolamine (PE), and cholesterol (Kulkarni, Betageri, & Singh, 1995). PEG–lipid conjugates, exposing the distal end able to interact with a selective ligand, or the preformed ligand–PEG–lipids were added during the preparation to obtain a targeted PEGylated liposome used in the studies reported in Table 3.

A moderate amount of cationic lipids was used when liposomes are prepared for brain gene therapy. When gene materials are mixed with lipid or preformed cationic liposomes, a spontaneous electrostatic interaction between the cationic lipid and the negatively charged plasmid DNA ensures the stable complex formation (Ruozi, Forni, Battini, & Vandelli, 2003). Biodegradability and toxicity of the cationic lipids (and consequently the related formulations) represent a very important critical point for the applicability to the treatment. In fact, several *in vitro* and *in vivo* experiments described a range of adverse effects associated with the use of cationic lipids and cationic liposomes such as inhibition of protein kinase C activity (Farhood, Bottega, Epand, & Huang, 1992; Zelphati & Szoka, 1996), induction of chromosome aberrations in human culture cells (Nuzzo et al., 1985), toxicity for nonphagocytic cells (Cortesi, Esposito, Menegatti, Gambari, & Nastruzzi, 1996; Lappalainen, Jääskeläinen, Syrjänen, Urtti, & Syrjänen, 1994; Parhham & Wetzig, 1993), downregulation of nitric oxide (NO), Tumor Necrosis Factor-α (TNF-α), and Prostaglandin E2 (PGE2) synthesis (Filion & Phillips, 1997), neurotoxicity after i.v. injection (Taniguchi, Yamamoto, Itakura, Miitchi, & Hayashi, 1988), and activation of complement via the alternative pathway (Chon, Cullis, & Devine, 1991). The toxic effect is mainly determined by the cationic nature of the vector that includes the structure of the hydrophilic group of cationic lipids, the linker between the headgroup and the hydrophobic parts along with the biodegradable pathway (Hongtao, Zhang, Wang, Cui, & Yanc, 2006). Cationic nanoparticles or nanogel has never been studied in this respect.

Preparation of the brain-targeted colloidal systems

Drug loading is performed during the preparation of the colloidal system, starting from the preformed polymer (in the case of polymeric Np) or from the lipids (in the case of liposomes). This allows the entrapment of the drug not only inside the colloidal system but also on their surface. In the case of PBCA Np, the polymerization reaction has also been conducted in the presence of the drug to be loaded; sometimes, in the case of nanogel, PBCA Np, and HSA Np, the drug is loaded simply mixing the drug solution and the preformed Np; obviously, in this case the drug is adsorbed on the Np surface.

After or before this step of drug loading there is a step dealing with the functionalization of the surface for brain targeting.

Preparation of polymeric Np

The most studied Np, made of PBCA, were prepared following the emulsion polymerization process; this method is based on the polymerization of a monomer at room temperature and at a constant stirring rate. This method has been widely applied (see Table 1) (Alyautdin et al., 1997; Ambruosi, Gelperina, et al., 2006; Kreuter *et al.*, 1995; Steiniger et al., 2004); as previously stated, several reports added the drug doxorubicin during the polymerization reaction (Ambruosi,

Khalansky, et al., 2006; Gulyaev et al., 1999; Petri et al., 2007). It has not been determined if this strategy modifies the drug. As can be seen in Table 2, this procedure ensures a greater loading than simple drug adsorption on the Np surface.

Thus, the preparation of brain-targeted Np follows several steps:

- Preparation of Np starting from butylcyanoacrylate monomer in an acidic polymerization medium containing dextran 70000 as stabilizer
- Neutralization (NaOH)
- Filtration (pore size 10 μM)
- Liophilization in presence of glucose (cryoprotectant)
- Loading of the drug (simple mixing with a drug solution)
- Loading of PS-80 (addition of the surfactant and simple mixing)
- Injection

The drug loading is always determined by the analysis of the drug remaining in the solution after Np preparation; either the kinetic of drug release from the Np or the release of drug that presumably happens during the coating with a surfactant has never been reported.

PEG-PHDCA Np (Brigger et al., 2002) and loperamide-loaded peptide-decorated Np (Tosi et al., 2007) (Table 1) were prepared following the method of nanoprecipitation (Fessi, Puisieux, Devissaguet, Ammoury, & Benita, 1989).

- Solution of PEG-PHDCA (or peptide-derivatized PLGA) in acetone
- The organic phase is poured in water containing Pluronic F-68 under magnetic stirring.
- Acetone was evaporated under reduced pressure.
- Ultracentrifugation
- The pellet was resuspended in a solution of glucose.
- Filtration on a sintered glass membrane
- Injection

Preparation of liposomes

Generally, three different multilamellar liposome vesicle (MLV) preparation techniques have been used.

Method 1: Thin layer evaporation (TLE) technique (Bangham, 1968).

- Lipids are dissolved in chloroform.
- A thin film is obtained by evaporating the organic solvent under reduced pressure in a rotary evaporator.
- The dried lipid film is hydrated with buffer solution and gently mixed to get unilamellar vesicles.

Method 2: Reverse phase evaporation (REV) methods (Szoka & Papahadjopoulos, 1978).

- Lipid mixture is dissolved in organic solvent (chloroform or diethyl ether) and mixed with the aqueous solution to obtain a water-in-oil emulsion.
- A gel is formed when the organic solvent is removed under reduced pressure.
- The gel evolved into a dispersion of liposomes upon agitation

Method 3: Frozen and thawed methods (FATMLV) (New, 1989).

- MLV suspension generally obtained by TLE methods is placed in a tube and dipped in a nitrogen bath for defined time.
- The tube is then removed and placed in a water bath at defined time and temperature.
- The sequence is repeated nine times.

In order to obtain liposome dispersions with a rather narrow particle size distribution and a small diameter (≤150 nm), a second step following the initial vesicle formation is required. Small unilamellar vesicles were obtained by

1. The extrusion technique (Berger, Sachse, Bender, Schubert, & Brandl, 2001) in which

MLV are extruded through a different pore size polycarbonate membrane for 19 times
2. Sonication technique (Lasch, Wesseig, & Brandl, 2003) using a bath or probe tip sonicators

Generally drugs are encapsulated during the hydration step: the excess of drug is removed by gel chromatography. Notwithstanding the general methodology of drug loading, some modifications can be applied such as in the encapsulation of doxorubicin which is performed via a pH gradient as well described by Mayer et al. (1986) and Bolotin et al. (1994). Gene material is instead added to the lipids (Shi et al., 2000; Xie et al., 2005) and then the liposomes are obtained using the methods previously described. The gene materials unbounded or bounded to the exterior of liposomes are routinely removed by nuclease treatment.

Surface functionalization of the colloidal systems

Nanoparticles

Before or after this step, there is the Np surface functionalization with CNS-targeting moieties.

Noncovalent strategy

Case of PS-80: A dispersion of PBCA Np in 1 mL of PBS was obtained by ultrasonication for 5 min and then PS-80 was added (1% m/v) followed by incubation for a period of 30 min. The quantity of PS-80 that it is adsorbed on Np surface and the kinetic of its release in plasma have never been determined.

Case of biotin–avidin: The egg white protein, avidin, is a 64000-Da homotetramer that has 4 biotin-binding sites per molecule: the avidin–biotin bond, even if it is not covalent, has an extremely high affinity; it is stable in the circulation, but it is labile at tissue depot sites (Bickel et al., 2001). Following this strategy, avidin is covalently coupled to the colloidal carrier, while the biotin is covalently linked to the vector (e.g., biotinylated antibodies). This strategy was applied on BSA Np (Michaelis et al., 2006), which were loaded with a drug after their preparation, or on unloaded PLA Np (Olivier, 2005). This methodology has been applied for brain-targeted liposomes in the case of liposomes containing therapeutic genes conjugated to multiple BBB and brain cell membrane-targeting agents and providing transport of the encapsulated gene across the BBB. In this case, a transportable peptide is conjugated to the surface of the liposome with avidin–biotin technology (Pardridge, 2002).

Covalent strategy

The targeting peptide H_2N–Gly-L-Phe-D-Thr-Gly-L-Phe-L-Leu–L-Ser–O-β-D-glucose-$CONH_2$ was covalently linked to the polyester PLGA before Np preparation; during the preparation of Np by nanoprecipitation, the hydrophilic peptide was located at the outside of the Np. In order to prepare the derivatized copolymer, the commercially available PLGA RG 503H is activated at its carboxylic end as an ester with *N*-hydroxysuccinimide; then the peptide is allowed to react with it by means of its aminic end, to form an amidic linkage with the polymer (Costantino et al., 2005).

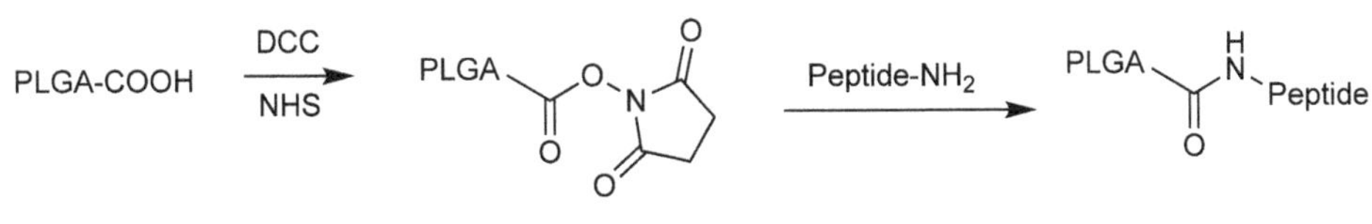

The carboxylic end of the polyester PLA has also been covalently derivatized following another strategy, namely the derivatization of the carboxy end of the polymer PEG-PLA copolymer with maleimide, which permits protein conjugation in nondenaturating conditions. Unloaded maleimide-functionalized PLA Np were then conjugated with thiolated mouse OX26 anti-rat TfR monoclonal antibodies (Olivier, 2005). This method has never been applied to drug-loaded Np; the possibility of a chemical reaction between maleimide and a suitable functional group present in the drug molecule could be hypothesized.

Liposomes

Three possible strategies could be applied to construct targeted liposomes:

1. Traditional methods in which a conjugate lipid–PEG–ligand is included in the thin film lipid formulation for hydration into liposomes (see Fig. 6);
2. Targeting ligands are coupled to the termini of functionalized PEG chains on liposomes;
3. Postinsertion technique by which a conjugate is micellized and incubated with preformed liposome to yield targeted liposomes.

Maleimide/thiolated ligands coupling

PEGylated immunoliposomes were frequently obtained by conjugation between thiolated antibody (8D3, 83-14, OX26) and maleimide moiety (DSPE-PEG-Mal) of the PEGylated liposomes (Huwyler et al., 1996; Shi et al., 2000; Shi, Boado et al., 2001; Zhang, Zhu, et al., 2002). In this case the drug should not have SH groups.

Tf liposomes loaded with HRP were obtained also by postinsertion technique. In this case, the Tf was tagged to the distal end of the PEG-chain via a maleimide–thiol coupling; Tf was modified with a thiol group using succinimidyl-acetylthioacetate (SATA) as described by Visser et al. (2004) and incubated by preformed micelles prepared using DSPE-PEG and DSPE-PEG-maleimide. Liposomes were separately obtained by PC and CHOL and mixed with micelles for 2 h at 40°C to obtain Tf-tagged liposomes (Visser et al., 2005) (see Fig. 6).

DCC coupling

Using the same technique (postinsertion) folate-liposome encapsulated doxorubicin were obtained. The conjugate lipid–PEG–folate was synthesized by dicyclohexylcarbodiimide (DCC) chemistry linking the amine group of lipid–PEG amine and the carboxyl of folic acid (Saul et al., 2003).

NHS/EDC coupling

Tf-coupled liposomes were obtained, as described by Maruyama et al. (1995), by conjugation of the ligand to the distal carboxylic end of PEG liposomes. The carboxy group of DSPE-PEG-COOH

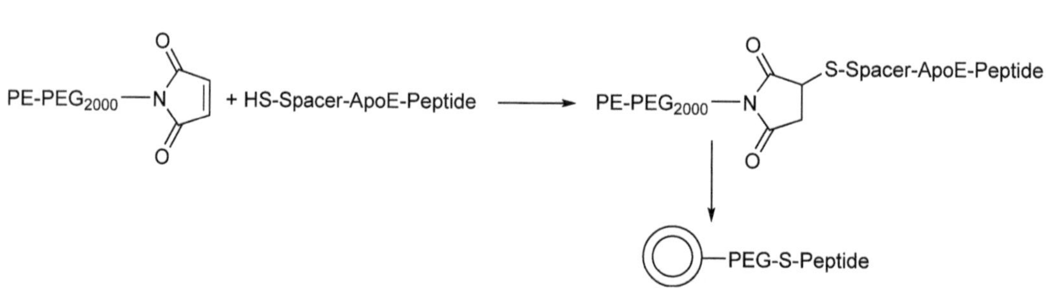

Fig. 6. Example of conjugation of thiolated protein (in this case ApoE) via PEG-maleimide phosphatidylethanolamine (PE), and preparation of functionalized liposomes.

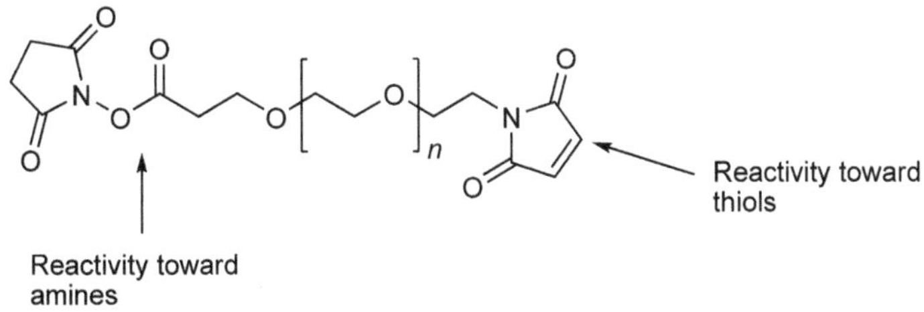

Fig. 7. Structure of the NHS-PEG-maleimide reagent.

was activated as an ester with NHS in the presence of EDC (Eavarone et al., 2000).

Albumin Np

In analogy with polyester copolymers, the covalent binding was also used to derivatize albumin-based Np. In this case two different pathways have been used:

1. Use of glutaraldehyde as a cross-linker and formation of a Schiff base with the ammine groups of the Ab.
2. In the case of albumin ApoA1 and B100 HSA Np, the Np were prepared by reaction with glutaraldehyde (no surface active agent is needed); then there was a transformation of the Np in sulfhydryl-reactive Np (by means of the reagents reported below NHS-PEG-maleimide); then thiolated apolipoproteins were added. The model drug loperamide was then loaded onto these Np simply shaking the preformed Np with a drug solution (Kreuter et al., 2007). In another case, Np were activated with NHS-PEG-Mal, then reacted with NeutrAvidin (or directly to thiolated ApoE), and subsequent linkage with biotinylated ApoE was formed (Michaelis et al., 2006).

References

Allemann, E., Leroux, J. C., Gurny, R., & Doelker, E. (1993). *In vitro* extended release properties of drug-loaded poly(D, L-lactic acid) nanoparticles produced by a salting-out procedure. *Pharmaceutical Research, 10*, 1732–1737.

Alyautdin, R. N., Petrov, V. E., Langer, K., Berthold, A., Kharkevich, D. A., & Kreuter, J. (1997). Delivery of loperamide across the blood-brain barrier with polysorbate 80-coated polybutylcyanoacrylate nanoparticles. *Pharmaceutical Research, 14*, 325–328.

Ambruosi, A., Gelperina, S., Khalansky, A., Tanski, S., Theisen, A., & Kreuter, J. (2006). Influence of surfactants, polymer and doxorubicin loading on the antitumor effect of poly (butyl cyanoacrylate) nanoparticles in a rat glioma model. *Journal of Microencapsulation, 23*, 582–592.

Ambruosi, A., Khalansky, A., Yamamoto, H., Gelperina, S., Begley, D. J., & Kreuter, J. (2006). Biodistribution of polysorbate 80-coated doxorubicin-loaded [^{14}C]-poly(butylcyanoacrylate) nanoparticles after intravenous administration to glioblastoma-bearing rats. *Journal of Drug Targeting, 14*, 97–105.

Ambruosi, A., Yamamoto, H., & Kreuter, J. (2005). Body distribution of polysorbate 80 and doxorubicin-loaded [^{14}C] poly(butyl cyanoacrylate) nanoparticles after i. v. administration in rats. *Journal of Drug Targeting, 13*, 535–542.

Anderson, K. E., Eliot, L. A., Stevenson, B. R., & Rogers, J. A. (2001). Formulation and evaluation of a folic acid receptor targeted oral vancomycin liposomal dosage form. *Pharmaceutical Research, 18*, 316–322.

Bangham, A. D. (1968). Membrane model with phospholipids. *Progress in Biophysics and Molecular Biology. 18*, 29–30.

Bangham, A. D., Standish, M. M., & Watkins, J. C. (1965). Diffusion of univalent ions across the lamellae of swollen phospholipids. *Journal of Molecular Biology, 13*, 238–252.

Batrakova, E. V., & Kabanov, A. V. (2008). Pluronic block copolymers: Evolution of drug delivery concept from inert nanocarriers to biological response modifiers. *Journal of Controlled Release, 130*, 98–106.

Bawa, R. (2008). Nanoparticle-based therapeutics in humans: A survey. *Nanotechnology Law and Business, 5*, 135–155.

Bazile, D. V., Ropert, C., Huve, P., Verrecchia, T., Marlard, M., Frydman, A., et al. (1992). Body distribution of fully biodegradable [^{14}C]-poly(lactic acid) nanoparticles coated with albumin after parenteral administration to rats. *Biomaterials, 13*, 1093–1102.

Berger, N., Sachse, A., Bender, J., Schubert, R., & Brandl, M. (2001). Filter extrusion of liposomes using different devices: Comparison of liposome size, encapsulation efficiency, and process characteristics. *International Journal of Pharmaceutics, 223*, 55–68.

Bickel, U., Yoshikawa, T., & Pardridge, W. M. (2001). Delivery of peptides through the blood brain barrier. *Advanced Drug Delivery Reviews*, *46*, 247–279.

Bolotin, E. M., Cohen, R., Bar, L. K., Emanuel, S. N., Lasic, D. D., & Barenholz, Y. (1994). Ammonium sulphate gradients for efficient and stable remote loading of amphipathic weak bases into liposomes and ligandosomes. *Journal of Liposome Research*, *4*, 455–479.

Brigger, I., Morizet, J., Aubert, G., Chacun, H., Terrier-Lacombe, M. J., Couvreur, P., et al. (2002). Poly(ethylene glycol)-coated hexadecylcyanoacrylate nanospheres display a combined effect for brain tumor targeting. *Journal of Pharmacology and Experimental Therapeutics*, *303*, 928–936.

Brigger, I., Morizet, J., Laudani, L., Aubert, G., Appel, M., Velasco, V., et al. (2004). Negative preclinical results with stealth nanospheres-encapsulated Doxorubicin in an orthotopic murine brain tumor model. *Journal of Controlled Release*, *100*, 29–40.

Calvo, P., Gouritin, B., Chacun, H., Desmaele, D., D'Angelo, J., Noel, J. P., et al. (2001). Long-circulating PEGylated polycyanoacrylate nanoparticles as new drug carrier for brain delivery. *Pharmaceutical Research*, *18*, 1157–1166.

Casy, A. F., & Parfitt, R. T. (1986). *Opioid analgesics. Chemistry and receptors*. New York: Plenum

Champion, J. A., Katare, Y. K., & Mitragotri, S. (2007). Particle shape: A new design parameter for micro-and nanoscale drug delivery carriers. *Journal of Controlled Release*, *121*, 3– 9.

Chellat, F., Merhi, Y., Moreau, A., & Yahia, L. H. (2005). Therapeutic potential of nanoparticulate systems for macrophage targeting. *Biomaterials*, *26*, 7260–7275.

Chon, A., Cullis, P. R., & Devine, D. V. (1991). The role of surface charge in the activation of the classical and alternative pathways of complement by liposomes. *Journal of Immunology*, *146*, 4234–4241.

Cortesi, R., Esposito, E., Menegatti, E., Gambari, R., & Nastruzzi, C. (1996). Effect of cationic liposomes composition on *in vitro* cytotoxicity and protective effect on carried DNA. *International Journal of Pharmaceutics*, *139*, 69–78.

Costantino, L., Gandolfi, F., Tosi, G., Rivasi, F., Vandelli, M. A., & Forni, F. (2005). Peptide-derivatized biodegradable nanoparticles able to cross the blood-brain barrier. *Journal of Controlled Release*, *108*, 84–96.

de Boer, A. G., van der Sandt, I. C.J., & Gaillard, P. J. (2003). The role of drug transporters at the blood-brain barrier. *Annual Review of Pharmacology and Toxicology*, *43*, 629–656.

Discher, B. M., Won, J. J., Ege, D. S., Lee, J., Bates, F. S., Discher, D. E., et al. (1999). Polymersomes: Tough vesicles made from diblock copolymers. *Science*, *284*, 1143–1146.

Discher, D. E., & Eisenberg, A. (2002). Poliymer vesicles. *Science*, *297*, 967–973.

Dong, Y., & Feng, S. S. (2004). Methoxy poly(ethylene glycol)-poly(lactide) (MPEG-PLA) nanoparticles for controlled delivery of anticancer drugs. *Biomaterials*, *25*, 2843–2849.

Eavarone, D. A., Yu, X., & Bellamkonda, R. V. (2000). Targeted drug delivery to C6 glioma by transferring-coupled liposomes. *Journal of Biomedical Materials Research*, *51*, 10–14.

Farhood, H., Bottega, R., Epand, R. M., & Huang, L. (1992). Effect of cationic cholesterol derivatives on gene transfer and protein kinase C activity. *Biochimica et Biophysica ACTA*, *1111*, 239–246.

Fenart, L., Casanova, A., Dehouck, B., Duhem, C., Slupek, S., Cecchelli, R., et al. (1999). Evaluation of the effect of charge and lipid coating on ability of 60-nm nanoparticles to cross an *in vitro* model of blood brain barrier. *Journal of Pharmacology and Experimental Therapeutics*, *291*, 1017–1022.

Feng, S. S., Mu, L., Win, K. Y., & Huang, G. (2004). Nanoparticles of biodegradable polymers for clinical administration of paclitaxel. *Current Medicinal Chemistry*, *11*, 413–424.

Fernandez-Urrusuno, R., Fattal, E., Feger, J., Couvreur, P., & Thérond, P. (1997). Evaluation of hepatic antioxidant systems after intravenous administration of polymeric nanoparticles. *Biomaterials*, *18*, 511–517.

Fernandez-Urrusuno, R., Fattal, E., Porquet, D., Feger, J., & Couvreur, P. (1995). Evaluation of liver toxicological effects induced by polyalkylcyanoacrylate nanoparticles. *Toxicology and Applied Pharmacology*, *130*, 272–279.

Fessi, H., Puisieux, F., Devissaguet, J. P., Ammoury, N., & Benita, S. (1989). Nanocapsule formation by interfacial polymer deposition following solvent displacement. *International Journal of Pharmaceutics*, *55*, R1–R4.

Filion, M. C., & Phillips, N. C. (1997). Anti-inflammatory activity of cationic lipids. *British Journal of Pharmacology*, *122*, 551–557.

Fishbein, I., Chorny, M., Rabinovich, L., Banai, S., & Gati, I. (2000). Nanoparticulate delivery system of a tyrphostin for the treatment of restenosis. *Journal of Controlled Release*, *65*, 221–229.

Friden, P. M., Walus, L. R., Musso, G. F., Taylor, M. A., Malfroy, B., & Starzyk, R. M. (1991). Anti-transferrin receptor antibody and antibody-drug conjugates cross the blood brain barrier. *Proceedings of the National Academy of Sciences of the United States of America*, *88*, 4771–4775.

Friese, A., Seiller, E., Quack, G., Lorenz, B., & Kreuter, J. (2000). Increase of the duration of the anticonvulsive activity of a novel NMDA receptor antagonist using poly(butylcyanoacrylate) nanoparticles as a parenteral controlled release system. *European Journal of Pharmaceutics and Biopharmaceutics*, *49*, 103–109.

Garcia-Garcia, E., Andrieux, K., Gil, S., & Couvreur, P. (2005). Colloidal carriers and blood-brain barrier (BBB) translocation: A way to deliver drugs to the brain? *International Journal of Pharmaceutics*, *298*, 274–292.

Garcia-Garcia, E., Gil, S., Andrieux, K., Desmaele, D., & Nicolas, V. (2005). A relevant *in vitro* rat model for the evaluation of blood-brain barrier translocation of nanoparticles. *Cellular and Molecular Life Science*, *62*, 1–9.

Gaspar, R., Préat, V., Opperdoes, F. R., & Roland, M. (1992). Macrophage activation by polymeric nanoparticles of polyalkylcyanoacrylates: Activity against intracellular *Leishmania donovani* associated with hydrogen peroxide production. *Pharmaceutical Research*, *9*, 782–787.

Gaumet, M., Gurny, R., & Delie, F. (2007). Fluorescent biodegradable PLGA particles with narrow size distributions:

Preparation by means of selective centrifugation. *International Journal of Pharmaceutics*, *342*, 222–230.

Gaumet, M., Vargas, A., Gurny, R., & Delie, F. (2008). Nanoparticles for drug delivery: The need for precision in reporting particle size parameters. *European Journal of Pharmaceutics and Biopharmaceutics*, *69*, 1–9.

Gelperina, S. E., Khalansky, A. S., Skidan, I. N., Smirnova, Z. S., Bobruskin, A. I., Severin, S. E., et al. (2002). Toxicological studies of doxorubicin bound to polysorbate 80-coated polybutylcyanoacrylate nanoparticles in healthy rats and rats with intracranial glioblastoma. *Toxicology Letters*, *126*, 131–141.

Geng, Y., Dalhaimer, P., Cai, S., Tsai, R., Tewari, M., Minko, T., et al. (2007). Shape effects of filaments versus spherical particles in flow and drug delivery. *Nature Nanotechnology*, *2*, 249–255.

Gessner, A., Olbrich, C., Schroeder, W., Kaiser, O., & Muller, R. H. (2001). The role of plasma proteins in brain targeting: Species dependent protein adsorption patterns on brain specific lipid drug conjugate (LDC) nanoparticles. *International Journal of Pharmaceutics*, *214*, 87–91.

Gessner, A., Waicz, R., Lieske, A., Paulke, B. R., Mader, K., & Muller, R. H. (2000). Nanoparticles with decreasing surface hydrophobicities: Influence on plasma protein adsorption. *International Journal of Pharmaceutics*, *196*, 245–249.

Gipps, E. M., Groscurth, P., Kreuter, J., & Speiser, P. P. (1987). The effects of polyalkylcyanoacrylate nanoparticles on human normal and malignant mesenchymal cells *in vitro*, *International Journal of Pharmaceutics*, *40*, 23–31.

Gonatas, N. K., Stieber, A., Hickey, W. F., Herbert, S. H., & Gonatas, J. O. (1984). Endosomes and Golgi vesicles in adsorptive and fluid phase endocytosis. *Journal of Cell Biology*, *99*, 1379–1390.

Grislain, L., Couvreur, P., Lenaerts, V., Roland, M., Deprez-Decampeneere, D., Speiser, P., et al. (1983). Pharmacokinetics and distribution of a biodegradable drug-carrier. *International Journal of Pharmaceutics*, *15*, 335–345.

Gulyaev, A. E., Gelperina, S. E., Skidan, I. N., Antropov, A. S., Kivman, G. Y., & Kreuter, J. (1999). Significant transport of doxorubicin into the brain with polysorbate 80-coated nanoparticles. *Pharmaceutical Research*, *16*, 1564–1569.

Harris, J. M., & Chess, R. B. (2003). Effect of PEGylation on pharmaceuticals. *Nature Reviews Drug Discovery*, *2*, 214–221.

Hawkins, M. J., Soon-Shiong, P., & Desai, N. (2008). Protein nanoparticles as drug carriers in clinical medicine. *Advanced Drug Delivery Reviews*, *60*, 876–885.

Hongtao, L. V., Zhang, S., Wang, B., Cui, S., & Yanc, J. (2006). Toxicity of cationic lipids and cationic polymers in gene delivery. *Journal of Controlled Release*, *114*, 100–109.

Horisawa, E., Hirota, T., Kawazoe, S., Yamada, J., Yamamoto, H., Takeuchi, H., et al. (2002). Prolonged anti-inflammatory action of DL-lactide/glycolide copolymer nanospheres containing betamethasone sodium phosphate for an intra-articular delivery system in antigen-induced arthritic rabbit. *Pharmaceutical Research*, *19*, 403–410.

Huang, C. Y., & Lee, Y. D. (2006). Core-shell type of nanoparticles composed of poly[(*n*-butyl cyanoacrylate)-*co*-(2-octyl cyanoacrylate)] copolymers for drug delivery application: Synthesis, characterization and *in vitro* degradation. *International Journal of Pharmaceutics*, *325*, 132–139.

Huwyler, J., Wu, D., & Pardridge, W. M. (1996). Brain drug delivery of small molecules using immunoliposomes. *Proceedings of the National Academy of Sciences of the United States of America*, *93*, 14164–14169.

Huwyler, J., Yan, J., & Pardridge, W. M. (1997). Receptor mediated delivery of daunomycin using immunoliposomes: Pharmacokinetics and tissue distribution in the rat. *Journal of Pharmacology and Experimental Therapeutics*, *282*, 1541–1546.

Jallouli, Y., Paillard, A., Chang, J., Sevin, E., & Betbeder, D. (2007). Influence of surface charge and inner composition of porous nanoparticles to cross the blood brain barrier *in vitro*. *International Journal of Pharmaceutics*, *344*, 103–109.

Kabanov, A. V., & Batrakova, E. V. (2004). New technologies for drug delivery across the blood brain barrier. *Current Pharmaceutical Design*, *10*, 1355–1363.

Kante, B., Couvreur, P., Dubois-Krack, G., De Meester, C., Guiot, P., Roland, M., et al. (1982). Toxicity of polyalkylcyanoacrylate nanoparticles I: Free NPs. *Journal of Pharmaceutical Sciences*, *72*, 786–790.

Kattan, J., Droz, J. P., Couvreur, P., Marino, J. P., Boutan-Laroze, L. A., Rougier, P., et al. (1992). Phase I clinical trial and pharmacokinetic evaluation of doxorubicin carried by polyisohexylcyanoacrylate nanoparticles. *Investigational New Drugs*, *10*, 191–199.

Kaur, I. P., Bhandari, R., Bhandari, S., & Kakkar, V. (2008). Potential of solid lipid nanoparticles in brain targeting. *Journal of Controlled Release*, *127*, 97–109.

Kerns, E., & Di, L. (2008). *Drug-like properties: Concepts, structure design and methods* (pp. 122–136). New York: Academic Press.

Kim, H. R., Andrieux, K., Gil, S., Taverna, M., Chacun, H., Desmaële, D., et al. (2007). Translocation of poly(ethylene glycol-co-hexadecyl)cyanoacrylate nanoparticles into rat brain endothelial cells: Role of apolipoproteins in receptor-mediated endocytosis. *Biomacromolecules*, *8*, 793–799.

Koziara, J. M., Lockman, P. R., Allen, D. D., & Mumper, R. J. (2003). In situ blood-brain barrier transport of nanoparticles. *Pharmaceutical Research*, *20*, 1772–1778.

Kratzer, I., Wernig, K., Panzenboeck, U., Bernhart, E., & Reicher, H. (2007). Apolipoprotein A-I coating on protamine oligonucleotide nanoparticles increases particle uptake and transcytosis in an *in vitro* model of the blood-brain barrier. *Journal of Controlled Release*, *117*, 301–311.

Kreuter, J. (2001). Nanoparticulate systems for brain delivery of drugs. *Advanced Drug Delivery Reviews*, *47*, 65–81.

Kreuter, J., Alyautdin, R. N., Kharkevich, D. A., & Ivanov, A. A. (1995). Passage of peptides through the blood-brain barrier with colloidal polymer particles (nanoparticles). *Brain Research*, *674*, 171–174.

Kreuter, J., Hekmatara, T., Dreis, S., Vogel, T., Gelperina, S., & Langer, K. (2007). Covalent attachment of apolipoprotein

A-I and B-100 to albumin nanoparticles enables drug transport into the brain. *Journal of Controlled Release, 118*, 54–58.

Kreuter, J., Petrov, V. E., Kharkevich, D. A., & Alyautdin, R. N. (1997). Influence of the type of surfactant on the analgesic effects induced by the peptide dalargin after its delivery across the blood-brain barrier using surfactant-coated nanoparticles. *Journal of Controlled Release, 49*, 81–87.

Kreuter, J., Ramge, P., Petrov, V., Hamm, S., Gelperina, S. E., Engelhardt, B., et al. (2003). Direct evidence that polysorbate 80-coated poly(butylcyanoacrylate) nanoparticles deliver drugs to the CNS via specific mechanisms requiring prior binding of drug to the nanoparticles. *Pharmaceutical Research, 20*, 409–416.

Kreuter, J., Shamenkov, D., Petrov, V., Ramge, P., Cychutek, K., Koch-Brandt, C., et al. (2002). Apolipoprotein-mediated transport of nanoparticle-bound drugs across the blood-brain barrier. *Journal of Drug Targeting, 10*, 317–325.

Kubiak, C., Couvreur, P., Manil, L., & Clausse, B. (1989). Increased cytotoxicity of nanoparticle-carried adriamycin *in vitro* and potentiation by verapamil and amiodarone. *Biomaterials, 10*, 553–556.

Kulkarni, S. B., Betageri, G. V., & Singh, M. (1995). Factors affecting microencapsulation of drugs in liposomes. *Journal of Microencapsulation, 12*, 229–246.

Kumagai, A. K., Eisenberg, J. B., & Pardridge, W. M. (1987). Absorptive-mediated endocytosis of cationized albumin and a beta-endorphin-cationized albumin chimeric peptide by isolated brain capillaries. Model system of blood brain barrier transport. *Journal of Biological Chemistry, 262*, 15214–15219.

Lappalainen, K., Jääskeläinen, I., Syrjänen, K., Urtti, A., & Syrjänen, S. (1994). Comparison of cell proliferation and toxicity assays using two liposomes. *Pharmaceutical Research, 11*, 1127–1131.

Lasch, J., Wesseig, V., & Brandl, M. (2003). Preparation of liposomes, In V. Torchilin & V. Wessing (Eds.), Liposomes: A practical approach (2nd ed., pp. 3–16). Oxford: Oxford University Press.

Lee, R. J., & Low, P. S. (1995). Folate-mediated tumor cell targeting of liposome-entrapped doxorubicin *in vitro*. *Biochimica et Biophysica ACTA, 1233*, 134–144.

Lemarchand, C., Gref, R., Passirani, C., Garcion, E., Petri, B., Müller, R., et al. (2006). Influence of polysaccharide coating on the interactions of nanoparticles with biological systems. *Biomaterials, 27*, 108–118.

Lherm, C., Müller, R. H., Puisieux, F., & Couvreur, P. (1992). Alkylcyanoacrylate drug carriers II. Cytotoxicity of cyanoacrylate nanoparticles with different alky chain length. *International Journal of Pharmaceutics, 84*, 13–22.

Li, S. (1999). Hydrolytic degradation characteristics of aliphatic polyesters derived from lactic and glycolic acids. *Journal of Biomedical Materials Research (Applied Biomaterials), 48*, 342–353.

Liu, L., Guo, K., Lu, J., Venkatraman, S. S., Luo, D., Moochhala, S. et al. (2008). Biologically active core/shell nanoparticles self-assembled from cholesterol-terminated PEG-TAT for drug delivery across the blood-brain barrier. *Biomaterials, 29*, 1509–1517.

Lockman, P. R., Koziara, J. M., Mumper, R. J., & Allen, D. D. (2004). Nanoparticle surface charges alter blood-brain barrier integrity and permeability. *Journal of Drug Targeting, 12* (9–10), 635–641.

Lockman, P. R., Koziara, J., Roder, K. E., Paulson, J., Abbruscato, T. J., Mumper, R. J., et al. (2003). *In vivo* and *in vitro* assessment of baseline blood brain barrier parameters in the presence of novel nanoparticles. *Pharmaceutical Research, 20*, 705–713.

Lockman, P. R., Oyewumi, M. O., Koziara, J. M., Roder, K. E., Mumper, R. J., & Allen, D. D. (2003). Brain uptake of thiamine-coated nanoparticles. *Journal of Controlled Release, 93*, 271–282.

Lode, J., Fichtner, I., Kreuter, J., Berndt, A., Diederichs, J. E., & Reszka, R. . (2001). Influence of surface-modifying surfactants on the pharmacokinetic behavior of 14C-poly(methylmethacrylate) nanoparticles in experimental tumor models. *Pharmaceutical Research, 18*, 1613–1619.

Lu, W., Sun, Q., Wan, J., She, Z., & Jiang, X. (2006). Cationic Albumin-conjugated pegylated nanoparticles allow gene delivery into brain tumors via intravenous administration. *Cancer Research, 66*, 11878–11887.

Lu, W., Wan, J., She, Z., & Jiang, X. (2007). Bain delivery property and accelerated blood clearance of cationic albumin conjugated pegylated nanoparticle. *Journal of Controlled Release, 118*, 38–53.

Lu, W., Zhang, X., Tan, Y. Z., Hu, K. L., Jiang, X. G., & Fu, S. K. (2005). Cationic albumin-conjugated pegylated nanoparticles as novel drug carrier for brain delivery. *Journal of Controlled Release, 107*, 428–448.

Lukowski, G., Müller, R. H., Müller, B. W., & Dittgen, M. (1992). Acrylic acid copolymer nanoparticles for drug delivery: I. Characterization of the surface properties relevant for *in vivo* organ distribution. *International Journal of Pharmaceutics, 84*, 23–31.

Maassen, S., Fattal, E., Müller, R. H., & Couvreur, P. (1993). Cell cultures for the assessment of toxicity and uptake of polymeric particulate drug carriers. *STP Pharma Sciences, 3*, 11–22.

Manjunath, K., & Venkateswarlu, V. (2005). Pharmacokinetics, tissue distribution and bioavailability of clozapine solid lipid nanoparticles after intravenous and intraduodenal administration. *Journal of Controlled Release, 107*, 215–228.

Maruyama, K., Takizawa, T., Yuda, T., Kennelm, S. J., Huang, L., & Iwatsuru, M. (1995). Targetability of novel immunoliposomes modified with amphipathic poly (ethylene glycol)s conjugated of their distal terminals to monoclonal antibodies. *Biochimica et Biophysica ACTA, 1234*, 74–80.

Mayer, L. D., Bally, M. B., & Cullis, P. R. (1986). Uptake of adriamycin into large unilamellar vesicles in response to a pH gradient. *Biochimica et Biophysica ACTA, 857*, 123–126.

Menei, P., Daniel, V., Montero-Menei, C., Brouillard, M., Pouplard-Barthelaix, A., & Benoit, J. P. (1993). Biodegradation and brain tissue reaction to poly(D,L-lactide-co-glycolide) microspheres. *Biomaterials, 14*, 470–478.

Michaelis, K., Hoffmann. M. M., Dreis. S., Herbert. E., Alyautdin, R. N., Michaelis, M., et al. (2006). Covalent linkage of apolipoprotein E to albumin nanoparticles strongly enhances drug transport into the brain. *Journal of Pharmacology and Experimental Therapeutics, 317*, 1246–1253.

Moghimi, S. M., & Szebeni, J. (2003). Stealth liposomes and long circulating nanoparticles: Critical issues in pharmacokinetics, opsonization and protein-binding properties. *Progress in Lipid Research, 42*, 463–478.

Mora, M., Sagrista, M. L., Trombetta, D., Bonina, F. P., De Pasquale, A., & Sajia, A. (2002). Design and characterization of liposomes containing long-chain N-AcylPEs for brain delivery: Penetration of liposomes incorporating GM_1 into the rat brain. *Pharmaceutical Research, 19*, 1430–1438.

Müller, R. H., Lherm, C., Herbert, J., & Couvreur, P. (1990). *In vitro* model for the degradation of alkylcyanoacrylate nanoparticles. *Biomaterials, 11*, 590–595.

Müller, R. H., Lherm, C., Herbort, J., Blunk, T., & Couvreur, P. (1992). Alkylcyanoacrylate drug carriers: I. Physicochemical characterization of nanoparticles with different alkyl chain length. *International Journal of Pharmaceutics, 84*, 1–11.

Nam, Y. S., Park, Y., Han, S. H., & Chang, I. S. (2002). Intracellular drug delivery using poly(D,L-lactide-co-glycolide) nanoparticles derivatized with a peptide from a transcriptional activator protein of HIV-1. *Biotechnology Letters, 14*, 2093–2098.

New, R. R.C. (1989). Preparation of liposomes. In R. R.C. New (Ed.), *Liposomes. A practical approach* (pp. 33–104). London: Oxford University Press.

Nicholas, A. P., McInnis, C., Gupta, K. B., Snow, W. W., Love, D. F., Mason, D. W., et al. (2002). The fate of biodegradable microspheres injected into rat brain. *Neuroscience Letters, 323*, 85–88.

Nuzzo, F., Sala, F., Biondi, O., Casati, A., Cestaro, B., & De Carli, L. (1985). Liposomes induce chromosome aberrations in human cultured cells. *Experimental Cell Research, 157*, 397–408.

Olivier, J. C. (2005). Drug transport to brain with targeted nanoparticles. *NeuroRx, 2*, 108–119.

Olivier, J. C., Fenart, L., Chauvet, R., Pariat, C., Cecchelli, R., & Couet, W. (1999). Indirect evidence that drug brain targeting using polysorbate-80 coated polybutylcyanoacrylate nanoparticles is related to toxicity. *Pharmaceutical Research, 16*, 1836–1842.

Owens, D. E., III, & Peppas, N. A. (2006). Opsonization, biodistribution, and pharmacokinetics of polymeric nanoparticles. *International Journal of Pharmaceutics, 307*, 93–102.

Pang, Z., Lu, W., Gao, H., Hu, K., Chen, J., Zhang, C., et al. (2008). Preparation and brain delivery property of biodegradable polymersomes conjugated with OX26. *Journal of Controlled Release, 128*, 120–127.

Panyam, J., Zhou, W. Z., Prabha, S., Sahoo, S. K., & Labhasetwar, V. (2002). Rapid endo-lysosomal escape of poly(D,L-lactide-co-glycolide) nanoparticles: Implications for drug and gene delivery. *FASEB Journal, 16*, 1217–1226.

Panzenboeck, U., Balasz, A., Sovic, A., Hrzenjac, A., Levak-Frank, S., Wintersperger, A., et al. (2002). ABCA1 and scavenger receptor class B, type I, are modulators of reverse sterol transport at an *in vitro* blood brain barrier constituted of porcine brain capillary endothelial cells. *Journal of Biological Chemistry, 277*, 42781–42789.

Pardridge, W. M. (1995). Vector-mediated peptide drug delivery to the brain. *Advanced Drug Delivery Reviews, 15*, 109–146.

Pardridge, W. M. (2001). *Brain drug targeting: The future of brain drug development* (pp. 1–370). Cambridge: Cambridge University Press.

Pardridge, W.M. (2002). Non-invasive gene targeting to the brain. US-Patent 6372250.

Parhham, M. J., & Wetzig, H. (1993). Toxicity screening of liposomes. *Chemistry and Physics of Lipids, 64*, 263–274.

Peracchia, M. T., Gref, R., Minamitake, Y., Domb, A., Lotan, N., & Langer, R. (1997). PEG-coated nanospheres from amphiphilic diblock and multiblock copolymers: Investigation of their drug encapsulation and release characteristics. *Journal of Controlled Release, 46*, 223–231.

Petri, B., Bootz, A., Khalansky, A., Hekmatara, T., Muller, R., Uhl, R., et al. (2007). Chemotherapy of brain tumors using doxorubicin bound to surfactant-coated poly(butyl cyanoacrylate) nanoparticles: Revisiting the role of surfactants. *Journal of Controlled Release, 117*, 51–58.

Polt, R., & Palian, M. M. (2001). Glycopeptide analgesics. *Drugs Future, 26*, 561–576.

Qian, Z. M., Li, H., Sun, H., & Ho, K. (2002). Targeted drug delivery via the transferrin receptor mediated endocytosis pathway. *Pharmacological Reviews, 54*, 561–587.

Reddy, H. L., Sharma, R. K., Chuttani, K., Mishra, A. K., & Murthy, R. R. (2004). Etoposide-incorporated tripalmitin nanoparticles with different surface charge: Formulation, characterization, radiolabeling, and biodistribution studies. *AAPS Journal, 6*, 1–10, Article 23.

Roser, M., Fischer, D., & Kissel, T. (1998). Surface-modified biodegradabile albumin nano- and microspheres. II: Effect of surface charge on *in vitro* phagocytosis and biodistribution in rats. *European Journal of Pharmaceutics and Biopharmaceutics, 46*, 255–263

Ruozi, B., Forni, F., Battini, R., & Vandelli, M. A. (2003). Cationic liposomes for gene transfection. *Journal of Drug Targeting, 11*, 407–414.

Sahoo, S. K., Panyam, J., Prabha, S., & Labhasetwar, V. (2002). Residual polyvinyl alcohol associated with poly (image,image-lactide-co-glycolide) nanoparticles affects their physical properties and cellular uptake. *Journal of Controlled Release, 82*, 105–114.

Santra, S., Yang, H., Dutta, D., Stanley, J. T., Holloway, P. H., Tan, W., et al. (2004). TAT conjugated, FITC doped silica nanoparticles for bioimaging applications. *Chemical Communications, 24*, 2810–2811.

Santra, S., Yang, H., Stanley, J. T., Holloway, P. H., Moudgil, B. M., Walter, G., et al. (2005). Rapid and effective labelling of brain tissue using TAT-conjugated CdS:Mn/ZnS quantum dots. *Chemical Communications, 25*, 3144–3146.

Sauer, I., Dunay, I. R., Weisgraber, K., Bienert, M., & Dathe, M. (2005). An apolipoprotein E-derived peptide mediates uptake of sterically stabilized liposomes into brain capillary endothelial cells. *Biochemistry, 44*, 2021–2029.

Sauer, I., Nikolenko, H., Keller, S., Ajaj, K. A., Bienert, M., & Dathe, M. (2006). Dipalmitoylation of a cellular uptake-mediating apolipoprotein E-derived peptide as a promising modification for stable anchorage in liposomal drug carriers. *Biochimica et Biophysica ACTA*, *1758*, 552–561.

Saul, J. M., Annapragada, A., Natarajan, J. V., & Bellamkonda, R. V. (2003). Controlled targeting of liposomal doxorubicin via the folate receptor *in vitro*. *Journal of Controlled Release*, *92*, 49–67.

Schlageter, K. E., Molnar, P., Lapin, G. D., & Groothuis, D. R. (1999). Microvessel organization and structure in experimental brain tumors: Microvessel populations with distinctive structural and functional properties. *Microvascular Research*, *58*, 312–328.

Schroeder, U., Schroeder, H., & Sabel, B. A. (2000). Body distribution of 3H-labelled dalargin bound to poly(butylcyanoacrylate) nanoparticles after their i.v. injection in mice. *Life Sciences*, *66*, 495–502.

Schwarze, S. R., Ho, A., Vocero-Akbani, A., & Dowdy, S. F. (1999). *In vivo* protein transduction: Delivery of a biologically active protein into the mouse. *Science*, *285*, 1569–1572.

Senior, J. (1987). Fate and behaviour of liposomes *in vivo*: A review of controlling factors. *Critical Reviews in Therapeutic Drug Carrier Systems*, *3*, 123–193.

Shi, N., Boado, R. J., & Pardridge, W. M. (2001). Receptor-mediated gene targeting to tissues *in vivo* following intravenous administration of pegylated immunoliposomes. *Pharmaceutical Research*, *18*(8), 1091–1095.

Shi, N., & Pardridge, W. M. (2000). Noninvasive gene targeting to the brain. *Proceedings of the National Academy of Sciences of the United States of America*, *97*, 7567–7572.

Shi, N., Zhang, Y., Zhu, C., Boado, R. J., & Pardridge, W. M. (2001). Brain specific expression of an exogenous gene after i.v. administration. *Proceedings of the National Academy of Sciences of the United States of America*, *98*, 12754–12759.

Smith, Q. R. (1993). Drug delivery to the brain and the role of carrier mediated transport. In L. R. Drewes & A. L. Betz (Eds.), *Frontiers in cerebral vascular biology: Transport and its regulation* (pp. 83–93). New York: Plenum.

Sonavane, G., Tomoda, K., & Makino, K. (2008). Biodistribution of colloidal gold nanoparticles after intravenous administration: Effect of particle size. *Colloids and Surfaces B: Biointerfaces*, *66*, 274–280.

Soni, V., Kohli, D. V., & Jain, S. K. (2005). Transferrin coupled liposomes as drug delivery carriers for brain targeting of 5-fluorouracil. *Journal of Drug Targeting*, *13*, 245–250.

Soni, V., Kohli, D. V., & Jain, S. K. (2008). Transferrin-conjugated liposomal system for improved delivery of 5-fluorouracil to brain. *Journal of Drug Targeting*, *16*, 73–78.

Soppimath, K. S., Aminabhavi, T. M., Kulkarni, A. R., & Rudzinski, W. E. (2001). Biodegradable polymeric nanoparticles as drug delivery devices. *Journal of Controlled Release*, *70*, 1–20.

Steiniger, S. C.J., Kreuter, J., Khalansky, A. S., Skidan, I. N., Bobruskin, A. T., Smirnova, Z. S., et al. (2004). Chemotherapy of glioblastoma in rats using doxorubicin-loaded nanoparticles. *International Journal of Cancer*, *109*, 759–767.

Szoka, F., & Papahadjopoulos, D. (1978). Procedure for preparation of liposomes with large internal aqueous space and high capture by reverse-phase evaporation. *Proceedings of the National Academy of Sciences of the United States of America*, *75*, 4194–4198.

Tabata, Y., & Ikada, Y. (1988). Effect of the size and surface charge of polymer microspheres on their phagocytosis by macrophage. *Biomaterials*, *9*, 356–362.

Takasato, Y., Rapaport, S. I., & Smith, Q. R. (1984). An in situ brain perfusion technique to study cerebrovascular transport in the rat. *American Journal of Physiology*, *247*, M484–M493.

Taniguchi, K., Yamamoto, Y., Itakura, K., Miitchi, H., & Hayashi, S. J. (1988). Assessment of ocular irritability of liposomes preparations. *Journal of Pharmacobio-Dynamics*, *11*, 607–611.

Thole, M., Nobmann, S., Huwyler, J., Bartmann, A., & Fricker, G. (2002). Uptake of cationized albumin coupled liposomes by cultured porcine brain microvessel endothelial cells and intact brain capillaries. *Journal of Drug Targeting*, *10*, 337– 344.

Tosi, G., Costantino, L., Rivasi, F., Ruozi, B., Leo, E., Vergoni, A. V., et al. (2007). Targeting the central nervous system: *In vivo* experiments with peptide-derivatized nanoparticles loaded with loperamide and Rhodamine-123. *Journal of Controlled Release*, *122*, 1–9.

Triguero, D., Buciak, J., & Pardridge, W. M. (1990). Capillary depletion method for quantification of blood brain barrier transport of circulating peptides and plasma proteins. *Journal of Neurochemistry*, *54*, 1882–1888.

Tseng, Y. C., Tabata, Y., Hyon, S. H., & Ikada, Y. (1990). *In vitro* toxicity of 2-cyanoacrylate polymers by cell culture method. *Journal of Biomedical Materials Research*, *24*, 1355–1367.

Ubrich, N., Bouillot, P., Pellerin, P., Hoffmann, M., & Maincent, P. (2004). Preparation and characterization of propranolol hydrochloride nanoparticles: A comparative study. *Journal of Controlled Release*, *97*, 291–300.

Umezawa, F., & Eto, Y. (1988). Liposome targeting to mouse brain: Mannose as a recognition marker. *Biochemical and Biophysical Research Communications*, *153*, 1038–1044.

Vajkoczy, P., & Menger, M. D. (2004). Vascular microenvironment in gliomas. *Cancer Treatment and Research*, *117*, 249–262.

Vauthier, C., Dubernet, C., Fattal, E., Pinto-Alphandary, H., & Couvreur, P. (2003). Poly(alkylcyanoacrylates) as biodegradable materials for biomedical applications, *Advanced Drug Delivery Reviews*, *55*, 519–548.

Vauthier, C., Labarre, D., & Ponchel, G. (2007). Design aspects of poly(alkylcyanoacrylate) nanoparticles for drug delivery. *Journal of Drug Targeting*, *15*, 641–663.

Vergoni, A. V., Tosi, G., Tacchi, R., Vandelli, M. A., Bertolini, A., Costantino, L. (in press). Nanoparticles as drug delivery agents specific for CNS: in vivo biodistribution. *Nanomedicine*, in press.

Vinogradov, S. V., Batrakova, E. V., & Kabanov, A. V. (2004). Nanogels for oligonucleotide delivery to the brain. *Bioconjugate Chemistry*, *15*, 50–60.

Vinters, H. V., & Ho, H. W. (1988). Effects of isobutyl 2-cyanoacrylate polymer on cultured cells derived from murine cerebral microvessels. *Toxicology. In Vitro*, *2*, 37–41.

Visser, C. C., Stefanovich, S., Voorwinden, L. H., Gaillard, P. J., Crommelin, A. J.A., Danhof, M., et al. (2004). Validation of transferrin receptor for drug delivery to brain capillary endothelial cells *in vitro*. *Journal of Drug Targeting*, *12*, 145–150.

Visser, C. C., Stefanovic, S., Voorwinden, L. H., van Bloois, L., Gaillard, P. J., Danhof, M., et al. (2005). Targeting liposomes with protein drugs to the blood-brain barrier *in vitro*. *European Journal of Pharmaceutical Science*, *25*, 299–305

Vonarbourg, A., Passirani, C., Saulnier, P., & Benoit, J. P. (2006). Parameters influencing the stealthiness of colloidal drug delivery systems. *Biomaterials*, *27*, 4356–4373.

Wang, S., Lee, R. J., Cauchon, G., Gorenstein, D. G., & Low, P. S. (1995). Delivery of antisense oligodeoxyribonucleotides against the human epidermal growth factor receptor into cultured KB cells with liposomes conjugated to folate via polyethylene glycol, *Proceedings of the National Academy of Sciences of the United States of America*, *92*, 3318–3322.

Wei, L., Tanb, Y., Hua, K., & Jiang, X. (2005). Cationic albumin conjugated pegylated nanoparticle with its transcytosis ability and little toxicity against blood-brain barrier, *International Journal of Pharmaceutics*, *295*, 247–260.

Weitman, S. D., Lark, R. H., Coney, L. R., Fort, D. W., Frasca, V., Zurawski, V. R., et al. (1992). Distribution of the folate receptor GP38 in normal and malignant cell lines and tissues. *Cancer Research*, *52*, 3396–3401.

Xiang, J. J., Tang, J. Q., Zhu, S. G., Nie, X. M., Lu, H. B., Shen, S. R., et al. (2003). IONP-PLL: A novel non-viral vector for efficient gene delivery. *Journal of Gene Medicine*, *5*, 803–817.

Xiang, J. J., Zhu, S. G., & Li, G. Y. (2001). Using iron oxide nanoparticles as gene carrier. *Chinese Journal of Cancer*, *20*, 1–6.

Xie, Y., Ye, L., Zhang, X., Cui, W., Lou, J., Nagai, T., et al. (2005). Transport of nerve growth factor encapsulated into liposomes across the blood–brain barrier: *In vitro* and *in vivo* studies. *Journal of Controlled Release*, *105*, 106–119.

Zelphati, O., & Szoka, F. C. (1996). Mechanism of oligonucleotide release from cationic liposomes. *Proceedings of the National Academy of Sciences of the United States of America*, *93*, 11493–11498.

Zhang, X., Xie, J., Li, S., Wang, X., & Hou, X. (2003). The study on brain targeting of the amphotericin B liposomes. *Journal of Drug Targeting*, *11*, 117–122.

Zhang, Y., Calon, F., Zhu, C., Boado, R., & Pardridge, W. M. (2003). Intravenous non-viral gene therapy causes normalization of striatal tyrosine hydroxylase and reversal of motor impairment in expertimental parkinsonism. *Human Gene Therapy*, *14*, 1–12.

Zhang, Y., Lee, H. J., Boado, R. J., & Pardridge, W. M. (2002). Receptor-mediated delivery of an antisense gene to human brain cancer cells. *Journal of Gene Medicine*, *4*, 183–194.

Zhang, Y., Schlachetzki, F., & Pardridge, W. M. (2003). Global non viral gene transfer to the primate brain following intravenous administration. *Molecular Therapy*, *7*, 11–18.

Zhang, Y., Schlachetzki, F., Zhang, Y. F., Boado, R. J., & Pardridge, W. M. (2004). Normalization of striatal tyrosine hydroxylase and reversal of motor impairment in experimental parkinsonism with intravenous non viral gene therapy and a brain specific promoter. *Human Gene Therapy*, *15*, 339–350.

Zhang, Y., Zhang, Y. F., Bryant, J., Charles, A., Boado, R. J., & Pardridge, W. M. (2004). Intravenous RNA interference gene therapy targeting the human epidermal growth factor receptor prolongs survival in intracranial brain cancer. *Clinical Cancer Research*, *10*, 3667–3677.

Zhang, Y., Zhu, C., & Pardridge, W. M. (2002). Antisense gene therapy of brain cancer with an artificial virus gene delivery system. *Molecular Therapy*, *6*, 67–72.

SECTION II

Application of Nanotechnology in Brain Disease

H.S. Sharma (Ed.)
Progress in Brain Research, Vol. 180
ISSN: 0079-6123

CHAPTER 4

Applications of nanotechnology in molecular imaging of the brain

Martina A. McAteer* and Robin P. Choudhury

Department of Cardiovascular Medicine, John Radcliffe Hospital, Headington, Oxford, UK

Abstract: Rapid advances in the field of nanotechnology promise revolutionary improvements in the diagnosis and therapy of neuroinflammatory disorders. An array of iron oxide nano- and microparticle agents have been developed for *in vivo* molecular magnetic resonance imaging (mMRI) of cerebrovascular endothelial targets, such as vascular cell adhesion molecule-1 (VCAM-1), E-selectin, and the glycoprotein receptor GP IIb/IIIa expressed on activated platelets. Molecular markers of glioma cells, such as matrix metalloproteinase-2 (MMP-2), and markers for brain tumor angiogenesis, such as alpha (v) beta (3) integrin ($\alpha_v\beta_3$), have also been successfully targeted using nanoparticle imaging probes. This chapter provides an overview of targeted, iron oxide nano- and microparticles that have been applied for *in vivo* mMRI of the brain in experimental models of multiple sclerosis (MS), brain ischemia, cerebral malaria (CM), brain cancer, and Alzheimer's disease. The potential of targeted nanoparticle agents for application in clinical imaging is also discussed, including multimodal and therapeutic approaches.

Keywords: Molecular magnetic resonance imaging; Ultra small particles of iron oxide; microparticles of iron oxide; nanoparticle contrast agents; vascular cell adhesion molecule-1; activated platelets; tumor angiogenesis; glioma

Introduction

Rapid expansion in the field of nanotechnology has led to the development of an array of nanoparticle contrast agents for application in medicine. The functionalization of the surface of nanoparticles with targeting ligands has led to substantial interest in the diagnostic and therapeutic potential of targeted nanoparticles for *in vivo* molecular imaging of the central nervous system (CNS). The most successful approaches have involved the synthesis of nanoparticles that combine target specificity with the capacity to carry a substantial payload of paramagnetic gadolinium chelates or superparamagnetic iron oxide agents. The targeted delivery of iron oxide nanoparticles to specific molecules of interest has been a focus of recent research and this objective has been accomplished by covalent conjugation of ligands, such as antibodies, peptides, or small-molecule peptidomimetics to the nanoparticle surface.

In this chapter, we highlight a number of methods available for the conjugation of ligands to the surface of functionalized iron oxide agents and review recent advances in their application to molecular imaging of

*Corresponding author.
Tel.: +44 (0) 1865 234647; Fax: +44 (0) 1865 234681
E-mail: martina@well.ox.ac.uk,
robin.choudhury@cardiov.ox.ac.uk

DOI: 10.1016/S0079-6123(08)80004-0

the brain. We focus in particular on molecular magnetic resonance imaging (mMRI), directed toward five main areas: (1) detection of macrophage infiltration in multiple sclerosis (MS) lesions, brain ischemia, and brain tumors; (2) vascular endothelial cell adhesion molecule expression in brain inflammation; (3) activated platelet-rich thrombi in cerebral malaria (CM) and ischemic stroke; (4) brain tumor detection and tumor angiogenesis; (5) β-amyloid plaques in Alzheimer's disease.

Multifunctional nanoprobes with combined diagnostic and therapeutic capabilities, and multimodal nanoprobes that can be simultaneously analyzed using a combination of MRI and positron emission tomography (PET) or single photon emission computed tomography (SPECT), or MRI and optical imaging techniques such as near-infra red fluorescence (NIRF) are also highlighted.

Molecular imaging modalities

Molecular imaging modalities include PET, SPECT, targeted ultrasound, mMRI, and optical fluorescence. Multimodal methods that combine two or more imaging modalities such as PET-MRI and SPECT-MRI are also under development (Beyer & Townsend, 2006; Catana et al., 2006; Pichler, Judenhofer, & Wehrl, 2008; Schlemmer, Pichler, Krieg, & Heiss, 2008; Somer, Benatar, O'Doherty, Smith, & Marsden, 2007). PET imaging uses exogenously administered radionuclides, such as 2-[$_{18}$F] fluoro-2-deoxy-D-glucose (FDG), which emit a signal that can be quantified by an external detector. PET has a much higher sensitivity (picomolar range) than MRI or SPECT and much better tissue penetration than ultrasound or optical imaging. However, the disadvantages of PET are its poor spatial resolution, substantial radiation exposure, and expense. In addition, despite the high sensitivity of FDG PET, its specificity is limited and therefore it is not used by itself to diagnose conditions such as Alzheimer's disease (Swerdlow, 2007). Contrast-enhanced ultrasound techniques using microbubbles have also been applied to image cerebrovascular inflammation (Klibanov, 2007; Linker et al., 2005; Raymond et al., 2008; Reinhardt et al., 2005) and tumor angiogenesis (Ellegala et al., 2003), but they are inherently limited by the need for acoustic windows and ultrasound cannot reliably penetrate the human skull.

By contrast, MRI provides exquisite spatial resolution and soft tissue contrast, without ionizing radiation, making MRI the imaging modality of choice for many brain pathologies and driving the need for mMRI contrast. The principal challenge of mMRI, compared to other imaging techniques such as PET and optical fluorescence, is its inherently low sensitivity. However, the use of paramagnetic (Gd-based) or superparamagnetic (iron oxide-based) nanoparticle MR contrast agents can overcome this limitation.

MRI contrast agents

Gadolinium-based agents

Gadolinium (Gd) is a lanthanide metal ion with seven unpaired electrons that has been demonstrated to enhance proton relaxation because of its high magnetic moment and water coordination (Runge & Gelblum, 1991). Paramagnetic Gd chelates shorten T_1 longitudinal relaxation times, producing positive contrast on T_1-weighted MR images. Gadolinium diethylene triamine pentaacetic acid (Gd-DTPA) is a low molecular weight (MW 590 Da) contrast agent routinely used clinically to visualize defects in the blood–brain barrier (BBB) of the CNS, caused by inflammatory lesions or tumors. Due to its small size, Gd-DTPA rapidly diffuses across disrupted BBB into the extracellular space, thus characterizing one feature that occurs relatively late in the inflammatory pathway (Grossman et al., 1988). However, Gd-enhanced MRI at best correlates moderately with clinical manifestations in MS and is not a strong predictor of cumulative impairment or disability (Kappos et al., 1999; Lee et al., 1999; Miller, 2004; Siva, 2006).

Conventional extracellular Gd chelates confer relatively low sensitivity (in the micromolar range), which is usually inadequate for application to mMRI. The extremely short intravascular half-lives and rapid renal excretion of Gd chelates further limit their use for most mMRI applications. To address these limitations, several novel

Gd nanoparticles have been synthesized that carry a substantial payload of Gd chelates. Examples include liposomes with amphipathic Gd chelates embedded in the outer membrane (Sipkins et al., 1998), perfluorocarbon lipid emulsions (Yu et al., 2000), micelles (Amirbekian et al., 2007; Lipinski et al., 2006), and lipoproteins (Frias, Williams, Fisher, & Fayad, 2004; Glickson et al., 2008).

Iron oxide-based agents

Superparamagnetic iron oxide-based contrast agents have superior contrast effects compared to Gd, due to their inherent amplification created by thousands of iron atoms (Mendonca Dias & Lauterbur, 1986; Renshaw, Owen, McLaughlin, Frey, & Leigh, 1986). Superparamagnetic iron oxide agents consist of a magnetite (Fe_3O_4) and/or maghemite (Fe_2O_3) crystalline core surrounded by a low molecular weight carbohydrate (usually dextran or carboxydextran) or polymer coat. They can be classified according to their size as ultrasmall superparamagnetic particles of iron oxide (USPIO) (20–50 nm diameter), superparamagnetic particles of iron oxide (SPIO) (60 to ~250 nm), and microparticles of iron oxide (MPIO) (from 0.9 μm upward). Monocrystalline iron oxide nanoparticles (MION) are a subset of USPIO (10–30 nm diameter), which have a single crystal core as opposed to the multiple iron oxide crystals of SPIO, and a much longer half-life (Thorek, Chen, Czupryna, & Tsourkas, 2006). Cross-linked iron oxide (CLIO) nanoparticles are modified MION with cross-linked (caged) dextran chains on their surface coat (McCarthy & Weissleder, 2008). These offer greater stability under harsh conditions without a change in size, blood half-life, or loss of dextran coat. Nanoparticles with biodegradable, high-affinity surface coats are under development (Arias, Gallardo, Gomez-Lopera, Plaza, & Delgado, 2001; Butoescu, Jordan, Petri-Fink, Hofmann, & Doelker, 2008; Ngaboni Okassa et al., 2005).

Iron oxide agents shorten T_2 and T_2^* relaxation times on T_2- and T_2^*-weighted MR images, creating low signal or negative contrast. They can also be detected by MRI with T_1, off resonance, and steady state free precession (SSFP) sequences (Sosnovik, Nahrendorf, & Weissleder, 2008). Several SPIO formulations have been developed, for example, Sinerem® (also known as Combidex®) for differentiating metastatic lymph nodes, GastroMARK for bowel imaging and Feridex® (same as Endorem®), a liver contrast agent approved for clinical use by the US Food and Drug Administration (FDA). USPIO are subject to phagocytotic clearance by Kupffer cells of the reticuloendothelial system and can be recycled using normal biochemical pathways for iron metabolism, including rapid turnover into body iron stores and incorporation into erythrocyte hemoglobin, since they are composed of biodegradable iron (Weissleder et al., 1990).

USPIO have become popular owing to their long blood half-life (up to 24 h) compared to Gd agents, which is a positive attribute for applications such as measurement of changes in cerebral perfusion. However, for mMRI this property is more of a hindrance since it may lead to high background contrast for an extended period. Furthermore, USPIO contrast effects are manifested in T_2^*-weighted images as indistinct areas of low signal that can be difficult to distinguish from the ordinary heterogeneity of normal tissue and other susceptibility artefacts. To overcome this limitation, several groups have developed "positive" contrast sequences for MRI of iron oxide particles, which allow for the generation of positive MR signal in volumes or regions containing iron particles (Briley-Saebo, Mani, Hyafil, Cornily, & Fayad, 2008; Lipinski, Briley-Saebo, Mani, & Fayad, 2008). A further drawback of USPIO is that they can be taken up nonspecifically by endothelial cells, leading to potential compromise to the specificity of molecular targeting (Briley-Saebo et al., 2004). Finally, although USPIO provide greater contrast than Gd, they may require sophisticated ligands that mediate internalization by endothelial cells in order to achieve adequate local concentrations (Nahrendorf et al., 2006a).

MPIO are superparamagnetic particles consisting of a Fe_2O_3 and/or Fe_3O_4 core surrounded by a polymer coat. For mMRI, MPIO create potent hypointense contrast effects on T_2^*-weighted images that extend to a distance roughly 50 times the physical diameter of the MPIO (Shapiro, Sharer, Skrtic, & Koretsky, 2006). The potency of the contrast effects derive from their high iron content and are significantly greater than that of USPIO. Shapiro and colleagues have reported the use of MPIO for *in vivo* detection of single cells (Shapiro, Sharer, Skrtic, & Koretsky, 2006) and cell tracking by MRI (Shapiro, Skrtic, & Koretsky, 2005). MPIO are less susceptible to nonspecific vascular egress or uptake by endothelial cells due to their size and incompressible nature (Briley-Saebo et al., 2006). Therefore, MPIO may be useful for the detection of endovascular molecular targets by mMRI. Unlike USPIO or Gd-based agents, it is not necessary to deliver very large numbers of MPIO to establish strong contrast effects. The nonbiodegradable coats of most commercially available MPIO makes them unsuitable for use in humans owing to potential toxicity, but the development of micron-sized biodegradable contrast particles for clinical use is underway (Chen et al., 2005; Hamoudeh & Fessi, 2006; Hemmingsson et al., 1987; Zhu et al., 2006).

Targeting of iron oxide-based contrast agents

The targeting of iron oxide-based contrast agents to specific sites of interest is accomplished by the conjugation of ligands to functional groups on the surface of the nanoparticle. The large surface area to volume ratio of nanoparticles permits multiple copies of a ligand to be attached per nanoparticle. This multivalent targeting increases the binding affinity of nanoparticles dramatically, allowing each nanoparticle to bind to multiple copies of receptors, expressed on the cell surface (Montet, Montet-Abou, Reynolds, Weissleder, & Josephson, 2006). Targeting ligands include antibodies or their immunospecific fragments F (ab), aptamers, peptides, or small-molecule peptidomimetics emerging from phage display or small-molecule screens. Phage display technology, introduced by Smith (1985) involves the generation of libraries of bacteriophage, viruses that infect bacterial cells, using standard recombinant DNA technology, with each phage expressing a different synthetic peptide sequence on its coat. After several consecutive rounds of affinity selection, DNA of the bound phage is sequenced, to produce novel antibody or peptide ligands.

A range of iron oxide nanoparticles and MPIO, functionalized with reactive surface groups, is available for covalent conjugation of protein, antibodies, or small peptide ligands. Nanoparticles functionalized with reactive carboxylic acid or amine surface groups are most commonly used. Carboxylic acid functionalized nanoparticles, unlike amine groups, require activation prior to conjugation, using either carbodiimide, most commonly 1-ethyl-3-(3-dimethylaminopropyl) (EDC), or a combination of EDC and *N*-hydroxyl succinimide ester (NHS). Amine functionalized nanoparticles can be directly conjugated to reactive ligands through reductive amination of aldehydes. The surface of amine-functionalized nanoparticles can also be modified by coupling cross-linking reagents, such as succinimidyl iodoacetate (SIA), *N*-succinimidyl-3-(2-pyridyldithio) propionate (SPDP), or succinimidyl-4-(*N*-maleimidomethyl)cyclohexane-1-carboxylate (SMCC), to enable conjugation of ligands bearing sulfhydryl groups, such as those present in peptides and antibodies.

MPIO with a variety of reactive surface groups are available from commercial sources including Invitrogen, Bangs Laboratories and Miltenyi Biotec Ltd. McAteer et al. (2007, 2008) have recently applied MPIO (1 μm diameter) with reactive tosyl groups on their surface for direct covalent conjugation of monoclonal antibodies. Tosyl-activated MPIO have the advantage that, unlike reactive carboxylic acid surface groups, they do not require surface activation and can be used for direct covalent conjugation of protein, peptides, or antibodies (Fig. 1). The hydrophobic properties of tosyl-activated MPIO facilitate optimal antibody orientation since Fc regions of the antibody, which are generally more hydrophobic than the Fab portion, will adsorb to the hydrophobic surface of MPIO followed by rapid covalent bond formation. This leaves the Fab portion exposed, thus maximizing the binding potential of antibody-conjugated MPIO to the target protein. McAteer et al. (2007) have also applied tosyl activated MPIO-based constructs for *in vivo* mMRI of

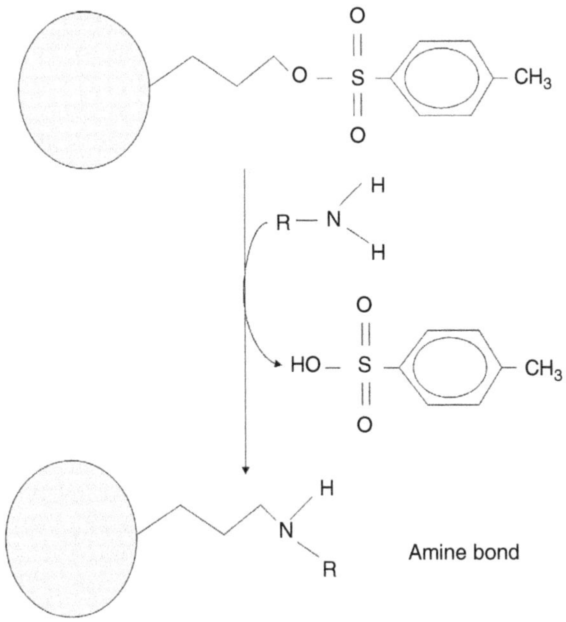

Fig. 1. Bioconjugation of ligands to tosyl activated MPIO. The hydroxy groups are activated by reaction with *p*-toluenesulfonyl chloride. The resulting sulfonyl ester can subsequently react covalently with proteins or other ligands containing amino or sulfhydryl groups.

VCAM-1 in acute brain inflammation . For labile temperature-sensitive ligands, MPIO with reactive epoxy groups are also available for direct covalent conjugation at low temperatures. Recently, cobalt-functional MPIO (1 μm diameter) have been used for direct covalent conjugation of MPIO to the histidine (His) tag of single-chain antibodies, targeted toward activated platelets in mouse models of CM (von Zur Muhlen et al., 2008c) and atherothrombosis (von Zur Muhlen et al., 2008b,d).

Molecular imaging of the brain

Macrophage infiltration in the CNS

Monocyte/macrophage infiltration in the CNS plays a key role in neuroinflammation and subsequent lesion development and brain injury in neurological diseases such as MS and stroke (Jander, Schroeter, & Saleh, 2007; Price, Warburton, & Menon, 2003; Stoll, Jander, & Schroeter, 1998). USPIO, administered intravenously *in vivo*, have been shown to detect activated macrophage infiltration in the CNS of experimental models including the experimental autoimmune encephalitis (EAE) rodent model of MS (Deloire et al., 2004; Dousset, Gomez, Petry, Delalande, & Caille, 1999a; Dousset et al., 1999b; Floris et al., 2004; Rausch, Hiestand, Baumann, Cannet, & Rudin, 2003), brain ischemia (Kleinschnitz et al., 2005; Rausch et al., 2001; Rausch, Baumann, Neubacher, & Rudin, 2002; Schroeter, Saleh, Wiedermann, Hoehn, & Jander, 2004), and brain tumors (Zimmer et al., 1995) using MRI. Most studies have used ferumoxtran-10 (Sinerem, Combidex, AMI-227) to visualize macrophages in the CNS (Brochet et al., 2006; Dousset et al., 1999a,b; Floris et al., 2004; Rausch et al., 2004), although MION agents (Xu et al., 1998) and the more recently developed USPIO agent, ferumoxytol (7228) (Floris et al., 2004), have also been investigated. In humans, ferumoxtran-10 has been used to detect macrophage infiltration in the brains of patients with MS (Dousset et al., 2006; Manninger et al., 2005), stroke (Nighoghossian et al., 2007; Saleh et al., 2004), and intracranial tumors

(Neuwelt et al., 2004). Recently, a novel USPIO agent, SHU555C, which is smaller than ferumoxtran-10 (25 vs. 30 nm), which has a shorter half-life (6–8 vs. 24–30 h), and is negatively charged, has been shown to be incorporated more efficiently by activated monocytes *in vitro* than ferumoxtran-10 (Metz et al., 2004). The SHU555C agent also produces *in vivo* hypointense contrast effects on MRI corresponding to macrophage infiltration in human MS lesions, which can both precede and persist longer than Gd contrast enhancement (Vellinga et al., 2008).

Similar differences in USPIO enhancement, distinct from the Gd enhancement of BBB permeability have been shown using ferumoxtran-10 in EAE (Dousset et al., 1999b; Floris et al., 2004) and more recently in human ischemic stroke (Saleh et al., 2004). Dousset et al. (2006) observed that USPIO enhancement corresponding to monocyte/macrophage infiltration in inflammatory lesions in humans with relapsing–remitting MS (RRMS) was detected in the presence or absence of Gd-enhanced BBB permeability (Dousset et al., 2006). It has been hypothesized that USPIO, due to their size (~30 nm diameter), cannot diffuse easily out of the vascular space but instead are actively taken up by circulating monocytes/macrophages across the BBB (Dousset et al., 2006; Manninger et al., 2005). It is also possible that USPIO may enter the CNS passively by transcytosis across brain endothelial cells (Xu et al., 1998) or via cellular incorporation outside the vasculature by activated microglia (Dousset et al., 1999c).

USPIO have been applied to track macrophage transmigration *in vivo* in the early stages of ischemia-reperfusion brain injury in a rat model following transient occlusion of the middle cerebral artery (Kim et al., 2008). Focal hypointense signal areas were detected on T_2^*-weighted MR images, corresponding to areas of iron-laden macrophages, in rat brain at days 3 and 4 postreperfusion. USPIO have also been used to track macrophage migration into specific regions of the brain at different stages of disease progression *in vivo* in a rat model of acute EAE (Baeten et al., 2008). Improved visualization of the extent and distribution of active macrophage infiltration was achieved by imaging earlier at disease onset rather than disease top and by increasing the time interval between USPIO administration and MRI to 24 h. Since hardly any hypointense MR signal intensity was detected at recovery, while the BBB is still disrupted (Floris et al., 2004), it is unlikely that USPIO passively diffuse into the CNS. Rather, it is hypothesized that the USPIO enhancement is due to cell-specific uptake.

To investigate monocyte infiltration *in vivo* in the CNS during an inflammatory response, Stroh et al. (2006) transfused spleen-derived monocytes labeled *ex vivo* with USPIO into splenectomized mice after middle cerebral artery occlusion and showed that USPIO-labeled monocytes were engrafted at the lesion border zone in the brain. Engberink et al. (2008) also used SPIO-labeled monocytes to investigate the temporal pattern of monocyte recruitment in the CNS in a rat model of neuroinflammation, experimentally induced photothrombosis (PT) (Engberink et al., 2008). *In vivo* MRI was performed at 24, 72, and 120 h postinjection. Following transfusion of SPIO-labeled monocytes, a significant decrease in signal intensity was observed in the PT lesion only after 72 h, whereas following administration of free USPIO, contrast enhancement was observed much earlier at 24 h and diminished thereafter. In theory, free USPIO are taken up in the blood stream by circulating monocytes. However, repetitive MRI directly after free USPIO administration showed that a decrease in signal intensity occurs rapidly after 2 h. This rapid USPIO enhancement is distinct from the much slower SPIO-labeled monocytes enhancement (>72 h), suggesting that USPIO enhancement does not primarily represent signal from peripherally labeled monocytes that migrate toward the inflammatory lesion. Therefore, SPIO-labeled monocytes may more accurately assess the time window of monocyte infiltration into the brain and aid the design of effective intervention treatments.

MPIO have been shown to detect single macrophages in mouse brain by *in vivo* MRI at clinical field strength of 1.5 T using a balanced steady state free precession (3D b-SSFP) imaging sequence with direct optical validation (Heyn et al., 2006a). MPIO have also been applied for *in vivo* cell tracking of cancer metastases from the single-cell stage in mouse brain by MRI at 1.5 T using MDA-MB-231BR cells labeled with MPIO

(0.9 μm) (Heyn et al., 2006b). Imaging at higher clinical field strength of 3 T significantly improved the number of single MPIO-labeled cells detected in mouse brain *in vivo*, with ~60% of MPIO-labeled cells not detectable using the previously validated 1.5 T protocol (Ramadan et al., 2008).

Endothelial adhesion molecule expression in vascular inflammation

The previous section details "nontargeted" contrast approaches that rely on nonspecific accumulation of iron particles through abnormally permeable blood vessels or their carriage into the vessel wall by macrophages. The following sections examine targeted mMRI, harnessing the signals used for leukocyte recruitment that occur earlier in the inflammatory process.

Early inflammation of the CNS is associated with enhanced expression of cell adhesion molecules on brain endothelial cells, which mediate cellular infiltration. Endothelial vascular cell adhesion molecule-1 (VCAM-1; CD106) and its ligand, $\alpha_4\beta_1$ integrin (also known as very late antigen-4, VLA-4), are key mediators of leukocyte recruitment and lesion development (Elices et al., 1990). VCAM-1 is not constitutively expressed on the cerebral vascular endothelium but is upregulated upon endothelial activation (Carlos et al., 1990). Selective inhibitors that bind to the α_4 subunit of $\alpha_4\beta_1$, blocking association with VCAM-1, have been shown to substantially reduce new or enlarging inflammatory lesions by MRI and clinical relapse in MS (Polman et al., 2006). Therefore, VCAM-1 is an attractive molecular imaging target of early cerebral vascular inflammation.

McAteer et al. (2007) have recently developed a novel, targeted MPIO probe that can identify VCAM-1 expression *in vivo* in mouse acute brain inflammation with exceptional conspicuity and at a time when pathology is undetectable by conventional MRI techniques. Commercially available autofluorescent MPIO (1 μm diameter), with reactive tosyl groups, were used for direct covalent conjugation of mouse monoclonal VCAM-1 antibodies (VCAM–MPIO). The capacity of VCAM–MPIO constructs for specific and quantitative binding was tested *in vitro* using activated mouse endothelial cells (sEND-1), stimulated with graded doses of tumor necrosis factor alpha (TNF-α). Differential interference confocal microscopy demonstrated sparse retention of VCAM–MPIO by unstimulated sEND-1 cells, reflecting low basal VCAM-1 expression, and a dose-dependent increase in VCAM–MPIO binding in response to TNF-α stimulation. Colocalization of VCAM–MPIO binding and VCAM-1 immunofluorescence on the cell surface of TNF-α stimulated cells was demonstrated by confocal microscopy.

The *in vivo* targeting ability of VCAM–MPIO was then tested in a mouse model of acute brain inflammation. In this model, acute inflammation was induced by stereotactic injection of the proinflammatory interleukin 1β (IL-1β) into the left corpus striatum. The right hemisphere of the brain received no injection and served as an internal control. After 3 h, VCAM–MPIO or negative isotype control IgG-MPIO (4×10^8 MPIO) were intravenously injected via a tail vein and allowed to circulate for 1.5–2 h prior to MRI. This allowed time for specific MPIO binding and clearance of unbound MPIO from the blood. To block VCAM-1 binding sites *in vivo*, a further group of mice were injected with VCAM-1 antibody 3 h after IL-1β injection and VCAM–MPIO administered 15 min later. *In vivo* MRI was performed at 7 T using a T_2^*-weighted 3D gradient-echo sequence, with a final isotropic resolution of 88 μm^3. VCAM–MPIO generated highly specific, potent hypointense contrast effects in the IL-1β activated hemisphere, which delineated the architecture of activated cerebral blood vessels, with minimal background contrast (Fig. 2). The specificity and potency of the contrast effects derived from a combination of targeted delivery of MPIO containing a high payload of iron oxide to sites of early vascular inflammation and rapid clearance of MPIO from the circulation which minimized background signal.

McAteer et al. (2008) have also constructed dual-ligand MPIO (4.5 μm diameter) targeting endothelial P-selectin and VCAM-1 adhesion molecules to mimic more closely leukocyte-binding pathways *in vivo*. Dual-ligand MPIO binding to arterial endothelium following *in vivo*

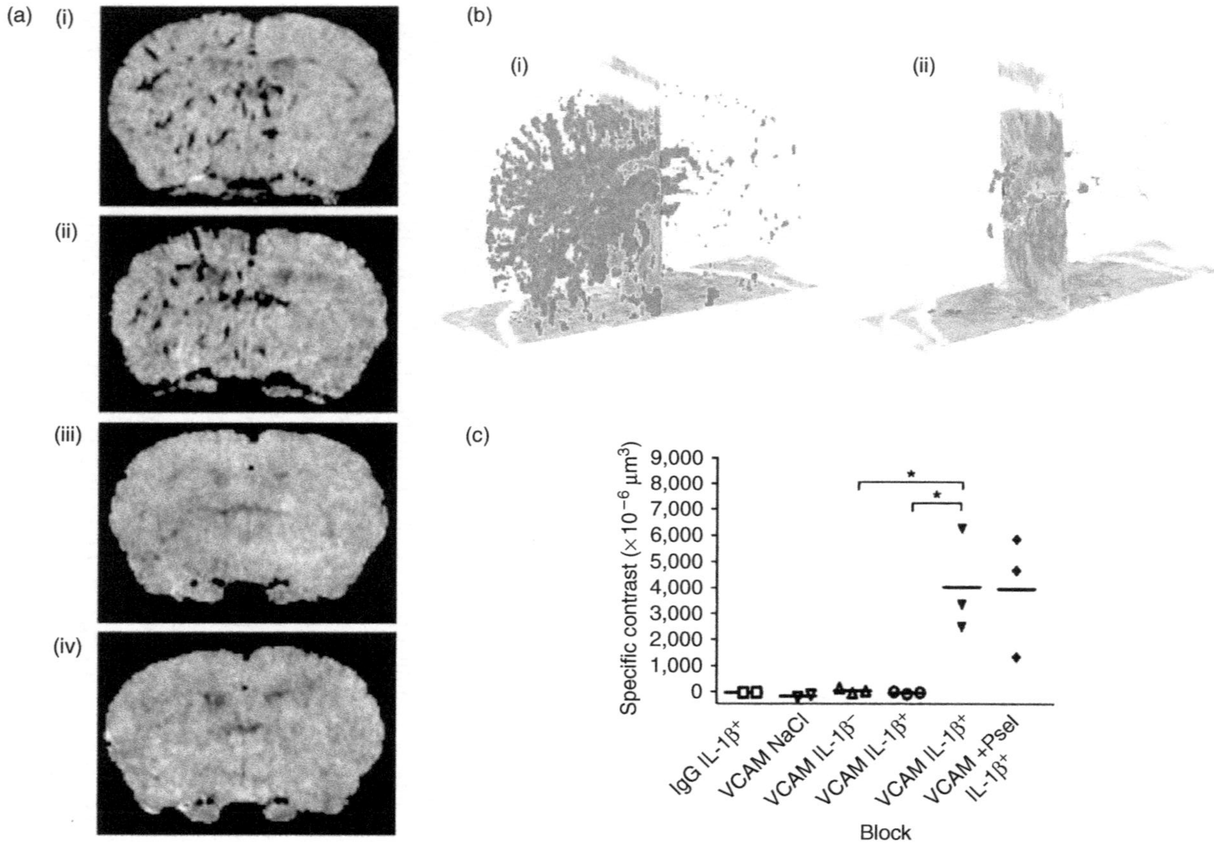

Fig. 2. (a) (i) Mouse injected intrastriatally with 1 ng of IL-1β in 1 µL of saline 3 h before i.v. injection of VCAM–MPIO (~4.5 mg Fe/kg). Intense low signal areas (black) on the left side of the brain reflect the specific MPIO retention on acutely activated vascular endothelium with almost absent contrast effect in the contralateral control hemisphere. (ii) Similar, unilateral MPIO contrast effects in a mouse injected as in (a) but with VCAM+P-selectin-MPIO. (iii) Absence of MPIO effects in a mouse injected as in (a) but with IgG-MPIO control. (iv) Absence of MPIO effects in a mouse injected with IL-1β into the striatum and with VCAM–MPIO intravenously after pretreatment with VCAM-1 antibody, which effectively blocked VCAM–MPIO binding. MRI data were obtained 1–2 h after MPIO. (b) (i) In each mouse, 41 contiguous images were segmented by an automated analysis of signal intensity histograms. VCAM–MPIO contrast effects delineated the architecture of cerebral vasculature in the IL-1β-stimulated hemisphere (left half of image) with almost total absence of binding on the contralateral, nonactivated side. The midlines are indicated by vertical sections. (ii) Preadministration of VCAM-1 antibody abolished VCAM–MPIO retention. (c) As compared with brains without IL-1β injection, specific contrast was increased >100-fold after administration of VCAM–MPIO. Dual conjugated MPIO targeting both VCAM-1 and P-selectin also bound specifically but did not further enhance contrast effects. Substitution of IgG-MPIO (IgG/IL-1β$^+$), sham intracerebral injection (VCAM/NaCl), no intracerebral injection (VCAM/IL-1β$^-$), and preblocking (VCAM/IL-1β$^+$ with block) were not associated with specific contrast effects. Bars indicate mean values for each group ($^*P = 0.02$) (McAteer et al., 2007).

intravenous (i.v.) injection was demonstrated by high resolution *ex vivo* MRI (9.4 T). Biodistribution studies showed that MPIO were sequestered by the liver and spleen after 24 h. Importantly, there was no evidence of tissue infarction, inflammation, or hemorrhage in any of the clearance organs.

Several generations of multimodal, magnetofluorescent nanoprobes have been developed for targeting endothelial VCAM-1, based on CLIO nanoparticles conjugated with Cy5.5, a NIRF molecule, to their surface (Cy5.5-CLIO). The first of these agents was a carboxylic acid functionalized Cy5.5-CLIO, conjugated to VCAM-1 or IgG monoclonal antibodies (Tsourkas et al., 2005). Anti-VCAM-1 targeted Cy5.5-CLIO were shown to specifically bind *in vivo* to the vasculature of TNF-α-stimulated mouse ear using intravital fluorescence microscopy (IVFM). Several novel peptides have also been generated by phage display that bind specifically to activated endothelium and are internalized, allowing progressive concentration by endothelial cells (Kelly et al., 2005; Nahrendorf et al., 2006b). These include a VCAM-1 specific, cyclic peptide sequence (CVHSPNKKC; termed VHS peptide) (Kelly et al., 2005) and a linear peptide, VHPKQHR, termed VCAM-1 internalizing peptide (VINP) (Nahrendorf et al., 2006b). By conjugating these peptide ligands to iron oxide nanoparticles, it was possible to image VCAM-1 expression in mouse atherosclerosis (Kelly et al., 2005; Nahrendorf et al., 2006b). This is an ingenious approach, but, in comparison to MPIO, also has several disadvantages. Firstly, cellular uptake is required, with potential for toxicity and ultimately loss of molecular specificity. Secondly it was necessary to wait 48 h after administration before imaging, and thirdly, the mode of cellular uptake is specific to VCAM-1 that the technology is not adaptable to other endovascular molecular targets.

E-selectin (CD62E, endothelial leukocyte adhesion molecule 1) is another important mediator of early rolling recruitment of leukocytes to the activated endothelium. sLe^X is a carbohydrate antigen associated with CD15 on the surface of leukocytes which binds to E-selectin expressed on activated endothelial cells. Sibson et al. (2004) successfully targeted E-selectin expression in rat brain *in vivo* using a Sialyl Lewis^x (sLe^X) mimetic moiety conjugated to Gd-DTPA (Gd-DTPA-B (sLe^X)A). Stereotactic injection of either IL-1β or TNF-α into the left striatum of Wistar rats was used to induce focal endothelial activation. Gd-DTPA-B(sLeX)A, administered systemically 3–4 h after intracerebral cytokine injection, produced hyperintense contrast effects on the activated brain endothelium on spin-echo T_1-weighted images, at a stage when no pathological changes are apparent with conventional MRI methods. The lack of enhancement with Gd-DTPA-BMA suggested that the response was specific and not related to leakage of the contrast agent across the BBB. Gd-DTPA-sLe^{x}A has also been used to detect early endothelial activation following transient focal ischemia in a mouse model of middle cerebral artery occlusion using *in vivo* MRI (9.4 T) (Barber et al., 2004). Sipkins, Gijbels, Tropper, Bednarski, Li, & Steinman (2000) have reported the detection of ICAM-1 upregulation on the cerebral microvasculature of mice with EAE by *ex vivo* MRI (9.4 T) using antibody-conjugated paramagnetic liposomes (ACPLs). However, the major limitation of low sensitivity Gd-based agents is the small quantity of Gd that can be delivered to an endothelial monolayer, which results in only modest contrast effects.

Ultrasound has shown promise for imaging endothelial molecules using targeted microbubbles (Villanueva et al., 1998), and recently, an enhanced technique, sensitive particle acoustic quantification (SPAQ) was applied to measure VCAM-1 and intercellular adhesion molecule-1 (ICAM-1) in the brains of rats with EAE (Reinhardt et al., 2005).

Targeted imaging of activated platelets in cerebral malaria

The pathogenesis of CM involves vascular inflammation, elevated levels of proinflammatory cytokines, and platelet aggregation in cerebral capillaries and venules (Wassmer, Combes, Candal, Juhan-Vague, & Grau, 2006). The platelet-specific, glycoprotein GP IIb/IIIa receptor

(CD41/CD61, also known as $\alpha_{IIb}\beta_3$ integrin) mediates the final common pathway of platelet aggregation via fibrinogen and is key to thrombus formation (Peter, Ahrens, Schwarz, Bode, & Ylanne, 2004). Immunohistochemistry for GP IIb/IIIa has demonstrated that platelet accumulation occurs in the microvasculature of patients with CM (Grau et al., 2003).

A single-chain antibody has recently been developed that specifically recognizes ligand-induced binding sites (LIBS) on GP IIb/IIIa receptors, which become exposed only upon activation by receptor–ligand binding, thus offering the opportunity to target activated platelet adhesion (Schwarz et al., 2004, 2006; Stoll et al., 2007). LIBS antibody, in contrast to nonactivation-specific GP IIb/IIIa antibodies such as Abciximab (Alonso et al., 2007), does not bind to circulating platelets but binds only to activated platelet aggregates or platelets adherent to damaged endothelium, for example, due to inflammation or plaque rupture.

von Zur Muhlen et al. (2008b) have recently constructed a novel GP IIb/IIIa receptor-targeted contrast agent consisting of LIBS single-chain antibodies conjugated to MPIO, for detection of activated platelets. Autofluorescent cobalt functionalized MPIO (1 μm diameter), conjugated to the histidine tag of LIBS single-chain antibodies, were shown to detect platelets in a mouse model of endovascular platelet aggregation using *ex vivo* MRI (11.7 T) von Zur Muhlen et al. (2008b). The LIBS–MPIO agent has also been applied for *in vivo* MRI detection of platelet aggregation associated with CM, at a time when no overt disease is detectable by conventional *in vivo* MRI (von Zur Muhlen et al., 2008c). LIBS–MPIO binding produced specific hypointense contrast effects on T_2-weighted MR images which delineated cortical vessels in CM-infected mice at day 6 following inoculation with *Plasmodium berghei* (Fig. 3). The dose of LIBS–MPIO (4×10^8 in 200 μL saline) was the same as that successfully used in our VCAM-1 brain imaging study (McAteer et al., 2007). Histological analysis confirmed the presence of LIBS–MPIO binding to endovascular platelet aggregates in CM-infected mice. Using LIBS–MPIO, it was also revealed that the proinflammatory cytokine TNF-α, but not IL-1β or lymphotoxin-α (LT-α), induced platelet adherence to cerebrovascular endothelium.

Thrombus detection in brain ischemia

Atherosclerotic plaque rupture and thrombus formation are key events resulting in ischemic stroke as well as myocardial infarction. Platelets have been shown to be involved in both the initiation of atherosclerotic lesion formation and the late stages of atherosclerotic events including plaque rupture and microembolism (Langer & Gawaz, 2008). Ruptured atherosclerotic plaques on the vessel wall and also vulnerable rupture-prone plaques are lined with activated platelets (Gawaz, Langer, & May, 2005; Massberg et al., 2002). In contrast to fibrin-rich thrombi, which form complex 3D reticular structure with high abundance of epitope, activated platelet thrombi may be partially occlusive and localized at the surface of the ruptured atherosclerotic plaque, presenting a challenge to targeted contrast delivery. However, the use of MPIO targeting platelet-rich thrombi has been demonstrated to overcome this challenge by von Zur Muhlen et al. (2008d), who successfully applied the LIBS–MPIO agent for *in vivo* MRI detection of small, platelet-rich thrombi in a mouse model of wall-adherent, carotid thrombosis. LIBS–MPIO reliably tracked a reduction in thrombus size in response to pharmacological thrombolysis treatment with urokinase (Fig. 4). The LIBS–MPIO agent has also been used to detect human platelet aggregates in explanted symptomatic carotid artery plaque specimens by *ex vivo* MRI (9.4 T) (von zur Muhlen et al., 2008d) and human platelet-rich clots *in vitro* using clinically relevant magnetic field strengths (3 T) (von Zur Muhlen et al., 2008a).

One potential limitation for *in vivo* clinical application of MPIO is the efficiency of MPIO–ligand binding affinity under shear stress, particularly under arterial flow conditions. To investigate this, von Zur Muhlen et al. (2008a) performed flow chamber experiments using human clots and confirmed that LIBS–MPIO remain bound to human platelets under venous and arterial flow conditions. Although the density of LIBS–MPIO binding was relatively

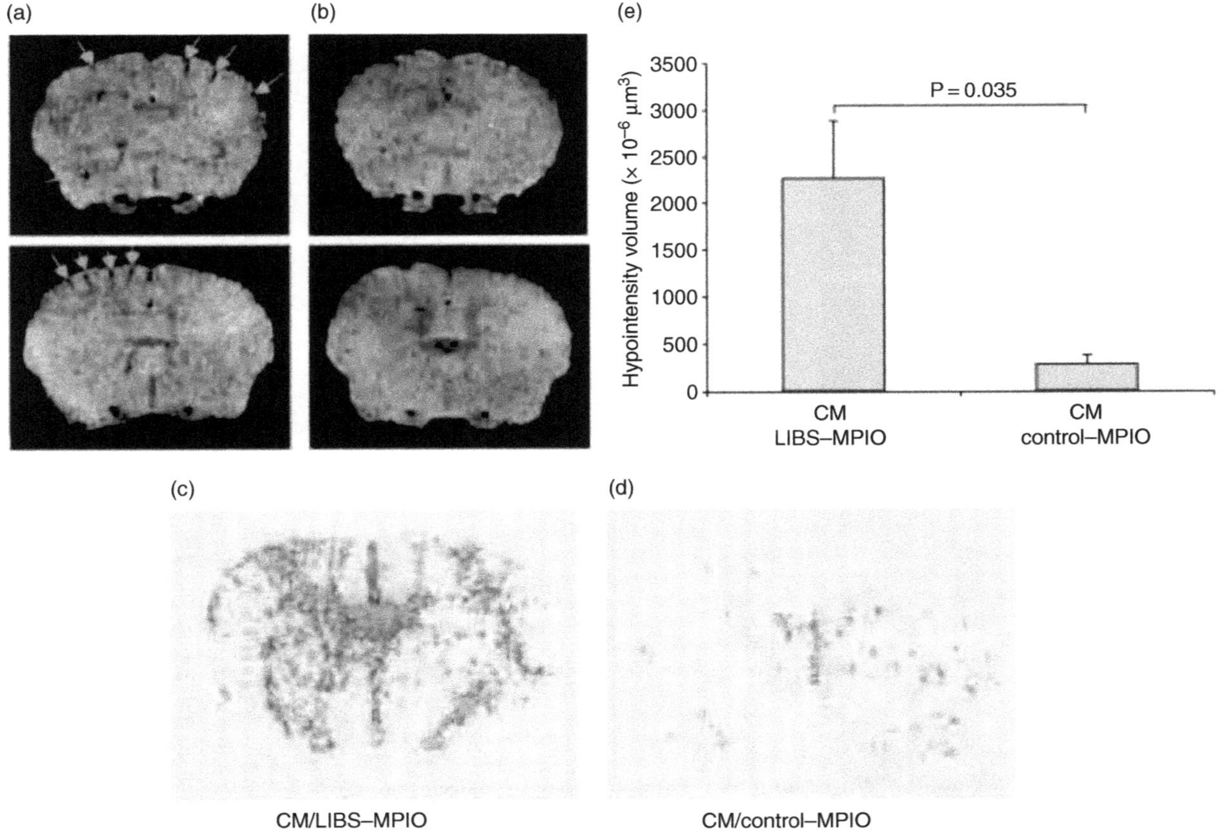

Fig. 3. (a and b) T_2-weighted 3D gradient-echo images from CM mice following i.v. injection of LIBS–MPIO (a) and control–MPIO (b) are presented in two representative slices at two different levels within the same brain. Areas of MPIO-induced signal appeared as dark signal voids (arrows) in cortical regions of the LIBS–MPIO-injected mouse, but not in the control–MPIO-injected mouse. (c and d) 3D reconstruction confirmed the cortical binding pattern in the LIBS–MPIO-injected mouse (c), whereas only modest background binding was evident in the control–MPIO animal (d). (e) Quantification of signal voids demonstrated a significant difference between the LIBS–MPIO and control–MPIO-injected animals (von Zur Muhlen et al., 2008c).

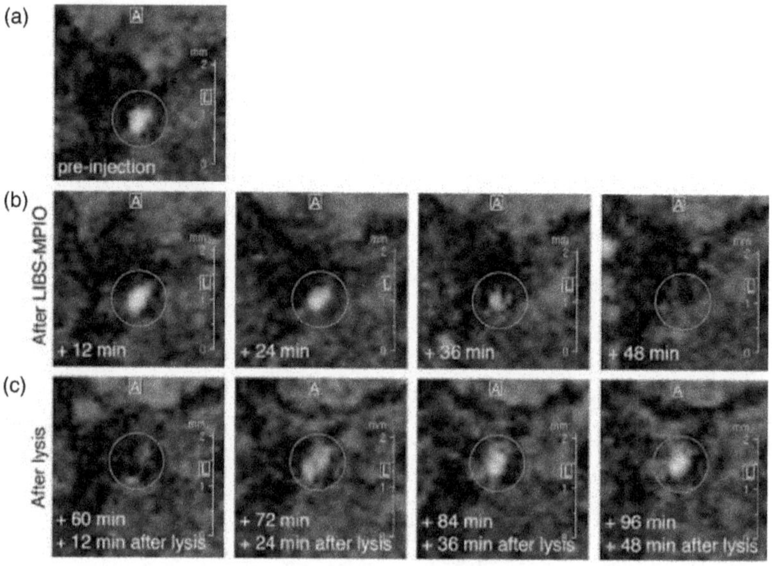

Fig. 4. Treatment with i.v. mouse urokinase. Enlarged transverse sections of the injured right carotid artery. (a) The vessel before injection, with a marked and progressive diminution in signal after LIBS–MPIO injection between 12 and 48 min (b). Thrombolysis was performed after the 48-min scan. Images obtained after i.v. application of urokinase show the reappearance of the vessel lumen signal over time indicated by (c) (von zur Muhlen et al., 2008d).

low, the potency of the contrast effects suggests that the use of targeted MPIO is a reasonable proposition for *in vivo* clinical application. Furthermore, the use of MPIO targeting platelet-rich thrombi may overcome the limitation of a previous *in vivo* MRI study using USPIO coupled to a cyclic arginine–glycine–aspartic acid (RGD) peptide, in which the *in vivo* spatial resolution ($0.2 \times 0.2 \times 1$ mm) and sensitivity were seen as potentially limiting for clinical application (Johansson, Bjornerud, Ahlstrom, Ladd, & Fujii, 2001). The use of single-chain antibodies as targeting ligands is also attractive for human application since they have low immunogenicity due to a lack of Fc regions. Single-chain antibodies can also be tailored in size for various purposes and produced in large quantities at low cost (Schwarz et al., 2006).

Contrast-enhanced ultrasound has also been shown to enable *in vivo* visualization of platelet-rich human thrombi in a rat model of carotid artery occlusion using microbubbles conjugated with a monoclonal antibody against GP IIb/IIIa, Abciximab, which is licensed for therapeutic use as a platelet aggregation inhibitor (Alonso et al., 2007). This highlights the feasibility of using therapeutic agents for targeting platelets and the potential for improving diagnosis and treatment of brain ischemia.

Gd-based agents have also been applied for the detection of fibrin-rich thrombi. In early studies, the targeting of mainly large, occlusive, fibrin-rich thrombi was achieved using monoclonal antibodies specific for fibrin and the sequential introduction of avidin and biotinylated perfluorocarbon particles (Lanza et al., 1996). However, this approach was limited by potential toxicity and immunogenicity. Subsequently, paramagnetic perfluorocarbon nanoparticles covalently bound to monoclonal antifibrin F(ab) fragments or to peptidomimetics were used to detect human thrombi (Yu et al., 2000). Due to their large size ($\sim$200 nm diameter), fibrin-targeted perfluorocarbon nanoparticles were unable to penetrate dense fibrin clots. However, a single layer of nanoparticle binding to the periphery of a fibrin clot was sufficient to enhance T_1 contrast substantially and detect fibrin-rich thrombi. This emphasizes both the technical limitations of contrast penetration to structures of potential interest and the potency of the lipid-encapsulated perfluorocarbon approach, whereby contrast enhancement can be detected from structures only one to two pixels thick.

A novel fibrin-targeted Gd-based agent EP-2104R (EPIX Pharmaceuticals, Lexington, MA, United States) has recently been applied to image human fibrin-rich thrombi with high specificity (Spuentrup et al., 2007a,b, 2008). EP-2104R is a small peptide that selectively and reversibly binds to fibrin. An earlier but very similar compound has also been described (Botnar et al., 2004). Sirol et al. (2005) demonstrated the superior contrast of EP-2104R for detection of chronic thrombus in the carotid artery of rabbits compared to standard Gd-DTPA and observed that the contrast enhancement lasted up to 8 weeks. EP-2104R has also been used in a pig model of cerebral venous sinus thrombosis to discriminate between thrombus, blood pool, and brain tissue (Stracke et al., 2007).

Molecular imaging of glioma

Malignant gliomas are the most common and malignant type of primary brain tumor (Wrensch, Minn, Chew, Bondy, & Berger, 2002). In its most aggressive form, glioblastoma multiforme, patients generally have a mean survival rate of less than 1 year (Behin, Hoang-Xuan, Carpentier, & Delattre, 2003). Surgical resection of brain tumor can improve survival rates but its effectiveness is limited by difficulties in visually distinguishing neoplastic and healthy brain tissue (Nitta & Sato, 1995). Gd-based contrast agents have been applied to aid preoperative localization of glioma using MRI. However, Gd chelates cannot accurately delineate tumor boundaries or quantify tumor volume due to rapid diffusion of Gd from the tumor site and difficulties in distinguishing glioma from surrounding tissue edema (Sun et al., 2008). Iron oxide nanoparticles have been shown to more accurately delineate brain tumors by *in vivo* MRI (Fleige et al., 2001; Moore, Marecos, Bogdanov, & Weissleder, 2000; Muldoon et al., 2006) due to enhanced internalization by glioma cells and prolonged clearance from the tumor site (Varallyay et al., 2002). The prolonged enhancement of USPIO may also be useful in comparing pre- and postoperative tumor burden and facilitate intraoperative MRI (Hunt, Bago, & Neuwelt, 2005).

Multimodal Cy5.5-CLIO have been applied for determining brain tumor margins preoperatively and during surgical resection in a rat model of gliosarcoma, implanted with green fluorescent protein (GFP)-expressing 9 L glioma cells (Kircher, Mahmood, King, Weissleder, & Josephson, 2003). Cy5.5-CLIO (15 mg Fe/kg body weight) were injected via tail vein and after 24 h, preoperative *in vivo* MRI was performed. The brain tumor and surrounding tissue were then surgically exposed and intraoperative NIRF imaging confirmed a good correlation between Cy5.5-CLIO and GFP fluorescence of true brain tumor cells.

The specificity of multimodal iron oxide nanoprobes for glioma cells has been improved by conjugating chlorotoxin (CTX) to the nanoparticle surface (Meng et al., 2007; Veiseh et al., 2005). CTX is a small 36-amino acid peptide, purified from the venom of the giant Israeli scorpion (*Leiurus quinquestriatus*). CTX has been shown to bind specifically to MMP-2 endopeptidase, which is preferentially upregulated on glioma cells and other tumors of neuroectodermal origin (Deshane, Garner, & Sontheimer, 2003) but is not expressed by normal brain tissue cells (Kachra et al., 1999). The specificity of multifunctional poly(ethylene glycol) (PEG)-coated iron oxide nanoparticles conjugated with CTX and the NIRF molecule, Cy5.5, has been demonstrated *in vitro* using 9 L glioma cells (Veiseh et al., 2005) and by *in vivo* MRI, using athymic (nu/nu) mice bearing 9 L xenograft tumors (Sun et al., 2008). Nanoparticle internalization into the cytoplasm of 9 L glioma cells was visualized by transmission electron microscopy and histological analysis of clearance organs found no acute toxic effects of these targeted nanoprobes. The potential of specifically targeting human glioma cells (U251) has also been demonstrated *in vitro* using multifunctional SPIO conjugated with fluorescein isothiocyanate (FITC) and CTX (SPIOFC) (Meng et al., 2007).

Light activated, multifunctional iron oxide nanoprobes have also been designed for targeted detection, diagnosis, and treatment of brain tumors (Reddy et al., 2006). The photodynamic therapy (PDT) agent, Photofrin®, a photosensitizer currently approved for clinical use in the USA, was encapsulated together with iron oxide nanoparticles within fluorescent-labeled (Alexa Fluor 594) PEGylated amine functionalized polyacrylamide (PAA) nanoparticles. The nanoparticles were targeted specifically to brain tumors by conjugating a vascular targeting peptide, F3, to the nanoparticle surface. F3 peptide is a 31-amino acid sequence present at the N-terminus of the human high mobility group protein 2 (HMGN2), identified by phage screens, that binds to tumor endothelial cells via the nucleolin receptor, which subsequently internalizes it to the nucleus. *In vitro* studies showed significant binding of F3-labeled nanoparticles to the surface of MDA-MB-435 breast cancer cells which, together with laser light, resulted in potent phototoxicity and over 90% cell death. Untargeted nanoparticles, with laser light, did not induce cell death.

The *in vivo* efficacy of F3-targeted nanoparticles was investigated using rats implanted with gliosarcoma (9 L) cells into the right forebrain. Using dynamic scanning MRI, F3-targeted nanoparticles produced significantly greater enhancement at the tumor site than in normal brain tissue, for a longer duration, compared to nontargeted nanoparticles or Photofrin alone (Fig. 5). F3-targeted nanoparticles also produced a significantly improved therapeutic outcome (higher animal survival rate and tumoral diffusion rate), with rats treated with F3-targeted agent having a median survival time of 33 days, compared to 13 days for Photofrin or untargeted nanoparticle-treated groups, and 7 days for the control group.

Overexpression of the tyrosine kinase cell surface receptor c-Met and its substrate, the hepatocyte growth factor (HGF), have been implicated in the progression of malignant glioblastomas (Koochekpour et al., 1997; Moriyama et al., 1998). Towner et al. (2008) reported the first *in vivo* visualization of c-Met overexpression in an intracerebral implantation C6 rat glioma model using a biotinylated Gd-DTPA-albumin agent coupled with a mouse monoclonal c-Met antibody. Specific binding of c-Met-targeted nanoparticles within gliomas was demonstrated by a decrease in T_1 relaxation time and a corresponding increase in MR signal intensity. Fluorescence microscopy of the biotinylated portion of the anti-c-Met probe within neoplastic and normal brain tissues confirmed binding specificity to glioma.

Tumor angiogenesis

Angiogenesis is a critical determinant of tumor proliferation and metastasis (Folkman, 2002). Once the tumor grows beyond 1–2 mm diameter, passive diffusion is no longer sufficient to support the viability of malignant cells, and neovascularization becomes essential (Folkman, 1996). The alpha (v) beta (3) ($\alpha_v\beta_3$) integrin is a well-recognized molecular marker of angiogenesis that is highly expressed during tumor growth and neovascularization (Barrett, Brechbiel, Bernardo, & Choyke, 2007). The amino acid sequence of RGD peptide targets $\alpha_v\beta_3$ integrin upregulation on activated endothelial cells of angiogenic vessels (Barrett et al., 2007). Cy5.5-labeled RGD peptides have been shown to bind specifically to tumor vasculature and tumor cells *in vivo* in an orthotopic glioblastoma brain tumor mouse model by *in vivo* NIRF imaging (Hsu et al., 2006). Targeted magneto-fluorescent Cy 5.5-labeled CLIO, conjugated to cyclic RGD peptides (cRGD-CLIO(Cy5.5)) have also been developed, which can detect $\alpha_V\beta_3$ expression on tumor cells *in vitro* and *in vivo* in implanted tumors, as determined by fluorescence and MRI (Montet et al., 2006). Multifunctional $\alpha_v\beta_3$-targeted liposomes containing Gd and fluorescein-PE, conjugated with RGD peptides (~700 RGD per liposome), have also been shown to detect activated tumor epithelium using both *in vivo* T_1-weighted MRI and *ex vivo* fluorescent microscopy (Mulder et al., 2005).

Hood et al. (2002) demonstrated the therapeutic application of cationic polymerized lipid-based nanoparticles coupled to a small organic $\alpha_v\beta_3$-targeting ligand ($\alpha_v\beta_3$-NP) for targeted delivery of an antiangiogenic gene to the tumor vasculature in tumor-bearing mice (Hood et al., 2002). Nanoparticles conjugated to a mutant *Raf-1* gene, which

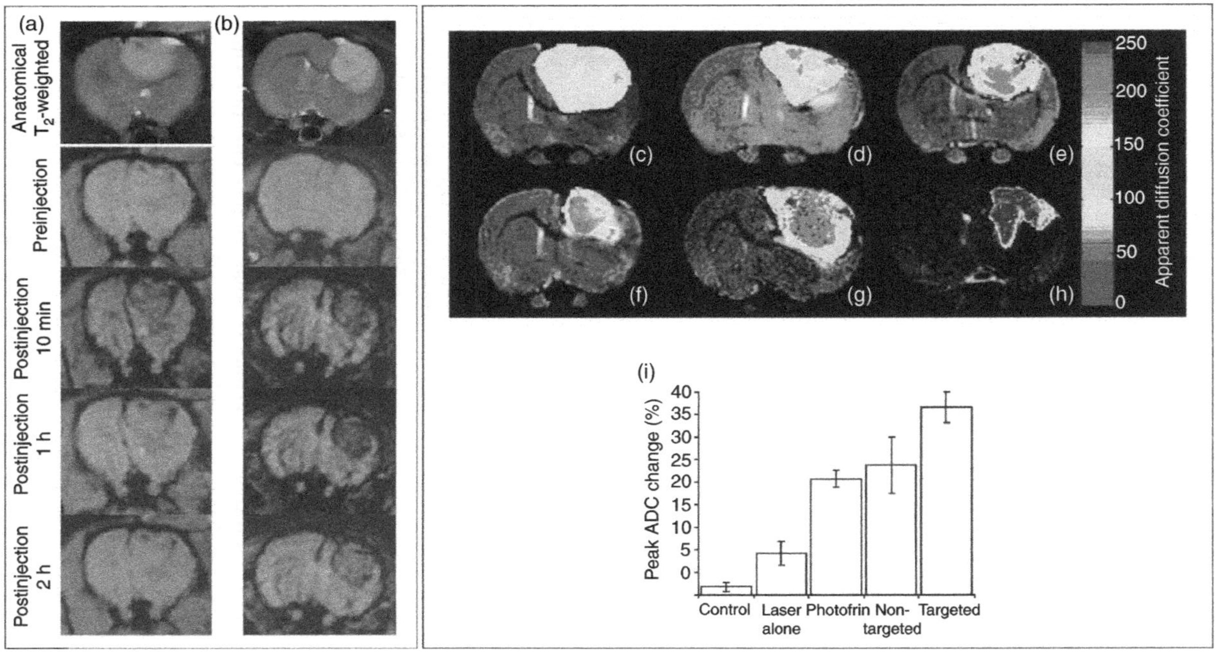

Fig. 5. Imaging and monitoring of therapeutic efficacy using multifunctional nanoparticles in 9 L brain tumors. (a) and (b), top; T2-weighted coronal images through the tumor of two different rats that reveal the anatomic extent of the tumor mass. Fast spin-echo images of the tumor following administration of (a) nontargeted nanoparticles and (b) F3-targeted nanoparticles obtained at the time points indicated. T2-weighted magnetic resonance images at day 8 after treatment from (c) a representative control i.c. 9 L tumor and tumors treated with (d) laser light only, (e) i.v. administration of Photofrin plus laser light, (f) nontargeted nanoparticles containing Photofrin plus laser light, and (g) targeted nanoparticles containing Photofrin plus laser light. The image shown in (h) is from the same tumor shown in (g), which was treated with the F3-targeted nanoparticle preparation but at day 40 after treatment. The color diffusion maps overlaid on top of T2-weighted images represent the apparent diffusion coefficient (ADC) distribution in each tumor slice shown. (i) Columns, mean peak percentage change in tumor apparent diffusion coefficient values for each of the experimental groups; bars, SE (Reddy et al., 2006).

blocks endothelial signaling and angiogenesis in response to multiple growth factors, were shown to result in apoptosis of tumor cells and tumor-associated endothelium, and sustained regression of established primary and metastatic tumors *in vivo*. A similar polymerized particle containing gadolinium and the $\alpha_v\beta_3$-targeting monoclonal antibody, LM609, has been used successfully to image angiogenic "hot spots" within tumor vasculature in tumor-bearing rabbits that were otherwise not detectable by conventional MRI (Sipkins et al., 1998). Gd-containing perfluorocarbon nanoparticles, targeted with $\alpha_v\beta_3$ integrin peptidomimetic antagonist, have also been used for *in vivo* MRI detection of neoangiogenesis in tumor-bearing rabbits (Winter et al., 2003) and mice (Schmieder et al., 2005). $\alpha_v\beta_3$-targeted paramagnetic nanoparticles generated increased T_1 signal intensity within tumors, predominantly in the tumor periphery at 2 h postinjection, which corresponded to areas of neoangiogenesis using histology (Winter et al., 2003). Despite their relatively large size (273 nm diameter), nanoparticles were able to penetrate into the leaky tumor neovasculature but did not appreciably migrate into the interstitium.

MicroPET imaging has demonstrated potential for imaging brain tumor angiogenesis. An *N*-4-[^{18}F] fluorobenzoyl-labeled cyclic RGD peptide ([^{18}F]FB-RGD) radiotracer has been used to visualize $\alpha_v\beta_3$-integrin positive brain tumors in orthotopically implanted U251T glioblastoma *nu/nu* mice, with virtually no uptake in the normal brain (Chen et al., 2004). Contrast-enhanced ultrasound (CEU) imaging of tumor angiogenesis has also been reported using microbubbles targeted to $\alpha_v\beta_3$. Microbubbles were conjugated with echistatin, a peptide derived from the venom of the viper *Echis carinatus* bearing the RGD motif with enhanced binding affinity for $\alpha_v\beta_3$. $\alpha_v\beta_3$-targeted microbubbles were applied to image angiogenesis in malignant glioma, induced by intracerebral implantation of U87MG human glioma cells in athymic rats (Ellegala et al., 2003). CEU signal from $\alpha_v\beta_3$-targeted microbubbles in tumors increased significantly from 14 to 28 days after implantation and was greatest at the periphery of tumors, where $\alpha_v\beta_3$ expression was most prominent, and correlated well with tumor microvascular blood volume. Dual-targeted perfluorocarbon-filled microbubbles, conjugated to vascular endothelial growth factor receptor 2 (VEGFR2) and $\alpha_v\beta_3$ antibodies, have also been shown to improve *in vivo* visualization of tumor angiogenesis in a human ovarian cancer xenograft tumor model in mice compared to single-ligand microbubbles (Willmann et al., 2008).

Gliosis

Gliosis is a fibrous proliferation of glial cells in injured areas of the CNS. Gliosis and neuronal loss is prevalent in glioma as well as in many other human neurological disorders including MS, viral encephalitis, Alzheimer's disease, traumatic brain injury, stroke, and cardiac arrest. Gliosis is traditionally detected by elevated glial fibrillary acidic protein (GFAP) levels in postmortem tissue samples using immunohistochemistry.

Novel SPIO probes have recently been developed for targeting the gene transcript of GFAP in glia and astrocytes *in vivo* using transcription MRI (tMRI) (Liu et al., 2008). This tMRI technique uses SPIO as labels for nucleic acid probes to report gene transcription in the brain *in vivo* at the mRNA level (Liu et al., 2007). SPIO were linked to a short DNA sequence complementary to cerebral mRNA of GFAP, found in glia and astrocyte (SPION-gfap) and a control sequence complementary to the mRNA of β-actin (SPION-β-actin). It was hypothesized that SPION probes could be distributed *via* the lymphatic system in mice with BBB disruption. Therefore, SPION-gfap was administered via eyedrop solution to the conjunctival sac, in three mouse models of acute neurological disorders: cortical spreading depression, minute intracranial puncture wound, and global cerebral ischemia. SPION-gfap produced areas of elevated signal in subtraction $R2^*$ maps using tMRI, which corresponded to areas of extensive gliosis in postmortem immunohistochemistry. Similarly, mice administered control SPION-β-actin exhibited foci of $R2^*$ elevation that matched β-actin-expressing endothelia in the vascular wall. Therefore, SPION-gfap and SPION-β-actin demonstrated specificity for reporting transcript imaging of glia and cells expressing smooth muscle actin in mouse brains *in vivo*.

β-amyloid plaques in Alzheimer's disease

Alzheimer's disease (AD) is a neurodegenerative disorder that is characterized by the accumulation of amyloid-β aggregates in the brain. PET has been shown to provide quantitative information on amyloid-β deposits in AD patients using a carbon-11 (C^{11})-labeled amyloid imaging PET tracer *N*-methy l[11C]2-(4′-methylaminophenyl-6-hydroxybenzathiazole), termed Pittsburgh Compound B (PIB) (Klunk et al., 2003, 2004; Nordberg, 2008). However, PIB has limited specificity in diagnosing and monitoring AD progression and is primarily a nonspecific marker of amyloid-β peptide-related cerebral amyloidosis (Lockhart et al., 2007). Iodinated benzothiazole derivatives radiolabeled with either C^{11} or I^{125}/I^{123} have also been developed for dual modal *in vivo* imaging of amyloid deposits in mice using PET or SPECT (Wang et al., 2004). However, PET and SPECT, due to their low spatial resolution, are unable to detect individual plaques and early-stage amyloid deposition, prior to irreversible damage.

MRI has been demonstrated to detect individual amyloid plaques *in vivo* (Jack et al., 2004, 2005). MicroMRI has been shown to detect amyloid-β plaques in AD transgenic mouse brains using Aβ1-40 conjugated with either Gd or MION after intracarotid coinjection with mannitol to transiently open the BBB (Fig. 6) (Wadghiri et al., 2003). Gd-Aβ1-40 peptide (PUT-Gd-Aβ), labeled with putrescine, a polyamine that increases BBB permeability, (Poduslo & Curran, 1996a,b), was also demonstrated to bind to amyloid plaques by *ex vivo* MRI. However, the use of full-length Aβ may have toxic side effects, since Aβ is known to seed amyloid deposition and actually promote plaque buildup (Jarrett, Berger, & Lansbury, 1993). Recently, a novel nontoxic Gd contrast agent coupled with a novel peptide K6Aβ1-30 homologous to Aβ that does not form amyloid aggregates was developed (Gd-DPTA-K6Aβ1-30) and enabled *in vivo* plaque detection in AD mouse model using microMRI (Sigurdsson et al., 2008). Antibody fragments F(ab′) (2) of a monoclonal antibody against fibrillar human Aβ42 that is polyamine (p)-modified has also demonstrated efficient targeting to amyloid plaques in AD mouse brains and may be a potential ligand for molecular imaging of amyloid plaques in AD with MRI or PET (Poduslo et al., 2007; Wengenack, Jack, Garwood, & Poduslo, 2008).

Conclusions and future perspectives

Molecular imaging using targeted iron oxide nano- and microparticle contrast agents is at the forefront in the advancement of *in vivo* diagnosis and monitoring of diseases of the brain. The key advantages offered by iron oxide contrast agents are their inherent amplification and surface-mediated multivalent affinity effects. Cellular MRI using iron oxide nanoparticles is a valuable tool to study the multiple aspects of neuroinflammation in MS and ischemic stroke, which may aid the development of effective treatments that limit monocyte/macrophage infiltration. Novel nano- and microparticle mMRI probes are rapidly evolving that specifically target cell adhesion molecules in neuroinflammation, activated platelets in CM and brain ischemia, glioma, and tumor angiogenesis. The potent contrast effects achieved using targeted MPIO may, with appropriate modification of the surface coat, provide a useful tool for earlier identification of neuroinflammation by MRI, which may accelerate accurate diagnosis and guide delivery of specific therapy. Furthermore, the identification of novel single-chain antibodies and peptide ligands using phage display or the use of synthetic glycoprotein ligands (van Kasteren, Kramer, Gamblin, & Davis, 2007) may overcome the issues of potential immunogenicity and steric constraints of monoclonal antibody ligands. In the future, a shift to clinical imaging at higher MRI field strengths (3 T)

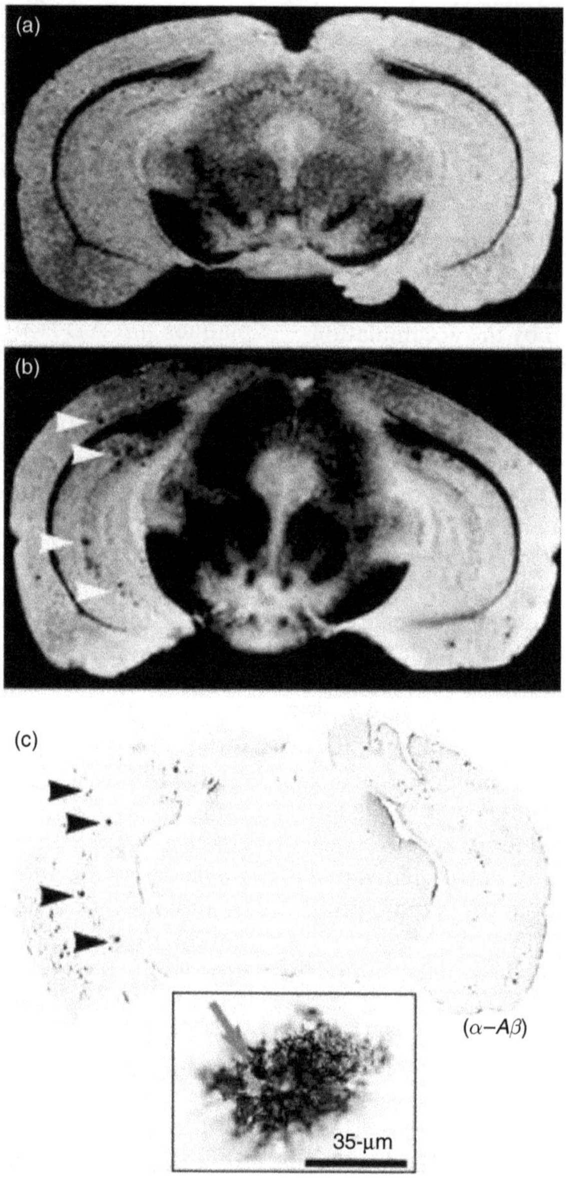

Fig. 6. Aβ plaques were detected with *ex vivo* μMRI after injection of MION-Aβ1-40 with mannitol. *Ex vivo* T_2-weighted SE coronal μMR images show 16-month-old control (a) and APP-transgenic (b) mouse brains. Both brains were extracted and prepared for imaging 6 h after carotid injection of MION-Aβ1-40 with 15% mannitol. Many μMRI lesions matched to Aβ plaques (arrowheads), as revealed by immunohistochemistry (c). High-power microscopic examination of the amyloid plaques, double-stained with an Aβ (6E10) antibody and a Mallory stain for iron (arrow), demonstrated the colocalization of MION with Aβ plaques (inset) (Wadghiri et al., 2003).

together with the introduction of combined modality methods and significant advances in mMRI contrast agents, such as multifunctional and multimodal nanoagents, offers great potential for the integration of diagnosis, treatment, and tracking of response to therapy.

Acknowledgment

Our work is funded by the Wellcome Trust.

References

Alonso, A., Della Martina, A., Stroick, M., Fatar, M., Griebe, M., Pochon, S., et al. (2007). Molecular imaging of human thrombus with novel abciximab immunobubbles and ultrasound. *Stroke*, *38*, 1508–1514.

Amirbekian, V., Lipinski, M. J., Briley-Saebo, K. C., Amirbekian, S., Aguinaldo, J. G., Weinreb, D. B., et al. (2007). Detecting and assessing macrophages in vivo to evaluate atherosclerosis noninvasively using molecular MRI. *Proceedings of the National Academy of Sciences of the United States of Amercia*, *104*, 961–966.

Arias, J. L., Gallardo, V., Gomez-Lopera, S. A., Plaza, R. C., & Delgado, A. V. (2001). Synthesis and characterization of poly(ethyl-2-cyanoacrylate) nanoparticles with a magnetic core. *Journal of Control Release*, *77*, 309–321.

Baeten, K., Hendriks, J. J., Hellings, N., Theunissen, E., Vanderlocht, J., Ryck, L. D., et al. (2008). Visualisation of the kinetics of macrophage infiltration during experimental autoimmune encephalomyelitis by magnetic resonance imaging. *Journal of Neuroimmunology*, *195*, 1–6.

Barber, P. A., Foniok, T., Kirk, D., Buchan, A. M., Laurent, S., Boutry, S., et al. (2004). MR molecular imaging of early endothelial activation in focal ischemia. *Annals of Neurology*, *56*, 116–120.

Barrett, T., Brechbiel, M., Bernardo, M., & Choyke, P. L. (2007). MRI of tumor angiogenesis. *Journal of Magnetic Resonance Imaging*, *26*, 235–249.

Behin, A., Hoang-Xuan, K., Carpentier, A. F., & Delattre, J. Y. (2003). Primary brain tumours in adults. *Lancet*, *361*, 323–331.

Beyer, T., & Townsend, D. W. (2006). Putting 'clear' into nuclear medicine: A decade of PET/CT development. *European Journal of Nuclear Medicine Molecular Imaging*, *33*, 857–861.

Botnar, R. M., Buecker, A., Wiethoff, A. J., Parsons, E. C., Jr., Katoh, M., Katsimaglis, G., et al. (2004). In vivo magnetic resonance imaging of coronary thrombosis using a fibrin-binding molecular magnetic resonance contrast agent. *Circulation*, *110*, 1463–1466.

Briley-Saebo, K., BjÃ¸rnerud, A., Grant, D., Ahlstrom, H., Berg, T., & Kindberg, G. M. (2004). Hepatic cellular distribution and degradation of iron oxide nanoparticles following single intravenous injection in rats: Implications for magnetic resonance imaging. *Cell and Tissue Research*, *316*, 315.

Briley-Saebo, K. C., Johansson, L. O., Hustvedt, S. O., Haldorsen, A. G., Bjornerud, A., Fayad, Z. A., et al. (2006). Clearance of iron oxide particles in rat liver: Effect of hydrated particle size and coating material on liver metabolism. *Investigative Radiology*, *41*, 560–571.

Briley-Saebo, K. C., Mani, V., Hyafil, F., Cornily, J. C., & Fayad, Z. A (2008). Fractionated Feridex and positive contrast: In vivo MR imaging of atherosclerosis. *Magnetic Resonance in Medicine*, *59*, 721–730.

Brochet, B., Deloire, M. S., Touil, T., Anne, O., Caille, J. M., Dousset, V., et al. (2006). Early macrophage MRI of inflammatory lesions predicts lesion severity and disease development in relapsing EAE. *Neuroimage*, *32*, 266–274.

Butoescu, N., Jordan, O., Petri-Fink, A., Hofmann, H., & Doelker, E. (2008). Co-encapsulation of dexamethasone 21-acetate and SPIONs into biodegradable polymeric microparticles designed for intra-articular delivery. *Journal of Microencapsulation*, *25*, 339–350.

Carlos, T. M., Schwartz, B. R., Kovach, N. L., Yee, E., Rosa, M., Osborn, L., et al. (1990). Vascular cell adhesion molecule-1 mediates lymphocyte adherence to cytokine-activated cultured human endothelial cells. *Blood*, *76*, 965–970.

Catana, C., Wu, Y., Judenhofer, M. S., Qi, J., Pichler, B. J., & Cherry, S. R. (2006). Simultaneous acquisition of multislice PET and MR images: Initial results with a MR-compatible PET scanner. *Journal of Nuclear Medicine*, *47*, 1968–1976.

Chen, H. H., Le Visage, C., Qiu, B., Du, X., Ouwerkerk, R., Leong, K. W., et al. (2005). MR imaging of biodegradable polymeric microparticles: A potential method of monitoring local drug delivery. *Magnetic Resonance in Medicine*, *53*, 614–620.

Chen, X., Park, R., Shahinian, A. H., Tohme, M., Khankaldyyan, V., Bozorgzadeh, M. H., et al. (2004). 18F-labeled RGD peptide: Initial evaluation for imaging brain tumor angiogenesis. *Nuclear Medicine and Biology*, *31*, 179–189.

Deloire, M. S., Touil, T., Brochet, B., Dousset, V., Caille, J. M., & Petry, K. G. (2004). Macrophage brain infiltration in experimental autoimmune encephalomyelitis is not completely compromised by suppressed T-cell invasion: In vivo magnetic resonance imaging illustration in effective anti-VLA-4 antibody treatment. *Multiple Sclerosis*, *10*, 540–548.

Deshane, J., Garner, C. C., & Sontheimer, H. (2003). Chlorotoxin inhibits glioma cell invasion via matrix metalloproteinase-2. *Journal of Biological Chemistry*, *278*, 4135–4144.

Dousset, V., Ballarino, L., Delalande, C., Coussemacq, M., Canioni, P., Petry, K. G., et al. (1999b). Comparison of ultrasmall particles of iron oxide (USPIO)-enhanced T2-weighted, conventional T2-weighted, and gadolinium-enhanced T1-weighted MR images in rats with experimental autoimmune encephalomyelitis. *AJNR. American Journal of Neuroradiology*, *20*, 223–227.

Dousset, V., Brochet, B., Deloire, M. S., Lagoarde, L., Barroso, B., Caille, J. M., et al. (2006). MR imaging of relapsing multiple sclerosis patients using ultra-small-particle iron oxide and compared with gadolinium. *AJNR. American Journal of Neuroradiology*, *27*, 1000–1005.

Dousset, V., Delalande, C., Ballarino, L., Quesson, B., Seilhan, D., Coussemacq, M., et al. (1999c). In vivo macrophage activity imaging in the central nervous system detected by magnetic resonance. *Magnetic Resonance in Medicine*, *41*, 329– 333.

Dousset, V., Gomez, C., Petry, K. G., Delalande, C., & Caille, J. M. (1999a). Dose and scanning delay using USPIO for central nervous system macrophage imaging. *Magma*, *8*, 185–189.

Elices, M. J., Osborn, L., Takada, Y., Crouse, C., Luhowskyj, S., Hemler, M. E., et al. (1990). VCAM-1 on activated endothelium interacts with the leukocyte integrin VLA-4 at a site distinct from the VLA-4/fibronectin binding site. *Cell*, *60*, 577–584.

Ellegala, D. B., Leong-Poi, H., Carpenter, J. E., Klibanov, A. L., Kaul, S., Shaffrey, M. E., et al. (2003). Imaging tumor angiogenesis with contrast ultrasound and microbubbles targeted to alpha(v)beta3. *Circulation*, *108*, 336–341.

Engberink, R. D., Blezer, E. L., Hoff, E. I., van der Pol, S. M., van der Toorn, A., Dijkhuizen, R. M., et al. (2008). MRI of monocyte infiltration in an animal model of neuroinflammation using SPIO-labeled monocytes or free USPIO. *Journal of Cerebral Blood Flow and Metabolism*, *28*, 841–851.

Fleige, G., Nolte, C., Synowitz, M., Seeberger, F., Kettenmann, H., & Zimmer, C. (2001). Magnetic labeling of activated microglia in experimental gliomas. *Neoplasia*, *3*, 489–499.

Floris, S., Blezer, E. L., Schreibelt, G., Dopp, E., van der Pol, S. M., Schadee-Eestermans, I. L., et al. (2004). Blood-brain barrier permeability and monocyte infiltration in experimental allergic encephalomyelitis: A quantitative MRI study. *Brain*, *127*, 616–627.

Folkman, J. (1996). New perspectives in clinical oncology from angiogenesis research. *European Journal of Cancer*, *32A*, 2534–2539.

Folkman, J. (2002). Role of angiogenesis in tumor growth and metastasis. *Seminar of Oncology*, *29*, 15–18.

Frias, J. C., Williams, K. J., Fisher, E. A., & Fayad, Z. A. (2004). Recombinant HDL-like nanoparticles: A specific contrast agent for MRI of atherosclerotic plaques. *Journal of American Chemical Society*, *126*, 16316–16317.

Gawaz, M., Langer, H., & May, A. E. (2005). Platelets in inflammation and atherogenesis. *Journal of Clinical Investigation*, *115*, 3378–3384.

Glickson, J. D., Lund-Katz, S., Zhou, R., Choi, H., Chen, I. W., Li, H., et al. (2008). Lipoprotein nanoplatform for targeted delivery of diagnostic and therapeutic agents. *Molecular Imaging*, *7*, 101–110.

Grau, G. E., Mackenzie, C. D., Carr, R. A., Redard, M., Pizzolato, G., Allasia, C., et al. (2003). Platelet accumulation in brain microvessels in fatal pediatric cerebral malaria. *Journal of Infectious Disease*, *187*, 461–466.

Grossman, R. I., Braffman, B. H., Brorson, J. R., Goldberg, H. I., Silberberg, D. H., & Gonzalez-Scarano, F. (1988). Multiple sclerosis: Serial study of gadolinium-enhanced MR imaging. *Radiology*, *169*, 117–122.

Hamoudeh, M., & Fessi, H. (2006). Preparation, characterization and surface study of poly-epsilon caprolactone magnetic microparticles. *Journal of Colloid and Interface Science*, *300*, 584–590.

Hemmingsson, A., Carlsten, J., Ericsson, A., Klaveness, J., Sperber, G. O., & Thuomas, K. A. (1987). Relaxation enhancement of the dog liver and spleen by biodegradable superparamagnetic particles in proton magnetic resonance imaging. *ACTA Radiologica*, *28*, 703–705.

Heyn, C., Ronald, J. A., Mackenzie, L. T., MacDonald, I. C., Chambers, A. F., Rutt, B. K., et al. (2006a). In vivo magnetic resonance imaging of single cells in mouse brain with optical validation. *Magnetic Resonance in Medicine*, *55*, 23–29.

Heyn, C., Ronald, J. A., Ramadan, S. S., Snir, J. A., Barry, A. M., MacKenzie, L. T., et al. (2006b). In vivo MRI of cancer cell fate at the single-cell level in a mouse model of breast cancer metastasis to the brain. *Magnetic Resonance in Medicine*, *56*, 1001–1010.

Hood, J. D., Bednarski, M., Frausto, R., Guccione, S., Reisfeld, R. A., Xiang, R., et al. (2002). Tumor regression by targeted gene delivery to the neovasculature. *Science*, *296*, 2404–2407.

Hsu, A. R., Hou, L. C., Veeravagu, A., Greve, J. M., Vogel, H., Tse, V., et al. (2006). In vivo near-infrared fluorescence imaging of integrin alphavbeta3 in an orthotopic glioblastoma model. *Molecular Imaging Biology*, *8*, 315–323.

Hunt, M. A., Bago, A. G., & Neuwelt, E. A. (2005). Single-dose contrast agent for intraoperative MR imaging of intrinsic brain tumors by using ferumoxtran-10. *AJNR. American Journal of Neuroradiology*, *26*, 1084–1088.

Jack, C. R., Jr., Garwood, M., Wengenack, T. M., Borowski, B., Curran, G. L., Lin, J., et al. (2004). In vivo visualization of Alzheimer's amyloid plaques by magnetic resonance imaging in transgenic mice without a contrast agent. *Magnetic Resonance in Medicine*, *52*, 1263–1271.

Jack, C. R., Jr., Wengenack, T. M., Reyes, D. A., Garwood, M., Curran, G. L., Borowski, B. J., et al. (2005). In vivo magnetic resonance microimaging of individual amyloid plaques in Alzheimer's transgenic mice. *Journal of Neuroscience*, *25*, 10041–10048.

Jander, S., Schroeter, M., & Saleh, A. (2007). Imaging inflammation in acute brain ischemia. *Stroke*, *38*, 642–645.

Jarrett, J. T., Berger, E. P., & Lansbury, P. T., Jr. (1993). The carboxy terminus of the beta amyloid protein is critical for the seeding of amyloid formation: Implications for the pathogenesis of Alzheimer's disease. *Biochemistry*, *32*, 4693–4697.

Johansson, L. O., Bjornerud, A., Ahlstrom, H. K., Ladd, D. L., & Fujii, D. K. (2001). A targeted contrast agent for magnetic resonance imaging of thrombus: Implications of spatial resolution. *Journal of Magnetic Resonance Imaging*, *13*, 615–618.

Kachra, Z., Beaulieu, E., Delbecchi, L., Mousseau, N., Berthelet, F., Moumdjian, R., et al. (1999). Expression of matrix metalloproteinases and their inhibitors in human brain tumors. *Clinical and Experimental Metastasis*, *17*, 555–566.

Kappos, L., Moeri, D., Radue, E. W., Schoetzau, A., Schweikert, K., Barkhof, F., et al. (1999). Predictive value of gadolinium-enhanced magnetic resonance imaging for relapse rate and changes in disability or impairment in multiple sclerosis: A meta-analysis. Gadolinium MRI Meta-analysis Group. *Lancet*, *353*, 964–969.

Kelly, K. A., Allport, J. R., Tsourkas, A., Shinde-Patil, V. R., Josephson, L., & Weissleder, R. (2005). Detection of vascular adhesion molecule-1 expression using a novel multimodal nanoparticle. *Circulation Research*, *96*, 327–336.

Kim, J., Kim, D. I., Lee, S. K., Kim, D. J., Lee, J. E., & Ahn, S. K. (2008). Imaging of the inflammatory response in reperfusion injury after transient cerebral ischemia in rats: Correlation of superparamagnetic iron oxide-enhanced magnetic resonance imaging with histopathology. *Acta Radiology*, *49*, 580–588.

Kircher, M. F., Mahmood, U., King, R. S., Weissleder, R., & Josephson, L. (2003). A multimodal nanoparticle for preoperative magnetic resonance imaging and intraoperative optical brain tumor delineation. *Cancer Research*, *63*, 8122–8125.

Kleinschnitz, C., Schutz, A., Nolte, I., Horn, T., Frank, M., Solymosi, L., et al. (2005). In vivo detection of developing vessel occlusion in photothrombotic ischemic brain lesions in the rat by iron particle enhanced MRI. *Journal of Cerebral Blood Flow and Metabolism*, *25*, 1548–1555.

Klibanov, A. L. (2007). Ultrasound molecular imaging with targeted microbubble contrast agents. *Journal of Nuclear Cardiology*, *14*, 876–884.

Klunk, W. E., Engler, H., Nordberg, A., Bacskai, B.J, Wang, Y., Price, J. C., et al. (2003). Imaging the pathology of Alzheimer's disease: Amyloid-imaging with positron emission tomography. *Neuroimaging Clinics of North Amercia*, *13*, 781–789, ix.

Klunk, W. E., Engler, H., Nordberg, A., Wang, Y., Blomqvist, G., Holt, D. P., et al. (2004). Imaging brain amyloid in Alzheimer's disease with Pittsburgh Compound-B. *Annals of Neurology*, *55*, 306–319.

Koochekpour, S., Jeffers, M., Rulong, S., Taylor, G., Klineberg, E., Hudson, E. A., et al. (1997). Met and hepatocyte growth factor/scatter factor expression in human gliomas. *Cancer Research*, *57*, 5391–5398.

Langer, H. F., & Gawaz, M. (2008). Platelet-vessel wall interactions in atherosclerotic disease. *Thrombosis and Haemostasis*, *99*, 480–486.

Lanza, G. M., Wallace, K. D., Scott, M. J., Cacheris, W. P., Abendschein, D. R., Christy, D. H., et al. (1996). A novel site-targeted ultrasonic contrast agent with broad biomedical application. *Circulation*, *94*, 3334–3340.

Lee, M. A., Smith, S., Palace, J., Narayanan, S., Silver, N., Minicucci, L., et al. (1999). Spatial mapping of T2 and gadolinium-enhancing T1 lesion volumes in multiple sclerosis: Evidence for distinct mechanisms of lesion genesis? *Brain*, *122*(Pt. 7), 1261–1270.

Linker, R. A., Reinhardt, M., Bendszus, M., Ladewig, G., Briel, A., Schirner, M., et al. (2005). In vivo molecular imaging of adhesion molecules in experimental autoimmune encephalomyelitis (EAE). *Journal of Autoimmunity*, *25*, 199–205.

Lipinski, M. J., Amirbekian, V., Frias, J. C., Aguinaldo, J. G., Mani, V., Briley-Saebo, K. C., et al. (2006). MRI to detect atherosclerosis with gadolinium-containing immunomicelles targeting the macrophage scavenger receptor. *Magnetic Resonance in Medicine*, *56*, 601–610.

Lipinski, M. J., Briley-Saebo, K. C., Mani, V., & Fayad, Z. A. (2008). "Positive contrast" inversion-recovery with ON[corrected]-resonant water suppression magnetic resonance imaging: A change for the better? *Journal of the Amercian College of Cardiology*, *52*, 492–494.

Liu, C. H., Huang, S., Cui, J., Kim, Y. R., Farrar, C. T., Moskowitz, M. A., et al. (2007). MR contrast probes that trace gene transcripts for cerebral ischemia in live animals. *FASEB Journal*, *21*, 3004–3015.

Liu, C. H., You, Z., Ren, J., Kim, Y. R., Eikermann-Haerter, K., & Liu, P. K. (2008). Noninvasive delivery of gene targeting probes to live brains for transcription MRI. *FASEB Journal*, *22*, 1193–1203.

Lockhart, A., Lamb, J. R., Osredkar, T., Sue, L. I., Joyce, J. N., Ye, L., et al. (2007). PIB is a non-specific imaging marker of amyloid-beta (Abeta) peptide-related cerebral amyloidosis. *Brain*, *130*, 2607–2615.

Manninger, S. P., Muldoon, L. L., Nesbit, G., Murillo, T., Jacobs, P. M., & Neuwelt, E. A. (2005). An exploratory study of ferumoxtran-10 nanoparticles as a blood-brain barrier imaging agent targeting phagocytic cells in CNS inflammatory lesions. *AJNR. Amercian Journal of Neuroradiology*, *26*, 2290–2300.

Massberg, S., Brand, K., Gruner, S., Page, S., Muller, E., Muller, I., et al. (2002). A critical role of platelet adhesion in the initiation of atherosclerotic lesion formation. *Journal of Experimental Medicine*, *196*, 887–896.

McAteer, M. A., Schneider, J. E., Ali, Z. A., Warrick, N., Bursill, C. A., von zur Muhlen, C., et al. (2008). Magnetic resonance imaging of endothelial adhesion molecules in mouse atherosclerosis using dual-targeted microparticles of iron oxide. *Arteriosclerosis, Thrombosis, and Vascular Biology*, *28*, 77–83.

McAteer, M. A., Sibson, N. R., von Zur Muhlen, C., Schneider, J. E., Lowe, A. S., Warrick, N., et al. (2007). In vivo magnetic resonance imaging of acute brain inflammation using microparticles of iron oxide. *Nature Medicine*, *13*, 1253–1258.

McCarthy, J. R., & Weissleder, R. (2008). Multifunctional magnetic nanoparticles for targeted imaging and therapy. *Advanced Drug Delivery Reviews*, *60*, 1241–1251.

Mendonca Dias, M. H., & Lauterbur, P. C. (1986). Ferromagnetic particles as contrast agents for magnetic resonance imaging of liver and spleen. *Magnetic Resonance in Medicine*, *3*, 328–330.

Meng, X. X., Wan, J. Q., Jing, M., Zhao, S. G., Cai, W., & Liu, E. Z. (2007). Specific targeting of gliomas with multifunctional superparamagnetic iron oxide nanoparticle optical and magnetic resonance imaging contrast agents. *Acta Pharmacologica Sinica*, *28*, 2019–2026.

Metz, S., Bonaterra, G., Rudelius, M., Settles, M., Rummeny, E. J., & Daldrup-Link, H. E. (2004). Capacity of human monocytes to phagocytose approved iron oxide MR contrast agents in vitro. *European Radiology*, *14*, 1851–1858.

Miller, D. H. (2004). Biomarkers and surrogate outcomes in neurodegenerative disease: Lessons from multiple sclerosis. *NeuroRx*, *1*, 284–294.

Montet, X., Montet-Abou, K., Reynolds, F., Weissleder, R., & Josephson, L. (2006). Nanoparticle imaging of integrins on tumor cells. *Neoplasia*, *8*, 214–222.

Moore, A., Marecos, E., Bogdanov, A., Jr., & Weissleder, R. (2000). Tumoral distribution of long-circulating dextran-coated iron oxide nanoparticles in a rodent model. *Radiology*, *214*, 568–574.

Moriyama, T., Kataoka, H., Kawano, H., Yokogami, K., Nakano, S., Goya, T., et al. (1998). Comparative analysis of expression of hepatocyte growth factor and its receptor, c-met, in gliomas, meningiomas and schwannomas in humans. *Cancer Letters*, *124*, 149–155.

Mulder, W. J., Strijkers, G. J., Habets, J. W., Bleeker, E. J., van der Schaft, D. W., Storm, G., et al. (2005). MR molecular imaging and fluorescence microscopy for identification of activated tumor endothelium using a bimodal lipidic nanoparticle. *FASEB Journal, 19*, 2008–2010.

Muldoon, L. L., Tratnyek, P. G., Jacobs, P. M., Doolittle, N. D., Christoforidis, G. A., Frank, J. A., et al. (2006). Imaging and nanomedicine for diagnosis and therapy in the central nervous system: Report of the eleventh annual Blood-Brain Barrier Disruption Consortium meeting. *AJNR. Amercian Journal of Neuroradiology, 27*, 715–721.

Nahrendorf, M., Jaffer, F. A., Kelly, K. A., Sosnovik, D. E., Aikawa, E., Libby, P., et al. (2006a). Noninvasive vascular cell adhesion molecule-1 imaging identifies inflammatory activation of cells in atherosclerosis. *Circulation, 114*, 1504–1511.

Nahrendorf, M., Jaffer, F. A., Kelly, K. A., Sosnovik, D. E., Aikawa, E., Libby, P., et al. (2006b) Noninvasive vascular cell adhesion molecule-1 imaging identifies inflammatory activation of cells in atherosclerosis. *Circulation, 114*, 1504–1511.

Neuwelt, E. A., Varallyay, P., Bago, A. G., Muldoon, L. L., Nesbit, G., & Nixon, R. (2004). Imaging of iron oxide nanoparticles by MR and light microscopy in patients with malignant brain tumours. *Neuropathology and Applied Neurobiology, 30*, 456–471.

Ngaboni Okassa, L., Marchais, H., Douziech-Eyrolles, L., Cohen-Jonathan, S., Souce, M., Dubois, P., et al. (2005). Development and characterization of sub-micron poly (D, L-lactide-co-glycolide) particles loaded with magnetite/ maghemite nanoparticles. *International Journal of Pharmaceutics, 302*, 187–196.

Nighoghossian, N., Wiart, M., Cakmak, S., Berthezene, Y., Derex, L., Cho, T. H., et al. (2007). Inflammatory response after ischemic stroke: A USPIO-enhanced MRI study in patients. *Stroke, 38*, 303–307.

Nitta, T., & Sato, K. (1995). Prognostic implications of the extent of surgical resection in patients with intracranial malignant gliomas. *Cancer, 75*, 2727–2731.

Nordberg, A. (2008). Amyloid plaque imaging in vivo: Current achievement and future prospects. *European Journal of Nuclear Medicine Molecular Imaging, 35*(Supplement 1), S46–S50.

Peter, K., Ahrens, I., Schwarz, M., Bode, C., & Ylanne, J. (2004). Distinct roles of ligand affinity and cytoskeletal anchorage in alphaIIbbeta3 (GP IIb/IIIa)-mediated cell aggregation and adhesion. *Platelets, 15*, 427–438.

Pichler, B. J., Judenhofer, M. S., & Wehrl, H. F. (2008). PET/ MRI hybrid imaging: Devices and initial results. *European Radiology, 18*, 1077–1086.

Poduslo, J. F., & Curran, G. L. (1996a). Increased permeability of superoxide dismutase at the blood-nerve and blood-brain barriers with retained enzymatic activity after covalent modification with the naturally occurring polyamine, putrescine. *Journal of Neurochemistry, 67*, 734–741.

Poduslo, J. F., & Curran, G. L. (1996b). Polyamine modification increases the permeability of proteins at the blood-nerve and blood-brain barriers. *Journal of Neurochemistry, 66*, 1599–1609.

Poduslo, J. F., Ramakrishnan, M., Holasek, S. S., Ramirez-Alvarado, M., Kandimalla, K. K., Gilles, E. J., et al. (2007). In vivo targeting of antibody fragments to the nervous system for Alzheimer's disease immunotherapy and molecular imaging of amyloid plaques. *Journal of Neurochemistry, 102*, 420–433.

Polman, C. H., O'Connor, P. W., Havrdova, E., Hutchinson, M., Kappos, L., Miller, D. H., et al. (2006). A randomized, placebo-controlled trial of natalizumab for relapsing multiple sclerosis. *New England Journal of Medicine, 354*, 899–910.

Price, C. J., Warburton, E. A., & Menon, D. K. (2003). Human cellular inflammation in the pathology of acute cerebral ischaemia. *Journal of Neurology, Neurosurgery, and Psychiatry, 74*, 1476–1484.

Ramadan, S.S, Heyn, C., Mackenzie, L. T., Chambers, A. F., Rutt, B. K., & Foster, P. J. (2008). Ex-vivo cellular MRI with b-SSFP: Quantitative benefits of 3T over 1.5 T. *Magma, 21*, 251–259.

Rausch, M., Baumann, D., Neubacher, U., & Rudin, M. (2002). In-vivo visualization of phagocytotic cells in rat brains after transient ischemia by USPIO. *NMR in Biomedicine, 15*, 278–283.

Rausch, M., Hiestand, P., Baumann, D., Cannet, C., & Rudin, M. (2003). MRI-based monitoring of inflammation and tissue damage in acute and chronic relapsing EAE. *Magnetic Resonance in Medicine, 50*, 309–314.

Rausch, M., Hiestand, P., Foster, C. A., Baumann, D. R., Cannet, C., & Rudin, M. (2004). Predictability of FTY720 efficacy in experimental autoimmune encephalomyelitis by in vivo macrophage tracking: Clinical implications for ultrasmall superparamagnetic iron oxide-enhanced magnetic resonance imaging. *Journal of Magnetic Resonance Imaging, 20*, 16–24.

Rausch, M., Sauter, A., Frohlich, J., Neubacher, U., Radu, E. W., & Rudin, M. (2001). Dynamic patterns of USPIO enhancement can be observed in macrophages after ischemic brain damage. *Magnetic Resonance in Medicine, 46*, 1018–1022.

Raymond, S. B., Treat, L. H., Dewey, J. D., McDannold, N. J., Hynynen, K., & Bacskai, B. J. (2008). Ultrasound enhanced delivery of molecular imaging and therapeutic agents in Alzheimer's disease mouse models. *PLoS one, 3*, e2175.

Reddy, G. R., Bhojani, M. S., McConville, P., Moody, J., Moffat, B. A., Hall, D. E., et al. (2006). Vascular targeted nanoparticles for imaging and treatment of brain tumors. *Clinical Cancer Research, 12*, 6677–6686.

Reinhardt, M., Hauff, P., Linker, R. A., Briel, A., Gold, R., Rieckmann, P., et al. (2005). Ultrasound derived imaging and quantification of cell adhesion molecules in experimental autoimmune encephalomyelitis (EAE) by Sensitive Particle Acoustic Quantification (SPAQ). *Neuroimage, 27*, 267–278.

Renshaw, P. F., Owen, C. S., McLaughlin, A. C., Frey, T. G., & Leigh, J. S., Jr. (1986). Ferromagnetic contrast agents: A new approach. *Magnetic Resonance in Medicine, 3*, 217–225.

Runge, V. M., & Gelblum, D. Y. (1991). Future directions in magnetic resonance contrast media. *Topics in Magnetic Resonance Imaging, 3*, 85–97.

Saleh, A., Schroeter, M., Jonkmanns, C., Hartung, H. P., Modder, U., & Jander, S. (2004). In vivo MRI of brain inflammation in human ischaemic stroke. *Brain, 127*, 1670–1677.

Schlemmer, H. P., Pichler, B. J., Krieg, R., & Heiss, W. D. (2008). An integrated MR/PET system: Prospective applications. Abdom Imaging, DOI: 10.1007/s00261-008-9450-2.

Schmieder, A. H., Winter, P. M., Caruthers, S. D., Harris, T. D., Williams, T. A., Allen, J. S., et al. (2005). Molecular MR imaging of melanoma angiogenesis with alphanubeta3-targeted paramagnetic nanoparticles. *Magnetic Resonance in Medicine, 53*, 621–627.

Schroeter, M., Saleh, A., Wiedermann, D., Hoehn, M., & Jander, S. (2004). Histochemical detection of ultrasmall superparamagnetic iron oxide (USPIO) contrast medium uptake in experimental brain ischemia. *Magnetic Resonance in Medicine, 52*, 403–406.

Schwarz, M., Meade, G., Stoll, P., Ylanne, J., Bassler, N., Chen, Y. C., et al. (2006). Conformation-specific blockade of the integrin GPIIb/IIIa: A novel antiplatelet strategy that selectively targets activated platelets. *Circulation Research, 99*, 25–33.

Schwarz, M., Rottgen, P., Takada, Y., Le Gall, F., Knackmuss, S., Bassler, N., et al. (2004). Single-chain antibodies for the conformation-specific blockade of activated platelet integrin alphaIIbbeta3 designed by subtractive selection from naive human phage libraries. *FASEB Journal, 18*, 1704–1706.

Shapiro, E. M., Skrtic, S., & Koretsky, A. P. (2005). Sizing it up: Cellular MRI using micron-sized iron oxide particles. *Magnetic Resonance in Medicine, 53*, 329–338.

Shapiro, E. M., Sharer, K., Skrtic, S., & Koretsky, A. P. (2006). In vivo detection of single cells by MRI. *Magnetic Resonance Medicine, 55*, 242–249.

Sibson, N. R., Blamire, A. M., Bernades-Silva, M., Laurent, S., Boutry, S., Muller, R. N., et al. (2004). MRI detection of early endothelial activation in brain inflammation. *Magnetic Resonance in Medicine, 51*, 248–252.

Sigurdsson, E. M., Wadghiri, Y. Z., Mosconi, L., Blind, J. A., Knudsen, E., Asuni, A., et al. (2008). A non-toxic ligand for voxel-based MRI analysis of plaques in AD transgenic mice. *Neurobiology of Aging, 29*, 836–847.

Sipkins, D. A., Cheresh, D. A., Kazemi, M. R., Nevin, L. M., Bednarski, M. D., & Li, K. C. (1998). Detection of tumor angiogenesis in vivo by alphaVbeta3-targeted magnetic resonance imaging. *Nature Medicine, 4*, 623–626.

Sipkins, D. A., Gijbels, K., Tropper, F. D., Bednarski, M., Li, K. C., & Steinman, L. (2000). ICAM-1 expression in autoimmune encephalitis visualized using magnetic resonance imaging. *Journal of Neuroimmunology, 104*, 1–9.

Sirol, M., Fuster, V., Badimon, J. J., Fallon, J. T., Moreno, P. R., Toussaint, J. F., et al. (2005). Chronic thrombus detection with in vivo magnetic resonance imaging and a fibrin-targeted contrast agent. *Circulation, 112*, 1594–1600.

Siva, A. (2006). The spectrum of multiple sclerosis and treatment decisions. *Clinical Neurology and Neurosurgery, 108*, 333–338.

Smith, G. P. (1985). Filamentous fusion phage: Novel expression vectors that display cloned antigens on the virion surface. *Science, 228*, 1315–1317.

Somer, E. J., Benatar, N. A., O'Doherty, M. J., Smith, M. A., & Marsden, P. K. (2007). Use of the CT component of PET-CT to improve PET-MR registration: Demonstration in soft-tissue sarcoma. *Physics in Medicine and Biology, 52*, 6991–7006.

Sosnovik, D. E., Nahrendorf, M., & Weissleder, R. (2008). Magnetic nanoparticles for MR imaging: Agents, techniques and cardiovascular applications. *Basic Research in Cardiology, 103*, 122–130.

Spuentrup, E., Botnar, R. M., Wiethoff, A. J., Ibrahim, T., Kelle, S., Katoh, M., et al. (2008). MR imaging of thrombi using EP-2104R, a fibrin-specific contrast agent: Initial results in patients. *European Radiology, 18*, 1995–2005.

Spuentrup, E., Katoh, M., Wiethoff, A. J., Buecker, A., Botnar, R. M., Parsons, E. C., et al. (2007a). Molecular coronary MR imaging of human thrombi using EP-2104R, a fibrin-targeted contrast agent: Experimental study in a swine model. *Rofo, 179*, 1166–1173.

Spuentrup, E., Katoh, M., Buecker, A., Fausten, B., Wiethoff, A. J., Wildberger, J. E., et al. (2007b). Molecular MR imaging of human thrombi in a swine model of pulmonary embolism using a fibrin-specific contrast agent. *Investigative Radiology, 42*, 586–595.

Stoll, P., Bassler, N., Hagemeyer, C. E., Eisenhardt, S. U., Chen, Y. C., Schmidt, R., et al. (2007). Targeting ligand-induced binding sites on GPIIb/IIIa via single-chain antibody allows effective anticoagulation without bleeding time prolongation. *Arteriosclerosis, Thrombosis, and Vascular Biology, 27*, 1206–1212.

Stoll, G., Jander, S., & Schroeter, M. (1998). Inflammation and glial responses in ischemic brain lesions. *Progress in Neurobiology, 56*, 149–171.

Stracke, C. P., Katoh, M., Wiethoff, A. J., Parsons, E. C., Spangenberg, P., & Spuntrup, E. (2007). Molecular MRI of cerebral venous sinus thrombosis using a new fibrin-specific MR contrast agent. *Stroke, 38*, 1476–1481.

Stroh, A., Zimmer, C., Werner, N., Gertz, K., Weir, K., Kronenberg, G., et al. (2006). Tracking of systemically administered mononuclear cells in the ischemic brain by high-field magnetic resonance imaging. *Neuroimage, 33*, 886–897.

Sun, C., Veiseh, O., Gunn, J., Fang, C., Hansen, S., Lee, D., et al. (2008). In vivo MRI detection of gliomas by chlorotoxin-conjugated superparamagnetic nanoprobes. *Small, 4*, 372–379.

Swerdlow, R. H. (2007). PET sheds light on Alzheimer's disease genetic risk. *Proceeding of the National Academy of Sciences of the United States of America, 104*, 18881–18882.

Thorek, D. L., Chen, A. K., Czupryna, J., & Tsourkas, A. (2006). Superparamagnetic iron oxide nanoparticle probes for molecular imaging. *Annual Review of Biomedical Engineering, 34*, 23–38.

Towner, R. A., Smith, N., Doblas, S., Tesiram, Y., Garteiser, P., Saunders, D., et al. (2008). In vivo detection of c-Met expression in a rat C6 glioma model. *Journal of Cell Molecular Medicine, 12*, 174–186.

Tsourkas, A., Shinde-Patil, V. R., Kelly, K. A., Patel, P., Wolley, A., Allport, J. R., et al. (2005). In vivo imaging of activated endothelium using an anti-VCAM-1 magnetooptical probe. *Bioconjugate Chemistry*, *16*, 576–581.

van Kasteren, S. I., Kramer, H. B., Gamblin, D. P., & Davis, B. G. (2007). Site-selective glycosylation of proteins: Creating synthetic glycoproteins. *Nature Protocols*, *2*, 3185–3194.

Varallyay, P., Nesbit, G., Muldoon, L. L., Nixon, R. R., Delashaw, J., Cohen, J. I., et al. (2002). Comparison of two superparamagnetic viral-sized iron oxide particles ferumoxides and ferumoxtran-10 with a gadolinium chelate in imaging intracranial tumors. *AJNR. Amercian Journal of Neuroradiology*, *23*, 510–519.

Veiseh, O., Sun, C., Gunn, J., Kohler, N., Gabikian, P., Lee, D., et al. (2005). Optical and MRI multifunctional nanoprobe for targeting gliomas. *Nano Letters*, *5*, 1003–1008.

Vellinga, M. M., Oude Engberink, R. D., Seewann, A., Pouwels, P. J., Wattjes, M. P., van der Pol, S. M., et al. (2008). Pluriformity of inflammation in multiple sclerosis shown by ultra-small iron oxide particle enhancement. *Brain*, *131*, 800–807.

Villanueva, F. S., Jankowski, R. J., Klibanov, S., Pina, M. L., Alber, S. M., Watkins, S. C., et al. (1998). Microbubbles targeted to intercellular adhesion molecule-1 bind to activated coronary artery endothelial cells. *Circulation*, *98*, 1–5.

von Zur Muhlen, C., von Elverfeldt, D., Choudhury, R. P., Ender, J., Ahrens, I., Schwarz, M., et al. (2008a). Functionalized magnetic resonance contrast agent selectively binds to glycoprotein IIb/IIIa on activated human platelets under flow conditions and is detectable at clinically relevant field strengths. *Molecular Imaging*, *7*, 59–67.

von Zur Muhlen, C., Peter, K., Ali, Z. A., Schneider, J. E., McAteer, M. A., Neubauer, S., et al. (2008b). Visualization of activated platelets by targeted magnetic resonance imaging utilizing conformation-specific antibodies against glycoprotein IIb/IIIa. *Journal of Vascular Research*, *46*, 6–14.

von Zur Muhlen, C., Sibson, N. R., Peter, K., Campbell, S. J., Wilainam, P., Grau, G. E., et al. (2008c). A contrast agent recognizing activated platelets reveals murine cerebral malaria pathology undetectable by conventional MRI. *Journal of Clinicial Investigation*, *118*, 1198–1207.

von zur Muhlen, C., von Elverfeldt, D., Moeller, J. A., Choudhury, R. P., Paul, D., Hagemeyer, C. E., et al. (2008d). Magnetic resonance imaging contrast agent targeted toward activated platelets allows in vivo detection of thrombosis and monitoring of thrombolysis. *Circulation*, *118*, 258–267.

Wadghiri, Y. Z., Sigurdsson, E. M., Sadowski, M., Elliott, J. I., Li, Y., Scholtzova, H., Tang, C. Y., et al. (2003). Detection of Alzheimer's amyloid in transgenic mice using magnetic resonance microimaging. *Magnetic Resonance in Medicine*, *50*, 293–302.

Wang, Y., Klunk, W. E., Debnath, M. L., Huang, G. F., Holt, D. P., Shao, L., et al. (2004). Development of a PET/SPECT agent for amyloid imaging in Alzheimer's disease. *Journal of Molecular Neuroscience*, *24*, 55–62.

Wassmer, S. C., Combes, V., Candal, F. J., Juhan-Vague, I., & Grau, G. E. (2006). Platelets potentiate brain endothelial alterations induced by Plasmodium falciparum. *Infection and Immunity*, *74*, 645–653.

Weissleder, R., Elizondo, G., Wittenberg, J., Rabito, C. A., Bengele, H. H., & Josephson, L. (1990). Ultrasmall superparamagnetic iron oxide: Characterization of a new class of contrast agents for MR imaging. *Radiology*, *175*, 489–493.

Wengenack, T. M., Jack, C. R., Jr., Garwood, M., & Poduslo, J. F. (2008). MR microimaging of amyloid plaques in Alzheimer's disease transgenic mice. *European Journal of Nuclear Medicine Molecular Imaging*, *35*(Supplement 1), S82–S88.

Willmann, J. K., Lutz, A. M., Paulmurugan, R., Patel, M. R., Chu, P., Rosenberg, J., et al. (2008). Dual-targeted contrast agent for US assessment of tumor angiogenesis in vivo. *Radiology*, *248*, 936–944.

Winter, P. M., Caruthers, S. D., Kassner, A., Harris, T. D., Chinen, L. K., Allen, J. S., et al. (2003). Molecular imaging of angiogenesis in nascent Vx-2 rabbit tumors using a novel alpha(nu)beta3-targeted nanoparticle and 1.5 tesla magnetic resonance imaging. *Cancer Research*, *63*, 5838–5843.

Wrensch, M., Minn, Y., Chew, T., Bondy, M., & Berger, M. S. (2002). Epidemiology of primary brain tumors: Current concepts and review of the literature. *Neuro-Oncology*, *4*, 278–299.

Xu, S., Jordan, E. K., Brocke, S., Bulte, J. W., Quigley, L., Tresser, N., et al. (1998). Study of relapsing remitting experimental allergic encephalomyelitis SJL mouse model using MION-46L enhanced in vivo MRI: Early histopathological correlation. *Journal of Neuroscience Research*, *52*, 549–558.

Yu, X., Song, S. K., Chen, J., Scott, M. J., Fuhrhop, R. J., Hall, C. S., et al. (2000). High-resolution MRI characterization of human thrombus using a novel fibrin-targeted paramagnetic nanoparticle contrast agent. *Magnetic Resonance in Medicine*, *44*, 867–872.

Zhu, D., White, R. D., Hardy, P. A., Weerapreeyakul, N., Sutthanut, K., & Jay, M. (2006). Biocompatible nanotemplate-engineered nanoparticles containing gadolinium: Stability and relaxivity of a potential MRI contrast agent. *Journal of Nanoscience and Nanotechnology*, *6*, 996–1003.

Zimmer, C., Weissleder, R., Poss, K., Bogdanova, A., Wright, S. C., Jr., & Enochs, W. S. (1995). MR imaging of phagocytosis in experimental gliomas. *Radiology*, *197*, 533–538.

H.S. Sharma (Ed.)
Progress in Brain Research, Vol. 180
ISSN: 0079-6123

CHAPTER 5

Development of iron chelator–nanoparticle conjugates as potential therapeutic agents for Alzheimer disease

Gang Liu[1], Ping Men[1], George Perry,[2,3] and Mark A. Smith[3,*]

[1]Department of Radiology, University of Utah, Salt Lake City, UT, USA
[2]College of Sciences, University of Texas at San Antonio, San Antonio, TX, USA
[3]Department of Pathology, Case Western Reserve University, Cleveland, OH, USA

Abstract: Oxidative stress is known to play a key role in the initiation and promotion of the neurodegeneration that characterizes the pathogenesis of Alzheimer disease (AD). An accumulation of redox active transition metals, including iron and copper, is likely a major generator of reactive oxidative species and other free radicals and is thought to induce a detrimental cycle of oxidative stress, amyloid-β aggregation, and neurodegeneration. As such, metal chelators may provide an alternative therapeutic approach to sequester redox active metals and prevent the onslaught of oxidative damage. Unfortunately, however, metal chelation approaches are currently limited in their potential, since many cannot readily pass the blood–brain barrier (BBB), due to their hydrophilicity, and many are neurotoxic at high concentrations. To circumvent such issues, here we describe the development of iron chelator–nanoparticle conjugation that allows delivery of target chelator to the brain in the absence of neurotoxicity. Such nanoparticle delivery of iron chelators will likely provide a highly advantageous mode of attack on the oxidative stress that plagues AD as well as other conditions characterized by excess metal accumulation.

Keywords: Alzheimer disease; amyloid-β aggregation; ion metal chelator; oxidative stress; nanoparticle delivery; neurodegeneration; prevention

Introduction

Alzheimer disease (AD) is a devastating neurodegenerative disease characterized by progressive and irreversible damage to thought, memory, and language (Smith, 1998). This disease is the most common form of dementia, and it is the fourth leading cause of death among people aged 65 and older (Minino, Heron, Murphy, & Kochanek, 2007; Yaari & Corey-Bloom, 2007). Currently, an estimated 24 million people suffer from dementia worldwide, and the affected population will exponentially increase to 42 million by year 2020 and to 81 million by year 2040 (Ferri et al., 2005) if no effective prevention and treatment become available (Hebert, Beckett, Scherr, & Evans, 2001;

* Corresponding author.
Tel.: +216-368-3670; Fax: +216-368-8964;
E-mail: mark.smith@case.edu

DOI: 10.1016/S0079-6123(08)80005-2

Scorer, 2001). Furthermore, AD places an incalculable emotional and physical drain on families and caretakers. As such, the disease poses a heavy economic and societal burden with associated annual cost of care over $100 billion and about $500 billion by year 2020 in the United States alone (Pardridge, 2007; Smith, 1998). The ravaging effects of the disease call for efforts to prevent, forestall, and reverse the disease. Unfortunately, despite much interest, currently available FDA-approved drugs, targeted mostly toward managing cognitive symptoms such as changes in memory and perception, provide only partial benefit to select patients and fall tremendously short as adequate means of therapeutic management (Liu et al., 2005; Marlatt et al., 2005). Obviously, there is an urgent need for a better understanding of AD development and more effective therapeutic agents.

Excess metal toxicity and amyloid-β aggregation

Although the etiology of AD is incompletely understood, oxidative stress resulting from various mechanisms may play a key role in the initiation and promotion of neurodegeneration in AD (Halliwell & Gutteridge, 1999; Markesbery, 1997; Perry, Castellani, Hirai, & Smith, 1998; Smith, Sayre, Monnier, & Perry, 1995). The central nervous system is particularly susceptible to oxidative damage compared with other tissues (Evans, 1993; Gutteridge, 1994), and oxidation reactions can be catalyzed by transition metals such as iron and copper (Halliwell & Gutteridge, 1999; Olanow, 1992). As such, the likelihood that an oxidation reaction will take place is probably increased by the regional concentrations of metals (Halliwell & Gutteridge, 1999; Olanow, 1992). In fact, some transition metals including iron and copper have been found in high concentrations in the brains of AD patients and accumulate specifically in the pathological lesions (Kong, Liochev, & Fridovich, 1992; Lovell, Robertson, Teesdale, Campbell, & Markesbery, 1998; Sayre, Perry, & Smith, 1999; Sayre et al., 2000; Smith, Harris, Sayre, & Perry, 1997), and these metals are suggested as key contributors to the altered redox state (Castellani et al., 2007; Liu, Men, Kudo, Perry, & Smith, 2009a). In addition, amyloid-β (Aβ) is also considered by many (Hardy, 2006; Hardy & Higgins, 1992), though not all (Castellani et al., 2006; Lee, Zhu, Nunomura, Perry, & Smith, 2006), as a major factor in AD pathogenesis. The neurotoxicity of Aβ may result from the formation of protease-resistant oligomeric and fibrillar forms of Aβ (Selkoe, 1991), and likewise, the aggregation and toxicity of Aβ is dependent on transition metals. Thus, chelation agents that selectively bind to, remove, and/or "redox silence" transition metals have long been considered an attractive therapeutic target for AD.

Metal chelation and nanoparticle delivery

Indeed, the thought that metal chelators are among the list of agents with potential to protect against metal-associated oxidation damage and to prevent and reverse Aβ aggregation (Crouch, Barnham, Bush, & White, 2006; Liu et al., 2005; Smith, 2006) is attracting renewed interest. Chelators provide a "three-pronged" mode of action. First, since iron and copper are suggested to play an important role in the self-assembly and neurotoxicity of Aβ (Atwood et al., 2004; Bush et al., 1994; Exley, 2006; Hayashi et al., 2007; Jobling et al., 2001; Nakamura et al., 2007), Aβ toxicity is expectedly attenuated by such chelators (Huang et al., 1999; Opazo et al., 2002; Rottkamp et al., 2001; Schubert & Chevion, 1995). In fact, the ability of Aβ to sequestrate redox metals likely explains conflicting *in vivo* and *in vitro* reports demonstrating Aβ as both oxidant (Behl, Davis, Cole, & Schubert, 1992) and antioxidant (Hayashi et al., 2007; Nakamura et al., 2007; Nunomura et al., 2001; Smith, Casadesus, Joseph, & Perry, 2002). Second, redox metals, as redox-active centers, lead to free radical generation (Bishop et al., 2002; Castellani et al., 2007; Sayre et al., 2000; Smith et al., 1997) and oxidative stress, which contribute to the initiation and promotion of neurodegeneration (Casadesus et al., 2004; Markesbery, 1997; Perry et al., 1998; Smith et al., 1995). Third, since oxidative stress, some of which is consequent to metal-mediated processes (Sayre et al., 2000), is associated with increased Aβ as a consequence of the coordinated upregulation of amyloid-β protein precursor (AβPP) (Yan et al., 1995) and β- and γ-secretases (Tamagno et al., 2008; Yang et al., 2003), it is also not surprising that treatment of AβPP-overexpressing transgenic mice, a model of AD that displays significant Aβ deposition and oxidative stress (Pappolla et al., 1998; Smith et al., 1998), with chelation agents results in less Aβ deposition (Adlard et al., 2008; Cherny et al., 2001).

Taken together, these data suggest chelation as a potential and powerful therapeutic approach to prevent and/or treat AD. Indeed, metal chelation compounds (Fig. 1), such as desferrioxamine (DFO), ethylenediaminetetraacetic acid (EDTA), and iodochlorhydroxyquin (clioquinol), have been used to treat patients with AD and provided significant clinical improvement (Casdorph, 2001; Crapper McLachlan et al., 1991; Regland et al., 2001; Ritchie et al., 2003). However, limitations concerning chelator bioavailability such as blood–brain barrier (BBB) penetration and toxic side-effects have hindered further investigation, limiting both the understanding of the pathologic role of metal dysregulation in AD as well as the evaluation of the efficacy and safety of chelation therapy. For instance, DFO, an iron chelator approved by the FDA for the treatment of iron overload, has been shown to slow progression of AD (Crapper McLachlan et al., 1991); however, it has serious side effects including neurotoxicity and neurological changes (Blake et al., 1985) and cannot penetrate the BBB due to its hydrophilic nature (Lynch, Fonseca, & Levine, 2000). While small molecular weight lipophilic chelators, like bi- or tridentate iron chelators, have the ability to penetrate the BBB, they also have considerable neurotoxicity (Hider, Epemolu, Singh, & Porter, 1994a) and do not remove iron from the brain despite effectively binding the metal (Crowe and Morgan, 1994). Thus, the use of these chelators is currently limited by their bioavailability and/or toxic side effects.

These obstacles, however, may be overcome by utilizing nanoparticle delivery systems. Currently, drug delivery systems using nanoparticles to target the brain have shown promise in both improved drug efficacy and reduced drug toxicity (Kreuter, 2001; Kreuter et al., 2002). A particular class of nanoparticles are able to cross the BBB by mimicking low-density lipoprotein (LDL) and enabling them to interact with the LDL receptor, resulting in their uptake by brain endothelial cells (Kreuter, 2001; Kreuter et al., 2002). Indeed, a recent report used nanoparticles to deliver rivastigmine, a drug

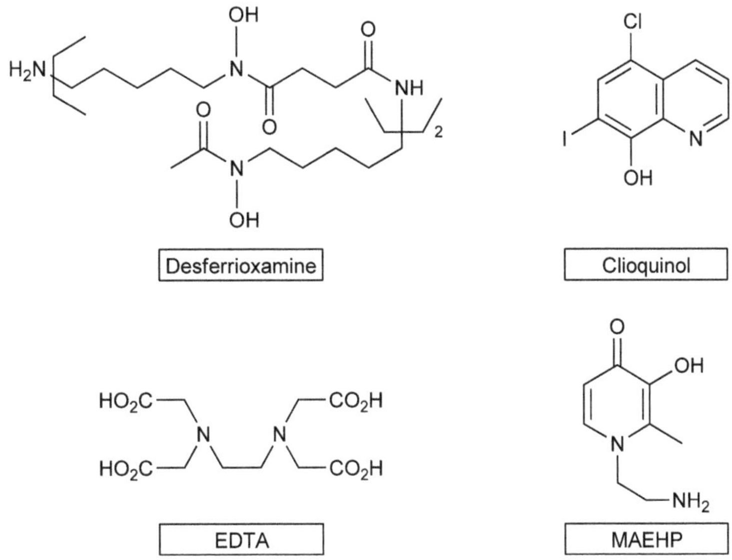

Fig. 1. Chemical structures of DFO, clioquinol, EDTA, and MAEHP.

for the treatment of AD, to show increased brain concentration of the drug in an animal model (Wilson et al., 2008). Significantly, our previous studies have suggested that nanoparticles covalently conjugated to chelators may have the potential to deliver chelators into the brain without altering metal-chelating capability (Liu et al., 2006).

Preparation and use of specific nanoparticle–chelator conjugates

As mentioned earlier, accumulating studies suggest that chelation agents could be a potential therapeutic approach to prevent and treat AD. Most importantly, the use of nanoparticles for target transport of chelators could provide more effective and safer therapeutics. In order to covalently conjugate them, however, chelators and nanoparticles should have functional groups that tend to react with each other, forming a chemical bond. For instance, Fig. 2 presents a prototypic method for conjugation of iron chelators and nanoparticles. The conjugation is achieved by simply forming amido bonds (Liu, Men, Perry, & Smith, 2007, 2009b). Briefly, monodispersed polystyrene particles with carboxyl functional groups on the surface were used to conjugate either DFO or 2-methyl-*N*-(2′-aminoethyl)-3-hydroxyl-4-pyridinone (MAEHP) (Fig. 1). Both

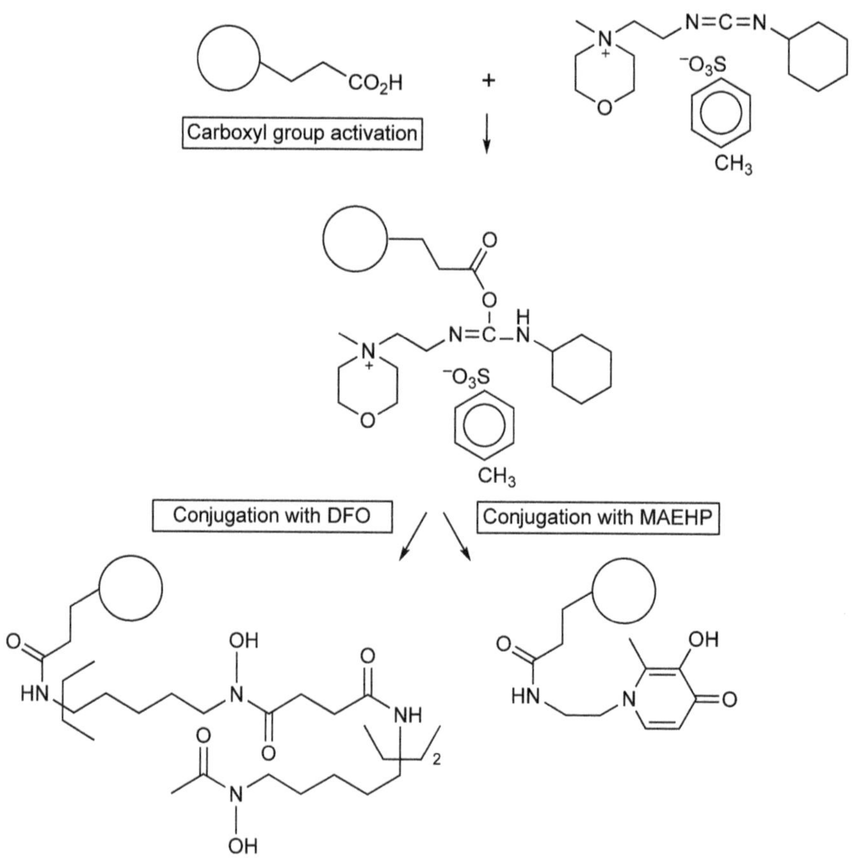

Fig. 2. Syntheses of chelator–nanoparticle conjugates (bottom left, Nano-DFO1 and bottom right, Nano-N2PY). First, reaction of carboxylic functionalized nanoparticles with CMC in MES buffer solution at room temperature. Then, conjugation of activated carboxylic nanoparticles with excessive DFO or MAEHP in MES at room temperature. (Modification with permission from Liu, 2007. Copyright Wiley-VCH Verlag GmbH & Co. KGaA. Reproduced with permission).

of these chelators contain a free primary amino group available for the conjugation. The carboxylated nanoparticles are washed with diluted NaOH solution, MES (2-[*N*-morpholino]ethane sulfonic acid) buffer and cold Milli-Q water and then centrifuged to separate the nanoparticles from the solutions. Carboxyl groups on the particles can then be activated by adding cold Milli-Q water containing *N*-cyclohexyl-*N*′-(2-morpholinoethyl)carbodiimide methyl-*p*-toluensulfonate (CMC) at room temperature. After removing the supernatant, CMC solution is added again along with MES buffer, and reacted for 30 min. The nanoparticles with activated carboxyl groups are washed with cold MES as quickly as possible, and added into MES buffer containing excess DFO or MAEHP for conjugation. After the mixture is vortexed well, it is followed by incubation for 30 min at room temperature with tilt rotation. The chelator–nanoparticle conjugates can then be washed with MES and PBS (phosphate-buffered saline, 10 mM, pH 7.4) buffer, and stored in PBS at 4°C. To confirm this conjugation method, chelator–nanoparticle conjugates spread on KCl crystal IR sample cards (Sigma-Aldrich, St. Louis, MO, USA) are routinely examined using a Fourier transform infrared spectroscopy (FTIR, Perkin-Elmer Spectrum 1000, Waltham, MA, USA). This method of conjugation yields greater than 70% of chelators with nanoparticles, estimated by measurements of the free chelator concentrations in the solutions before and after conjugation.

The conjugated chelators with nanoparticles should still retain their metal binding ability in order for them to be used for chelation. To determine this capability, reaction of chelator–nanoparticle conjugates with ferric iron were conducted. In brief, an aliquot of freshly prepared 1% ferric iron solution ($Fe(NO_3)_3$) in Milli-Q water was reacted with chelator–nanoparticle conjugates for 1 h at room temperature (Liu et al., 2007, 2009b). After separating the conjugates by centrifugation, the particles were washed thoroughly with EDTA solution (5 mM) and MES buffer to remove noncomplexed iron ions. To examine the metal-binding ability of the conjugates, Perls method was conducted with potassium ferrocyanide–HCl solution (3%) for ferric iron stain (Liu et al., 2009a; Sheehan & Hrapchak, 1980). After incubation, solution was removed and the iron chelator nanoparticle conjugates were washed with Milli-Q water for three times. As observed, the chelator–nanoparticle conjugates changed their color from white to blue, implying the presence of iron.

For further examination, the samples were dispersed in Milli-Q water, drop-cast onto carbon-coated copper grid for transmission electron microscopy (TEM) after air dry at room temperature. A TEM (Philips Tecnai 12 TEM, Eagle, FEI Company, Hillsboro, OR, USA) was used for examining the nanoparticles at 37,000× magnification. The TEM study shows ferric ferrocyanide granules on the nanoparticle surface (Fig. 3c). Figure 3b shows the nanoparticles without chelator conjugation which, compared with Fig. 3c, also indicates that the conjugation and metal binding did not affect the nanoparticle size. These studies demonstrate that the nanoparticle conjugation of chelators does not affect the chelator capability of complexing iron. These results are consistent with our previous studies, which show that DFO conjugated to particle still retains its hexadentate property, forming a 1:1 complex with iron (Liu et al., 2007, 2009b). Interestingly, the conjugation of a bidentate chelator, MAPHP [2-methyl-*N*-(3′-aminopropyl)-3-hydroxyl-4-pyridinone, an analog of MAEHP] with nanoparticles may assemble two such molecules (four oxygen binding donors) into a hexadentate chelator by utilizing two additional oxygen donors from amido groups (Fig. 3a) (Liu et al., 2009b). The hexadentate iron chelators possess advantages including kinetic stability, concentration independence of iron affinity, and low toxicity (Hider, Porter, & Singh, 1994b).

Effects of chelator–nanoparticle conjugate on Aβ aggregates and cytotoxicity

A large body of studies suggests that neurotoxicity of Aβ may result from the formation of protease-resistant oligomeric and fibrillar forms of Aβ (Selkoe, 1991), and blocking Aβ aggregation may provide a valuable therapeutic approach (Cohen & Kelly, 2003). Nanoparticles of C60 fullerene

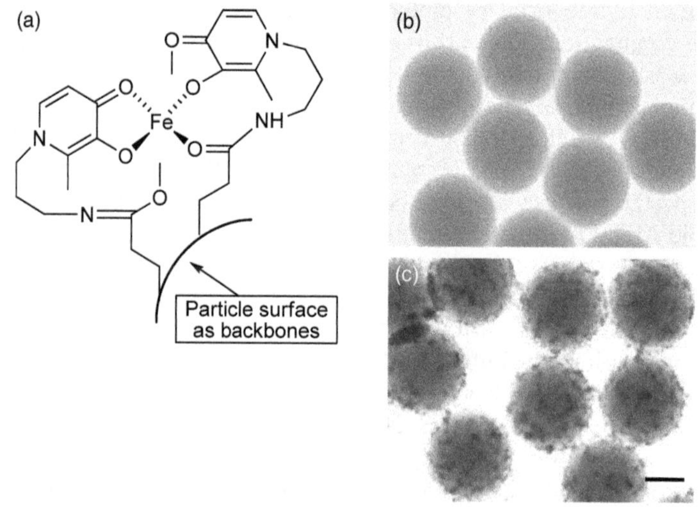

Fig. 3. (a) Four oxygen chelation sites in two MAPHPs and two oxygen chelation donors from amido groups, which assemble a hexadentate chelator through particle surface as the backbones. Nanoparticle TEM images: (b) Nanoparticles without chelator conjugation and iron binding, and (c) Nanoparticles with both reactions. (Modification with permission from Liu et al., 2009a,b.)

(Kim & Lee, 2003; Podolski et al., 2007) have been reported as an inhibitor for Aβ fibrillation *in vitro* studies. In contrast, TiO_2 nanoparticles have reportedly been able to promote Aβ fibrillar formation (Wu et al., 2008). It is possible that the surface properties of nanoparticles play a key role in manipulation of Aβ aggregation (Rocha et al., 2008). In our *in vitro* study, the ability of chelator–nanoparticle conjugates has been examined for prevention of Aβ aggregation. Aβ was incubated in the presence and absence of the conjugates in PBS (10 mM, pH 7.4) at a molar ratio of 1 (Aβ) to 2 (based on chelators) at 37°C (Liu et al., 2009a; Watanabe et al., 2002). After specific incubation periods, an aliquot of the solutions was mixed with Congo red solution (3 μM, PBS), and fibrillar formation in duplicate samples were examined using a UV–vis spectrophotometer. The solutions were allowed to stand at room temperature, and red color precipitates of Aβ aggregates were observed from Aβ solutions which did not contain chelator–nanoparticle conjugates.

Interestingly, results also show there is no precipitate of Aβ aggregates in PBS buffer containing Aβ and chelator–nanoparticle conjugates. Figure 4a shows Aβ aggregate precipitates stained with Congo red, which is imaged with fluorescence microscopy (AmScope, 400× magnification, Chino, CA, USA). Coincubation of Aβ with chelator–nanoparticle conjugates can completely prevent the formation of such precipitates (Fig. 4b). This study demonstrates that the chelator–nanoparticle conjugate can prevent Aβ aggregate formation *in vitro* under physiological conditions. It also suggests that the protective mechanism of the conjugates against Aβ-associated neurotoxicity may be due partially to inhibiting Aβ aggregation.

Chelator–nanoparticle conjugates have the ability to block Aβ aggregate formation, and their formation is suggested as a key contributor to neurodegeneration in AD. It would be very interesting to know whether such capability could lead to protection of neurons from Aβ-associated toxicity. To answer this question, the possibility that specific chelator conjugates such as Nano-N2PY and Nano-DFO1 can inhibit Aβ cytotoxicity was examined *in vitro* with human cortical neuronal cells. The cells were cultured with Aβ, chelator–nanoparticle conjugates or Aβ conjugates. In brief, cells (5,000/well) were placed in a 96-well microplate, and cultured in 100 μL of Dulbecco's modified Eagle's medium (DMEM) with 1% of fetal bovine serum (FBS) at 37°C in a 5% CO_2 atmosphere. The final concentrations of

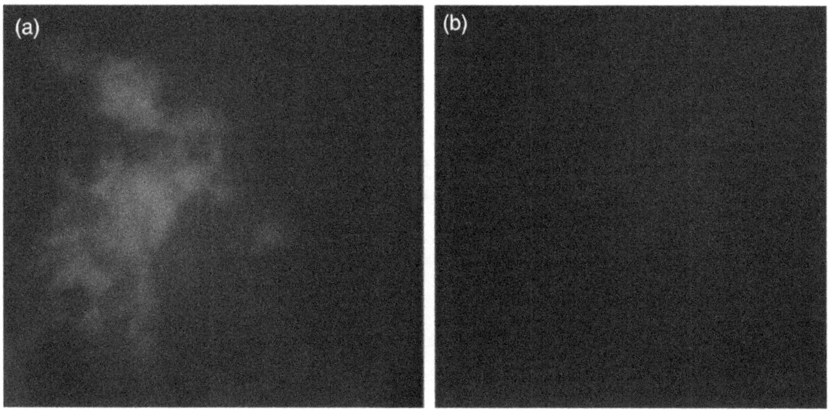

Fig. 4. Images of fluorescence microscopy: (a) Precipitates of Aβ aggregates readily form in PBS without chelator–nanoparticle conjugates. (b) No Aβ precipitates form in PBS-containing chelator–nanoparticle conjugates. (Modification with permission from Liu et al., 2009a.)

conjugates and Aβ in the culture medium were 2 μM (based on chelators) and 1 μM, respectively (Liu et al., 2009a; Watanabe et al., 2002). After three days culture, Aβ-induced cytotoxicity was investigated using a Cytotoxicity LDH Detection Kit with an ELx808 microplate reader (BioTek, Winooski, VT, USA) according to the manufacturer's instructions. Absorbance wavelength measured in this experiment was 490 nm with a reference at 630 nm.

Aβ, in fact, induces toxicity to the neurons compared to control cells, while chelator–nanoparticle conjugates significantly mitigate the Aβ-associated cytotoxicity to the cells (Fig. 5a). The chelator–nanoparticle conjugates at the given concentration do not cause significant toxicity to the neurons. Moreover, the neuronal cell proliferation was examined with a Cell Proliferation WST-1 Reagent assay. Briefly, the cells (2,000 cells/well) in a 96-well microplate were treated with Aβ (final concentration, 1 μM), chelator–nanoparticle conjugates (final 2 μM, based on chelators) or the conjugates/Aβ (final concentration, 2 (based on chelators)/1 μM), and incubated in DMEM with 10% FBS at 37°C and 5% CO_2 for 3 days following a provided protocol from Roche. Absorbance wavelength measured in this experiment was 450 nm with a reference at 630 nm.

After 3 days culture, cells treated with the chelator–nanoparticle conjugates/Aβ have similar proliferation values reflected by absorbance as the control cells, which are significantly higher than the Aβ-treated cells (Fig. 5b). Again, the results show that there are no significant differences of cell proliferation between control neurons and the neurons treated with chelator–nanoparticle conjugates alone. Thus, these studies demonstrate that the chelator–nanoparticle conjugates are effective at protecting neuronal cells against Aβ-associated cytotoxicity *in vitro* and have no significant adverse effects on cell growth/proliferation. Interestingly, these studies, in turn, may also provide evidence to support the Aβ hypothesis.

How might these nanoparticle conjugates enter the brain

To examine the ability of chelator–nanoparticle conjugates to enter the brain and leave the brain after complexing metals, 2-D PAGE analyses have been performed to evaluate the protein absorption patterns on free chelator–nanoparticle conjugates and chelator–nanoparticle conjugates with iron. Human plasma proteins absorbed on both conjugates alone and the conjugates with iron were examined. Briefly, the conjugates are overcoated with polysorbate 80 at room temperature, and the conjugates with iron (100 μL of each conjugate, 2.5%, w/v in

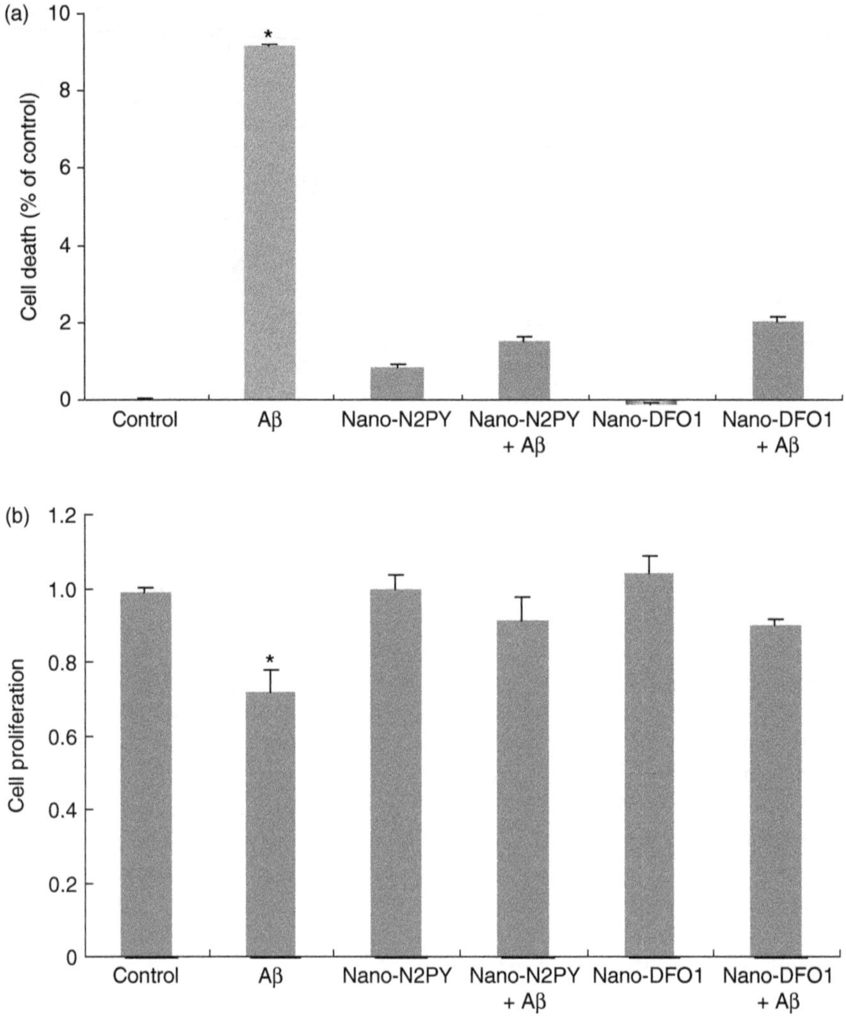

Fig. 5. (a) The cytotoxicity of chelator–nanoparticle conjugates, Aβ only, and the conjugates with Aβ when incubated with neuronal cells as measured by Cytotoxicity LDH Detection Assay. Values, as % of control, were represented as mean ± standard errors ($n = 4$–6; *significantly different from other groups at $p < 0.05$). (b) Effects of chelator–nanoparticle conjugates, Aβ only, and the conjugates with Aβ on cell proliferation of neuron cells as determined by WST-1 assay. Results were represented as mean ± standard errors ($n = 3$–5; *significantly different from other groups at $p < 0.05$).

PBS buffer) were incubated separately in 1 mL of citrated human plasma for 5 min at 37°C (Blunk, Hochstrasser, Sanchez, Muller, & Muller, 1993). After separation, the systems were washed four times with Milli-Q water. The adsorbed proteins were eluted from the conjugate surface with a protein-solubilizing solution (5% SDS, 5% dithioerythritol, 10% glycerol and 60 mM Tris, pH 6.8) (Blunk et al., 1993). The eluted proteins from the conjugates were analyzed with 2-D PAGE. In the first dimension (isoelectric focusing, IEF), the proteins are separated according to their isoelectric points (p*I*). The IEF was carried out in glass tubes of inner diameter 2.0 mm using 2.0% pH 3.5–10 ampholines for 9,600 V-h. In the second dimension of SDS-PAGE, the separation is based on molecular weight (MW). Each tube was equilibrated for

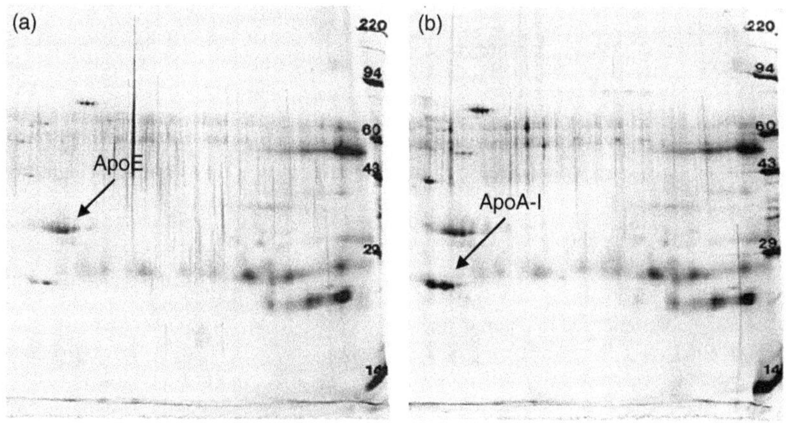

Fig. 6. 2-D PAGE images of plasma protein patterns on (a) chelator–nanoparticle conjugates coated with polysorbate 80 and (b) chelator–nanoparticle conjugates complexed with iron. (Modification with permission from Liu et al., 2006.)

10 min in 62.5 mM Tris, pH 6.8 buffer containing 2.3% SDS, 50 mM dithioerythritol and 10% glycerol, and sealed to the top of a stacking gel that is on the top of a 10% acrylamide slab gel ($145 \times 145 \times 0.75\ mm^3$). The SDS slab gel electrophoresis was performed for about 4 h at 12.5 mA/gel. After SDS-PAGE, the gels were dried between sheets of cellophane, and silver-stained (Blunk et al., 1993).

The polysorbate 80-coated chelator–particle conjugates have the ability to preferentially absorb ApoE on their surface (Fig. 6a), while the chelator–particle conjugates complexed with iron prefer to absorb ApoA (Fig. 6b). These results are very interesting because the selected ApoE absorption of free conjugates and the preferential ApoA absorption of conjugates with metals may allow them to mimic the ApoE and ApoA nanoparticles, respectively. This feature could facility the free conjugates to enter the brain and the conjugates with metals to leave the brain using LDL transport mechanisms (Alyautdin et al., 1998; Davson and Segal, 1996). Furthermore, our preliminary studies show that chelator–nanoparticle conjugates have the ability to penetrate a monolayer composed of bovine brain microvascular endothelial cells, an *in vitro* BBB model. Importantly, the chelator–nanoparticle conjugates also show the ability to cross BBB in our preliminary *in vivo* studies; the conjugate contents in the brain of mice have been found significantly higher after intravenous injection compared with controls. More *in vivo* and *in vivo* studies are warranted.

Conclusion

Chelators can easily be conjugated with nanoparticles that have the potential to transport the conjugated chelators across the BBB. The conjugation does not affect the chelation capability of the chelators, and in some cases, even improves the chelation coordination property. Our *in vitro* studies also show that chelator–nanoparticle conjugates can effectively inhibit Aβ aggregate formation and, thereby, protect human brain cells from Aβ-related toxicity. As such conjugates have the potential to cross the BBB and thereafter be actively transported out of the brain, this approach may offer great potential for AD therapeutics. Moreover, this novel approach of nanoparticle–chelator delivery could significantly improve the efficacy and reduce the toxicity of chelation therapy. Likewise, it could also provide a valuable tool to uncover the role of metals in AD pathogenesis. Further *in vitro* and *in vivo* studies toward clinical application, however, are needed.

Acknowledgments

The authors are very grateful for the financial support from the National Institutes of Health (NS052677 to GL and AG026151 to MAS) and the Alzheimer's Association.

References

Adlard, P. A., Cherny, R. A., Finkelstein, D. I., Gautier, E., Robb, E., Cortes, M., et al. (2008). Rapid restoration of cognition in Alzheimer's transgenic mice with 8-hydroxy quinoline analogs is associated with decreased interstitial Abeta. *Neuron, 59*(1), 43–55.

Alyautdin, R. N., Tezikov, E. B., Ramge, P., Kharkevich, D. A., Begley, D. J., & Kreuter, J. (1998). Significant entry of tubocurarine into the brain of rats by adsorption to polysorbate 80-coated polybutylcyanoacrylate nanoparticles: An in situ brain perfusion study. *Journal of Microencapsulation, 15*(1), 67–74.

Atwood, C. S., Perry, G., Zeng, H., Kato, Y., Jones, W. D., Ling, K. Q., et al. (2004). Copper mediates dityrosine cross-linking of Alzheimer's amyloid-beta. *Biochemistry, 43*(2), 560–568.

Behl, C., Davis, J., Cole, G. M., & Schubert, D. (1992). Vitamin E protects nerve cells from amyloid beta protein toxicity. *Biochemical and Biophysical Research Communication, 186* (2), 944–950.

Bishop, G. M., Robinson, S. R., Liu, Q., Perry, G., Atwood, C. S., & Smith, M. A. (2002). Iron: A pathological mediator of Alzheimer disease? *Developmental Neuroscience, 24*(2–3), 184–187.

Blake, D. R., Winyard, P., Lunec, J., Williams, A., Good, P. A., Crewes, S. J., et al. (1985). Cerebral and ocular toxicity induced by desferrioxamine. *Quarterly Journal of Medicine, 56*(219), 345–355.

Blunk, T., Hochstrasser, D. F., Sanchez, J. C., Muller, B. W., & Muller, R. H. (1993). Colloidal carriers for intravenous drug targeting: Plasma protein adsorption patterns on surface-modified latex particles evaluated by two-dimensional polyacrylamide gel electrophoresis. *Electrophoresis, 14*(12), 1382–1387.

Bush, A. I., Pettingell, W. H., Multhaup, G., d Paradis, M., Vonsattel, J. P., Gusella, J., et al. (1994). Rapid induction of Alzheimer A beta amyloid formation by zinc. *Science, 265* (5177), 1464–1467.

Casadesus, G., Smith, M. A., Zhu, X., Aliev, G., Cash, A. D., Honda, K., et al. (2004). Alzheimer disease: Evidence for a central pathogenic role of iron-mediated reactive oxygen species. *Journal of Alzheimer's Disease, 6*(2), 165–169.

Casdorph, H. R. (2001). EDTA chelation therapy: Efficacy in brain disorders. In E. M. Cranton (Ed.), *A textbook on EDTA chelation therapy* (pp. 142–163). Charlottesville: Hampton Roads Publishing Company, Inc.

Castellani, R. J., Lee, H. G., Zhu, X., Nunomura, A., Perry, G., & Smith, M. A. (2006). Neuropathology of Alzheimer disease: Pathognomonic but not pathogenic. *Acta Neuropathologica (Berlin), 111*(6), 503–509.

Castellani, R. J., Moreira, P. I., Liu, G., Dobson, J., Perry, G., Smith, M. A., et al. (2007). Iron: The Redox-active center of oxidative stress in Alzheimer disease. *Neurochemical Research, 32*(10), 1640–1645.

Cherny, R. A., Atwood, C. S., Xilinas, M. E., Gray, D. N., Jones, W. D., McLean, C. A., et al. (2001). Treatment with a copper-zinc chelator markedly and rapidly inhibits beta-amyloid accumulation in Alzheimer's disease transgenic mice. *Neuron, 30*(3), 665–676.

Cohen, F. E., & Kelly, J. W. (2003). Therapeutic approaches to protein-misfolding diseases. *Nature, 426*(6968), 905–909.

Crapper McLachlan, D. R., Dalton, A. J., Kruck, T. P., Bell, M. Y., Smith, W. L., Kalow, W., et al. (1991). Intramuscular desferrioxamine in patients with Alzheimer's disease. *Lancet, 337*(8753), 1304–1308.

Crouch, P. J., Barnham, K. J., Bush, A. I., & White, A. R. (2006). Therapeutic treatments for Alzheimer's disease based on metal bioavailability. *Drug News & Perspectives, 19*(8), 469–474.

Crowe, A., & Morgan, E. H. (1994). Effects of chelators on iron uptake and release by the brain in the rat. *Neurochemical Research, 19*(1), 71–76.

Davson, H., & Segal, M. B. (1996). *Physiology of the CSF and blood-brain barriers.* Boca Raton: CRC Press.

Evans, P. H. (1993). Free radicals in brain metabolism and pathology. *British Medical Bulletin, 49*(3), 577–587.

Exley, C. (2006). Aluminium and iron, but neither copper nor zinc, are key to the precipitation of beta-sheets of Abeta_{42} in senile plaque cores in Alzheimer's disease. *Journal of Alzheimer's Disease, 10*(2–3), 173–177.

Ferri, C. P., Prince, M., Brayne, C., Brodaty, H., Fratiglioni, L., Ganguli, M., et al. (2005). Global prevalence of dementia: A Delphi consensus study. *Lancet, 366*(9503), 2112–2117.

Gutteridge, J. M. (1994). Hydroxyl radicals, iron, oxidative stress, and neurodegeneration. *Annals of the New York Academy of Science, 738*, 201–213.

Halliwell, B., & Gutteridge, J. M.C. (1999). *Free radicals in biology and medicine.* New York: Oxford University Press.

Hardy, J. (2006). Alzheimer's disease: The amyloid cascade hypothesis: An update and reappraisal. *Journal of Alzheimer's Disease, 9*(3 Suppl.), 151–153.

Hardy, J. A., & Higgins, G. A. (1992). Alzheimer's disease: The amyloid cascade hypothesis. *Science, 256*(5054), 184–185.

Hayashi, T., Shishido, N., Nakayama, K., Nunomura, A., Smith, M. A., Perry, G., et al. (2007). Lipid peroxidation and 4-hydroxy-2-nonenal formation by copper ion bound to amyloid-beta peptide. *Free Radical Biology & Medicine, 43* (11), 1552–1559.

Hebert, L. E., Beckett, L. A., Scherr, P. A., & Evans, D. A. (2001). Annual incidence of Alzheimer disease in the United States projected to the years 2000 through 2050. *Alzheimer Disease and Associated Disorders, 15*(4), 169–173.

Hider, R. C., Epemolu, O., Singh, S., & Porter, J. B. (1994a). Iron chelator design. *Advances in Experimental Medicine and Biology, 356*, 343–349.

Hider, R. C., Porter, J. B., & Singh, S. (1994b). The design of therapeutically useful iron chelators. In R. J. Bergeron & G. M. Brittenham (Eds.), *The development of iron chelators for clinical use* (pp. 353–371). Boca Raton: CRC.

Huang, X., Atwood, C. S., Hartshorn, M. A., Multhaup, G., Goldstein, L. E., Scarpa, R. C., et al. (1999). The A beta peptide of Alzheimer's disease directly produces hydrogen peroxide through metal ion reduction. *Biochemistry, 38*(24), 7609–7616.

Jobling, M. F., Huang, X., Stewart, L. R., Barnham, K. J., Curtain, C., Volitakis, I., et al. (2001). Copper and zinc binding modulates the aggregation and neurotoxic properties of the prion peptide PrP106-126. *Biochemistry, 40*(27), 8073–8084.

Kim, J. E., & Lee, M. (2003). Fullerene inhibits beta-amyloid peptide aggregation. *Biochemical and Biophysical Research Communications, 303*(2), 576–579.

Kong, S., Liochev, S., & Fridovich, I. (1992). Aluminum(III) facilitates the oxidation of NADH by the superoxide anion. *Free Radical Biology & Medicine, 13*(1), 79–81.

Kreuter, J. (2001). Nanoparticulate systems for brain delivery of drugs. *Advanced Drug Delivery Reviews, 47*(1), 65–81.

Kreuter, J., Shamenkov, D., Petrov, V., Ramge, P., Cychutek, K., Koch-Brandt, C., et al. (2002). Apolipoprotein-mediated transport of nanoparticle-bound drugs across the blood-brain barrier. *Journal of Drug Targeting, 10*(4), 317–325.

Lee, H. G., Zhu, X., Nunomura, A., Perry, G., & Smith, M. A. (2006). Amyloid beta: The alternate hypothesis. *Current Alzheimer Research, 3*(1), 75–80.

Liu, G., Garrett, M. R., Men, P., Zhu, X., Perry, G., & Smith, M. A. (2005). Nanoparticle and other metal chelation therapeutics in Alzheimer disease. *Biochimica et Biophysica Acta, 1741*(3), 246–252.

Liu, G., Men, P., Harris, P. L., Rolston, R. K., Perry, G., & Smith, M. A. (2006). Nanoparticle iron chelators: A new therapeutic approach in Alzheimer disease and other neurologic disorders associated with trace metal imbalance. *Neuroscience Letters, 406*(3), 189–193.

Liu, G., Men, P., Kudo, W., Perry, G., & Smith, M. A. (2009a) Nanoparticle-chelator conjugates as inhibitors of amyloid-beta aggregation and neurotoxicity: A novel therapeutic approach for Alzheimer disease. *Neuroscience Letters, 455*, 187–190.

Liu, G., Men, P., Perry, G., & Smith, M. A. (2007). Nanoparticles for the treatment of Alzheimer's disease: Theoretical rationale, present status and future perspectives. In C. S. S. R. Kumar (Ed.), *Nanomaterials for medical diagnosis and therapy (nanotechnologies for the life sciences)* (pp. 644– 706). Hoboken: John Wiley & Sons.

Liu, G., Men, P., Perry, G., & Smith, M. A. (2009b) Metal chelators coupled with nanoparticles as potential therapeutic agents for Alzheimer's disease. *Journal of Nanoneuroscience, 1*, 42–55.

Lovell, M. A., Robertson, J. D., Teesdale, W. J., Campbell, J. L., & Markesbery, W. R. (1998). Copper, iron and zinc in Alzheimer's disease senile plaques. *Journal of the Neurological Sciences, 158*(1), 47–52.

Lynch, S. G., Fonseca, T., & Levine, S. M. (2000). A multiple course trial of desferrioxamine in chronic progressive multiple sclerosis. *Cellular and Molecular Biology (Noisy-le-grand), 46*(4), 865–869.

Markesbery, W. R. (1997). Oxidative stress hypothesis in Alzheimer's disease. *Free Radical Biology & Medicine, 23*(1), 134–147.

Marlatt, M. W., Webber, K. M., Moreira, P. I., Lee, H. G., Casadesus, G., Honda, K., et al. (2005). Therapeutic opportunities in Alzheimer disease: One for all or all for one? *Current Medicinal Chemistry, 12*(10), 1137–1147.

Minino, A. M., Heron, M. P., Murphy, S. L., & Kochanek, K. D. (2007). Deaths: Final data for 2004. *National Vital Statistics Reports, 55*(19), 1–119.

Nakamura, M., Shishido, N., Nunomura, A., Smith, M. A., Perry, G., Hayashi, Y., et al. (2007). Three histidine residues of amyloid-beta peptide control the redox activity of copper and iron. *Biochemistry, 46*(44), 12737–12743.

Nunomura, A., Perry, G., Aliev, G., Hirai, K., Takeda, A., Balraj, E. K., et al. (2001). Oxidative damage is the earliest event in Alzheimer disease. *Journal of Neuropathology and Experimental Neurology, 60*(8), 759–767.

Olanow, C. W. (1992). An introduction to the free radical hypothesis in Parkinson's disease. *Annals of Neurology, 32*(Suppl.), S2–S9.

Opazo, C., Huang, X., Cherny, R. A., Moir, R. D., Roher, A. E., White, A. R., et al. (2002). Metalloenzyme-like activity of Alzheimer's disease beta-amyloid. Cu-dependent catalytic conversion of dopamine, cholesterol, and biological reducing agents to neurotoxic H(2)O(2). *Journal of Biological Chemistry, 277*(43), 40302–40308.

Pappolla, M. A., Chyan, Y. J., Omar, R. A., Hsiao, K., Perry, G., Smith, M. A., et al. (1998). Evidence of oxidative stress and in vivo neurotoxicity of beta-amyloid in a transgenic mouse model of Alzheimer's disease: A chronic oxidative paradigm for testing antioxidant therapies in vivo. *American Journal of Pathology, 152*(4), 871–877.

Pardridge, W. M. (2007). Blood-brain barrier delivery. *Drug Discovery Today, 12*(1–2), 54–61.

Perry, G., Castellani, R. J., Hirai, K., & Smith, M. A. (1998). Reactive oxygen species mediate cellular damage in Alzheimer disease. *Journal of Alzheimer's Disease, 1*(1), 45–55.

Podolski, I. Y., Podlubnaya, Z. A., Kosenko, E. A., Mugantseva, E. A., Makarova, E. G., Marsagishvili, L. G., et al. (2007). Effects of hydrated forms of C60 fullerene on amyloid 1-peptide fibrillization in vitro and performance of the cognitive task. *Journal of Nanoscience and Nanotechnology, 7*(4–5), 1479–1485.

Regland, B., Lehmann, W., Abedini, I., Blennow, K., Jonsson, M., Karlsson, I., et al. (2001). Treatment of Alzheimer's disease with clioquinol. *Dementia and Geriatric Cognitive Disorders, 12*(6), 408–414.

Ritchie, C. W., Bush, A. I., Mackinnon, A., Macfarlane, S., Mastwyk, M., MacGregor, L., et al. (2003). Metal-protein attenuation with iodochlorhydroxyquin (clioquinol) targeting Abeta amyloid deposition and toxicity in Alzheimer disease: A pilot phase 2 clinical trial. *Archives of Neurology*, *60*(12), 1685–1691.

Rocha, S., Thunemann, A. F., Pereira Mdo, C., Coelho, M., Mohwald, H., & Brezesinski, G. (2008). Influence of fluorinated and hydrogenated nanoparticles on the structure and fibrillogenesis of amyloid beta-peptide. *Biophysical Chemistry*, *137*(1), 35–42.

Rottkamp, C. A., Raina, A. K., Zhu, X., Gaier, E., Bush, A. I., Atwood, C. S., et al. (2001). Redox-active iron mediates amyloid-beta toxicity. *Free Radical Biology & Medicine*, *30*(4), 447–450.

Sayre, L. M., Perry, G., Harris, P. L., Liu, Y., Schubert, K. A., & Smith, M. A. (2000). In situ oxidative catalysis by neurofibrillary tangles and senile plaques in Alzheimer's disease: A central role for bound transition metals. *Journal of Neurochemistry*, *74*(1), 270–279.

Sayre, L. M., Perry, G., & Smith, M. A. (1999). Redox metals and neurodegenerative disease. *Current Opinion in Chemical Biology*, *3*(2), 220–225.

Schubert, D., & Chevion, M. (1995). The role of iron in beta amyloid toxicity. *Biochemical and Biophysical Research Communication*, *216*(2), 702–707.

Scorer, C. A. (2001). Preclinical and clinical challenges in the development of disease-modifying therapies for Alzheimer's disease. *Drug Discovery Today*, *6*(23), 1207–1219.

Selkoe, D. J. (1991). The molecular pathology of Alzheimer's disease. *Neuron*, *6*(4), 487–498.

Sheehan, D. C., & Hrapchak, B. R. (1980). *Theory and practice of histotechnology*. St. Louis: CV Mosby.

Smith, M. A. (1998). Alzheimer disease. *International Review Neurobiology*, *42*, 1–54.

Smith, M. A. (2006). Oxidative stress and iron imbalance in Alzheimer disease: How rust became the fuss! *Journal of Alzheimer's Disease*, *9*(3 Suppl.), 305–308.

Smith, M. A., Casadesus, G., Joseph, J. A., & Perry, G. (2002). Amyloid-beta and tau serve antioxidant functions in the aging and Alzheimer brain. *Free Radical Biology & Medicine*, *33*(9), 1194–1199.

Smith, M. A., Harris, P. L., Sayre, L. M., & Perry, G. (1997). Iron accumulation in Alzheimer disease is a source of redox-generated free radicals. *Proceedings of the National Academy of Sciences, USA*, *94*(18), 9866–9868.

Smith, M. A., Hirai, K., Hsiao, K., Pappolla, M. A., Harris, P. L., Siedlak, S. L., et al. (1998). Amyloid-beta deposition in Alzheimer transgenic mice is associated with oxidative stress. *Journal of Neurochemistry*, *70*(5), 2212–2215.

Smith, M. A., Sayre, L. M., Monnier, V. M., & Perry, G. (1995). Radical AGEing in Alzheimer's disease. *Trends in Neurosciences*, *18*(4), 172–176.

Tamagno, E., Guglielmotto, M., Aragno, M., Borghi, R., Autelli, R., Giliberto, L., et al. (2008). Oxidative stress activates a positive feedback between the gamma- and beta-secretase cleavages of the beta-amyloid precursor protein. *Journal of Neurochemistry*, *104*(3), 683–695.

Watanabe, K., Nakamura, K., Akikusa, S., Okada, T., Kodaka, M., Konakahara, T., et al. (2002). Inhibitors of fibril formation and cytotoxicity of beta-amyloid peptide composed of KLVFF recognition element and flexible hydrophilic disrupting element. *Biochemical and Biophysical Research Communications*, *290*(1), 121–124.

Wilson, B., Samanta, M. K., Santhi, K., Kumar, K. P., Paramakrishnan, N., & Suresh, B. (2008). Poly(n-butylcyanoacrylate) nanoparticles coated with polysorbate 80 for the targeted delivery of rivastigmine into the brain to treat Alzheimer's disease. *Brain Research*, *1200*, 159–168.

Wu, W. H., Sun, X., Yu, Y. P., Hu, J., Zhao, L., Liu, Q., et al. (2008). TiO_2 nanoparticles promote beta-amyloid fibrillation in vitro. *Biochemical and Biophysical Research Communications*, *373*(2), 315–318.

Yaari, R., & Corey-Bloom, J. (2007). Alzheimer's disease. *Seminars in Neurology*, *27*(1), 32–41.

Yan, S. D., Yan, S. F., Chen, X., Fu, J., Chen, M., Kuppusamy, P., et al. (1995). Non-enzymatically glycated tau in Alzheimer's disease induces neuronal oxidant stress resulting in cytokine gene expression and release of amyloid beta-peptide. *Nature Medicine*, *1*(7), 693–699.

Yang, L. B., Lindholm, K., Yan, R., Citron, M., Xia, W., Yang, X. L., et al. (2003). Elevated beta-secretase expression and enzymatic activity detected in sporadic Alzheimer disease. *Nature Medicine*, *9*(1), 3–4.

SECTION III

Nanoparticles Therapy and Neuroregeneration

H.S. Sharma (Ed.)
Progress in Brain Research, Vol. 180
ISSN: 0079-6123

CHAPTER 6

Carbon nanotubes as substrates/scaffolds for neural cell growth

William Lee and Vladimir Parpura*

Department of Neurobiology, Center for Glial Biology in Medicine, Atomic Force Microscopy & Nanotechnology Laboratories, Civitan International Research Center, Evelyn F. McKnight Brain Institute, University of Alabama, Birmingham, AL, USA

Abstract: Carbon nanotubes (CNTs) due to their unique properties have sparked interest for their use in biomedical applications in recent years. In particular, the use of CNTs as substrates/scaffolds for neural cell growth has been an area of active research over the past decade. CNTs, either native or functionalized with various chemical groups, are biocompatible with neuronal cell adhesion and growth. Functionalized CNTs can modulate the neuronal growth in graded manner; positively charged CNTs promoted neurite outgrowth of hippocampal neurons in culture to a greater extent than when these cells were grown on neutral or negatively charged CNTs. Conductivity and mechanical properties of CNTs have been shown to affect neuronal morphology as well. Other neural cells, such as stem and glial cells, can also be successfully grown on CNT substrates. While currently the acute toxicity of CNTs is considered comparable to that of other forms of carbon, the long-term exposures limits need to be established in order to use these materials as neural prosthesis. Nonetheless, accumulating data support the use of CNTs as a biocompatible and permissive substrate/scaffold for neural cells and such application holds great potential in biomedicine.

Keywords: carbon nanotubes; glial cells; neural stem cells; neurite outgrowth; neurons

Introduction

The rapid expansion of nanotechnological applications in neuroscience holds great promise for both basic and clinical research, since this technology provides novel tools and approaches (reviewed in Silva, 2006). The driving force for implementation of nanoscale materials in biology is embedded within the notion that such materials could achieve the desired high-specificity interactions with the cellular constituents which otherwise could not be possible using materials engineered at a larger scale. Indeed, the nanomaterials and nanodevices available to date show a wide range of applications in neuroscience research, such as serving as biosensors, use in targeted drug delivery and as platforms for neural cell growth. Currently, out of the variety of available nanomaterials, carbon nanotubes (CNTs) arguably showed the most promise on the neuroscience scene (reviewed in Bekyarova, Haddon, & Parpura, 2005; Bekyarova, Ni, et al., 2005; Malarkey & Parpura, 2007).

* Corresponding author.
Tel.: (205) 996-7369;
E-mail: vlad@uab.edu

DOI: 10.1016/S0079-6123(08)80006-4

The first evidence for the nanosized carbon tubes was presented in 1952 (Radushkevich & Lukyanovich, 1952; also see Monthioux & Kuznetsov, 2006). However, the use of CNTs was unrealized for almost 40 years since this initial report. The description of synthesis and characterization of a variety of CNTs published in the early 1990s (Ijima, 1991; Ijima & Ichihashi, 1993) ignited interest by electronics, computer, and aerospace industries due to CNTs unique electrical and physical properties (see below). Additionally, the availability of chemistry to functionalize CNTs with biologically relevant molecules piqued interest of biologists, most notably neuroscientists. In this chapter we focus our discussion on a subset of biological applications of CNTs as scaffolds/substrates for adhesion and growth of neural cells. We begin with a primer (section "Structure, properties, and functionalization of CNTs") on the structure, properties, and functionalization of CNTs (for detailed description, see Bekyarova et al., 2005; Bekyarova, Ni, et al., 2005). We then discuss the use of CNTs as substrates/scaffolds for neuronal growth (section "CNTs as substrates/scaffolds for neuronal cell growth"), followed by the description of CNT utilization for growth of other neural cells in the central nervous system (CNS), which are glial and stem cells (section "CNTs as substrate for other neural cells"). We put an effort throughout the chapter to disclose the cell type used the in reported studies in an attempt to delineate findings obtained using primary neural cells from those made utilizing cell lines. We find this necessary since cell lines may emulate only a subset of the primary cell characteristics. Additionally, when information is available we also report on the brain regions from which primary cells of several types were isolated and also on the animal type/species used, since neural cells can show species- and region-specific morphological and functional variability.

Structure, properties, and functionalization of CNTs

CNTs are composed of graphene sheets rolled into a cylinder. These cylindrical structures have a hollow core with their ends being capped with a fullerene dome (Fig. 1). CNTs are classified according to the number of concentric graphene cylinders they contain. Single-walled CNTs (SWNT), double-walled CNTs (DWNT), or multi-walled carbon nanotubes (MWNT) contain a single graphene cylinder, two, or multiple concentric cylinders, respectively. MWNTs have an outer diameter that typically ranges from 2 to 100 nm, while the inner diameter varies between 1 and 3 nm. The smallest SWNTs reported to date have a diameter of 0.4 nm (Qin et al., 2000; Wang, Tang, Li, & Chen, 2000), although they are usually produced with a random distribution of diameters between 0.7 and 2 nm. SWNTs are mixture of single CNTs and hexagonally, closely packed CNT bundles held together by van der Waals forces. The length of synthesized CNTs is typically in the micrometer range. However, longer lengths of synthesized SWNT up to 4 cm have been reported (Zheng et al., 2004); this translates into an amazing length-to-diameter ratio of $\sim 3\times 10^7$:1. CNTs have an exceptional mechanical strength with a Young's modulus of ~1 TPa. They are chemically relatively inert and are not biodegradable. These physical properties make CNTs a durable nanomaterial for bioengineering, especially in applications where a sustained presence of the material is desirable, such as scaffolds for support of cellular growth.

CNTs exhibit unique electrical properties which enable them to be used as biosensors. The conductivity of CNTs is dependent on the structural arrangement of the hexagonal graphene lattice; they can be metallic or semiconductive. In metallic CNTs, the hexagonal lattice can be arranged in any of the three configurations: arm-chair, zig-zag, or chiral; all arm-chair CNTs are metallic. In semiconductive CNTs, the lattice can be arranged in a zig-zag or chiral configuration. In this chapter, we limit our discussion of CNTs electrical properties solely to the role they play in the growth pattern of neurons. For an in-depth discussion on CNT–neuron electrical interface, see a recent review published elsewhere (Sucapane et al., 2009).

The techniques that are commonly used to synthesize so-called as-prepared CNTs include electric arc, laser ablation, chemical vapor deposition (CVD), and high-pressure carbon monoxide

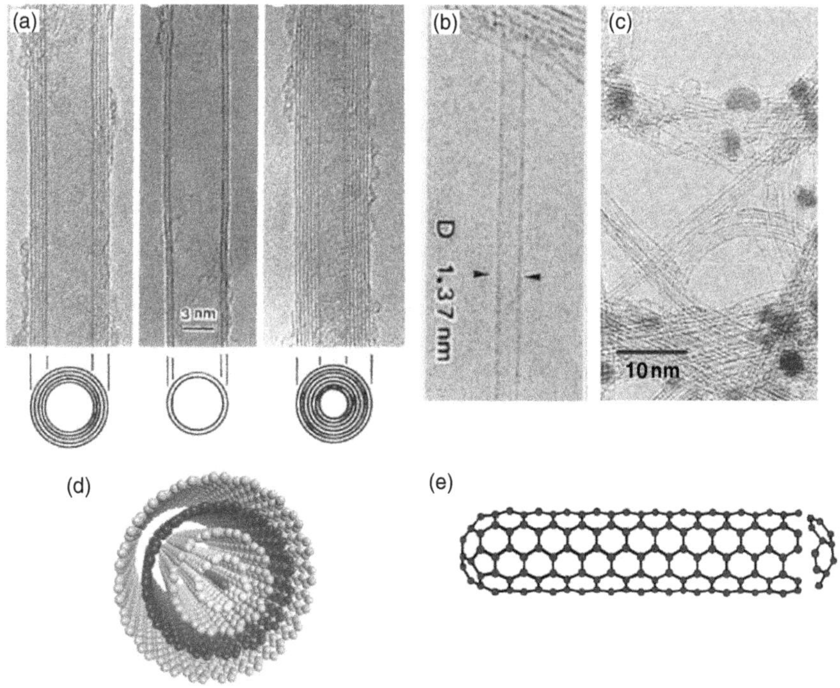

Fig. 1. CNTs are composed of graphene sheets rolled into a cylinder. Transmission electron microscopy (TEM) micrographs of (a) MWNTs and DWNTs (Ijima, 1991) and (b) SWNTs (Ijima & Ichihashi, 1993). (c) TEM micrograph showing bundles of SWNTs. The dark spots are catalyst particles used for nanotube growth. (d, e) Schematics of (d) MWNT and (e) SWNT with its ends capped with fullerene domes (separated from the tubular structure for clarity at the right end). Modified from Bekyarova, Ni, et al. (2005).

disproportionation (reviewed in Journet & Bernier, 1998). As-prepared CNTs typically contain a substantial fraction of metal catalyst, together with carbonaceous impurities which usually consist of amorphous carbon and nanoparticles; many purification procedures have been reported (referenced in Bekyarova, Ni, et al., 2005). Although synthesized CNTs are relatively chemically inert, they can be modified via noncovalent or covalent attachment of molecules. Naturally, covalently linked molecules have more stable association to CNTs. Noncovalent modification is the preferred method if the preservation of the electronic characteristics of the nanotubes is required. Molecules such as DNA, proteins, and lipids can be adsorbed to as-prepared CNTs to generate noncovalently functionalized CNTs. This functionalization can result in, for example, dispersion of SWNT bundles into individual SWNTs. In contrast, the process of covalently linking molecules to CNTs requires the conversion of as-prepared CNTs to reactive intermediates. This can be achieved by refluxing CNTs with strong oxidizing agents such as nitric acid; during this process a carboxyl group is added to CNTs. CNT-COOH can be then used for covalent functionalization with molecules of interest. Alternatively, the additional chemical reaction, which converts the carboxyl group to acyl chloride, can be implemented; the resulting intermediate product CNT-SOCl in turn can be further reacted with the compound of interest to complete the covalent functionalization process of CNTs. These approaches have been successfully applied to covalently functionalize CNTs with various compounds. A somewhat more challenging task, due to the hydrophobic nature of CNTs, has it been to disperse them in aqueous solutions, a prerequisite for many of biological applications. At present, this can be achieved by interacting CNTs with biological molecules that can adsorb to them, coating them with surfactants or by functionalization of CNTs with water-soluble polymers. Detailed reviews on biofunctionalization of CNTs are available elsewhere (Bekyarova et al., 2005; Bekyarova, Ni, et al., 2005).

CNTs as substrates/scaffolds for neuronal cell growth

There is an increasing body of work supporting the usage of CNTs as permissive substrates/scaffolds for cell adhesion and growth. The first study that explored the possibility of using CNTs as substrates for neuronal growth was done by Mattson and colleagues (Mattson, Haddon, & Rao, 2000). The researchers grew embryonic rat hippocampal neurons on nonfunctionalized, as-prepared, MWNTs that covered standard polyethyleneimine (PEI)-coated coverslips. As a control, neurons were seeded on glass coverslips coated with PEI alone. Neuronal morphological parameters of growth were assessed using fixed cells and scanning electron microscopy (SEM). Neurons cultured on MWNTs did not exhibit as elaborate branching of neurites as those cells cultured on PEI-coated coverslips, although neurons adhered to MWNTs and remained alive in culture for at least 8 days. Next, MWNTs were noncovalently functionalized by physiosorption of 4-hydroxynonenal (4HNE), a product of lipid peroxidation that can promote neurite outgrowth (for details on actions of 4HNE on neurons, see Mattson et al., 2000 and references therein). Neurons grown on MWNT-4HNE displayed an increased number, length, and branching of neurites when compared to neurons cultured on nonfunctionalized MWNTs. Taken together, this study demonstrated that MWNTs can serve as a permissive substrate for neuronal cell adhesion and growth and that modifying MWNTs with a biologically relevant molecule can be used to modulate neuronal growth and neurite outgrowth. These findings raised the possibility that MWNTs could be used as scaffolds in designing neural prostheses.

CNTs are not biodegradable, and as such they could be used as implants where long-term extracellular molecular cues for neurite outgrowth are necessary, such as in regeneration after spinal cord or brain injury. However, biological compounds, such as 4HNE would not be as stable as MWNTs. Additionally, if attached by physiosorption, molecules could dissociate from MWNTs thus reducing their local concentration. Consequently, it would be advantageous to modulate neuronal growth and neurite outgrowth by systematically varying properties of MWNTs themselves via stable functionalities covalently attached to MWNTs. This goal was achieved in a follow-up study (Hu, Ni, Montana, Haddon, & Parpura, 2004), where surface charge of MWNTs were systematically varied to control the outgrowth and branching pattern of neuronal processes. Here, rather than examining the neuronal growth patterns of fixed neurons, Hu and colleagues (Hu et al., 2004) examined the morphological parameters of rat neonatal hippocampal neurons in culture by imaging live cells that were loaded the vital dye, calcein. Their results using live cells were consistent with the initial finding by Mattson and colleagues using fixed cells; MWNTs were not as good as a substrate for neuron outgrowth when compared to PEI (Fig. 2). After this confirmatory observation, they compared neuronal growth on various MWNTs. Functionalization was carried out by addition of carboxyl groups, poly-*m*-aminobenzene sulfonic acid (PABS), or ethylenediamine (EN) to MWNTs. Because the extracellular media used to grow and investigate neurons required physiological pH of 7.35 and due to association constants of these functionalities, MWNT-COOH, MWNT-PABS, and MWNT-EN displayed negative, zwitterionic, and positive surface charge, respectively. Results showed the presence of more numerous growth cones, longer average neurite length, and elaborate neurite branching when measurements for neurons grown on positively charged MWNT-EN were compared to those of neurons grown on neutral or negatively charged MWNTs; branching of neurites showed graded dependency of MWNT charge in the order MWNT-EN > MWNT-PABS > MWNT-COOH (Fig. 3a).

Because PEI is positively charged at physiological pH, one approach to control neuronal growth and neurite outgrowth would be to make a composite PEI-CNT polymer, thus, "diluting"

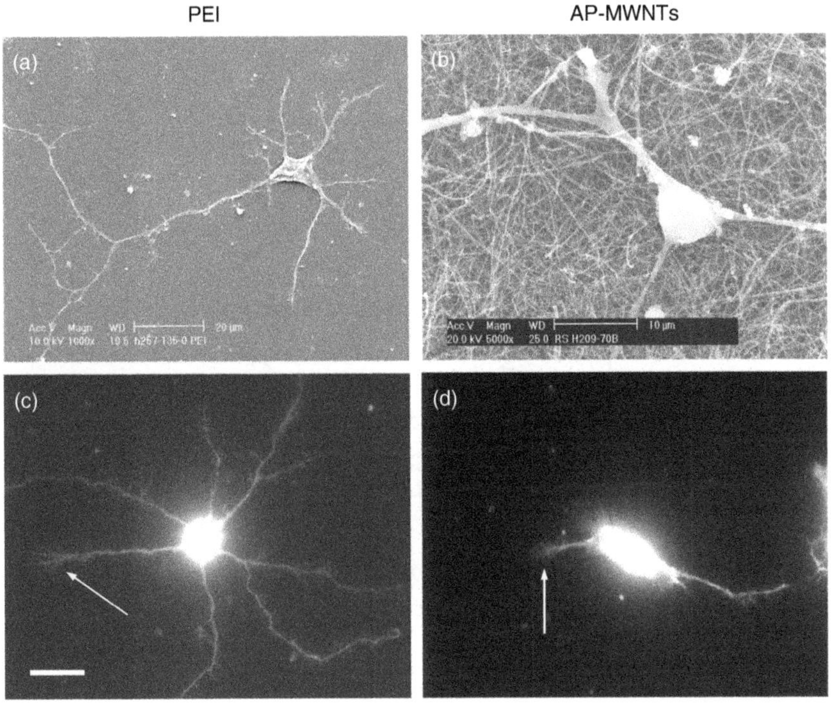

Fig. 2. MWNTs are permissive substrate for neurons. Both PEI (positive control, left column images) and as-prepared MWNTs (AP-MWNTs; right column images) support neuronal viability and permit neurite outgrowth. (a, b) SEM images of neurons grown on PEI (a) and AP-MWNTs (b). (c, d) Fluorescence images showing live neurons, which accumulate a vital stain, calcein. Arrows indicate growth cones. Scale bar: 20 μm, except 10 μm in B. Modified from Hu et al. (2004).

the PEI charge using neutral CNTs. Hu et al. (2005) synthesized such a polymer where branched PEI was grafted onto SWNTs. Since the SWNT's weight load in this copolymer was 18–19%, the positive charge of PEI was perhaps proportionally reduced. This graft copolymer, SWNT-PEI, was used to coat glass coverslips onto which neonatal rat hippocampal neurons were seeded and grown in culture. This manipulation of CNTs resulted in neurite outgrowth and branching intermediate to those displayed by neurons grown on as-prepared MWNTs or PEI alone as substrates (Fig. 3b). Some future experiments should be conducted to chemically control the percent load of the SWNTs in the copolymer and test the effects of that on neurite outgrowth and branching. The ability of such control may allow grading of the neuronal growth if graft copolymers would be used in neural prosthesis.

One strategy for treatment of injury sites in the CNS is to cause the extension of neurite length in selected neurites, thus increasing the chance of "bridging" the injured site rather that enhancing the generalized outgrowth. Ni et al. (2005) achieved selected neurite outgrowth by treating neonatal rat hippocampal neurons grown on PEI-coated coverslips with water-soluble SWNT graft copolymers. SWNTs were chemically functionalized with either PABS or polyethylene glycol (PEG) to form the corresponding graft copolymers, which at physiological extracellular pH would show zwitterionic or neutral charge, respectively. This covalent functionalization made SWNTs soluble/dispersible in aqueous media. The addition of SWNT graft copolymers to culturing media caused concentration-dependent effects on neurons. At a dosage of 1μg/mL, there was a decrease in the number of neurites per

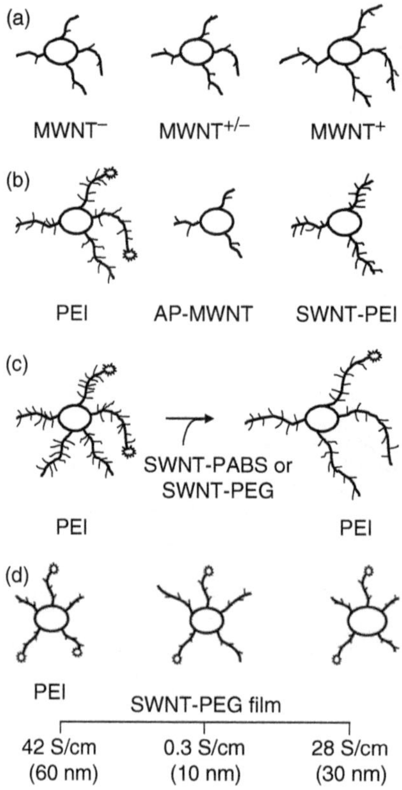

Fig. 3. The effects of CNTs on neurite outgrowth, growth cones, and body size. (a) Manipulating the charge carried by functionalized MWNTs utilized as a scaffold for neuronal growth can be used to control the outgrowth and branching pattern of neurites. Neurons grown on positively charged MWNTs have longer and more decorated neurites that neurons grown on negative or zwitterionic MWNTs. (b) SWNT-PEI graft copolymer substrate promotes neurite outgrowth and branching intermediate to those exhibited on PEI and AP-MWNTs. (c) Water-soluble SWNT-PABS and SWNT-PEG graft copolymers when added to the culturing medium of neurons grown on PEI were able to increase the length of selected neuronal processes and to reduce the number of growth cones. (d) The neurite outgrowth was significantly greater in neurons grown on the 10-nm-thick SWNT-PEG films. The average area of the neuron cell body grown on the 30-nm-thick SWNT films was enlarged. Neurons grown on the PEI had a significantly higher number of growth cones than those grown on 10- and 30-nm SWNT films, but not higher than those grown on 60-nm SWNT films. Conductivity of each film is listed in S/cm. (a) Modified from Hu et al. (2004). (b) Modified from Hu et al. (2005). (c) Modified from Ni et al. (2005). (d) Modified from Malarkey et al. (2009).

neuron in treated cells. Additionally, there was no significant change in the total neurite length per neuron, since neurons displayed a concurrent increase in the average length of neurites. Thus, neurons treated with SWNT-PABS or SWNT-PEG graft copolymers had sparser, but longer neurites, consistent with the enhancement of selected neurites outgrowth (Fig. 3c). This effect on neuronal growth involved the modulation of intracellular Ca^{2+} homeostasis, since SWNT-PEG acted as an inhibitor of depolarization-dependent influx of Ca^{2+} to the cytosol from the extracellular space. An increase in intracellular Ca^{2+} levels due to depolarization of neurons can regulate plasma membrane/vesicular recycling which has been implicated to play a role in the rate of neurite elongation. To further elucidate the mechanisms responsible for the effects of SWNTs on neurite outgrowth, Malarkey, Reyes, Zhao, Haddon, & Parpura (2008) examined how SWNTs could affect membrane recycling, using the same experimental conditions as the study of Ni et al. (2005).

An increase in total membrane area during neurite outgrowth occurs through incorporation of new membrane by vesicle fusion or exocytosis. Membrane can also be retrieved by the reverse process of endocytosis. Data pointed to a preferential inhibiting action of SWNT-PEG on regulated, depolarization-dependent, endocytosis. Thus, the exocytotic incorporation of vesicles into the plasma membrane is not balanced by the endocytotic retrieval in the presence of SWNTs, which could effectively cause the increase in neurite length observed by Ni et al. (2005). Based on these findings, it is possible that the reduction of the number of neurites seen in neurons treated with SWNT-PEG could be a compensatory mechanism to keep the cell surface/volume relatively constant. Taken together, these results suggest the exciting possibility that water-soluble SWNTs could be delivered locally to the site of CNS injury to enhance neurite outgrowth which might aid in the process of regenerating and rewiring synaptic connections lost due to injury. Of course, this new direction for possible therapeutic intervention will need to be assessed in experimental models of mammalian brain injury.

Biofunctionalization of CNTs can be carried out to achieve inhibitory effects on cellular growth (Liopo, Stewart, Hudson, Tour, & Pappas, 2006). Using NG108 cells, a mouse neuroblastoma × rat glioma hybridoma cell line, as a neuronal model in culture to explore CNT–neuron interaction, Liopo et al. showed that SWNT mats functionalized with 4-*tert*-butylphenyl (nonpolar/hydrophobic molecule) or 4-benzoic acid (negatively charged molecule at normal extracellular pH of ~7.4) reduced cell adhesion, proliferation, and cell survival of NG108 cells when compared to growth characteristics of these tumorigenic cells cultured on as-prepared SWNT mats and even more so when compared to their growth on polystyrene tissue culture plastic. Functionalized SWNTs with 4-benzoic acid had a greater effect on the reduction of cell adhesion and survival then SWNTs functionalized with 4-tert-butylphenyl. Altogether, the above studies suggest that CNTs are an accommodating source material for generating diverse substrates that can be used to modulate adhesion and growth of neurons (or arguably similar cells) and outgrowth of their processes in culture. The mechanism for the action of biofunctionalized CNTs in modulation of neuronal growth and outgrowth, with exception of 4HNE (see discussion on possible mechanisms in Mattson et al., 2000), appears to be simply the level of charge and/or polarity of functional chemical groups. Such functionalizations lack specificity that neural cells utilize in intercellular interactions. Consequently, it should be essential to design and engineer CNT substrates that are functionalized with biological molecules that possess ligand–receptor specificity.

A first step toward such a goal has been accomplished by Matsumoto et al. (2007) who functionalized CNTs using endogenous ligands in the CNS to assess the retention of ligand activity when conjugated to CNTs. They covalently linked neurotrophins, nerve growth factor (NGF) or brain-derived neurotrophic factor (BDNF), to MWNTs. Dorsal root ganglia (DRG) from chick embryos were dissociated to establish neuronal cultures that were grown in the standard laminin-coated well plates. The numbers of DRG neurons with neurite outgrowth longer than the cell body were counted. The addition of soluble NGF or BDNF prompted neurite outgrowth. The extent of such outgrowth could be matched when instead of soluble neutrotrophins, a dispersion of either MWNT-NGF or MWNT-BDNF was applied to the DRG neurons, indicating that neurotrophins covalently attached to CNTs retained their bioactivity. Future experiments will have to be carefully designed to assess the use of MWNT substrates/mats functionalized with neurotrophins for support of neurite outgrowth. It should be noted, however, that when soluble neurotrophin is added to neurons upon binding to its receptor, the resulting neurotrophin–receptor complex can be internalized into a cell. It is then credible that dispersible MWNT-neurotrophin could be internalized by an endocytotic process. However, a caveat to this prospective CNT use presents itself in the ability of CNTs, SWNT-PEG, to inhibit depolarization-dependent endocytosis in neurons (Malarkey et al., 2008). Indeed, future research will be required to assess this possible method of delivery of CNTs and their conjugates to the intracellular space.

The studies discussed above focused on investigating CNT–neuron interactions by varying the surface chemistries on CNT substrates. However, mechanical and electrical properties of CNT substrates such as surface roughness and conductivity, respectively, could also modulate neuronal growth and neurite outgrowth. A recent study examined how the passive conductivity of the substrate affects neurite outgrowth (Malarkey et al., 2009). Malarkey et al. generated retainable films of SWNT-PEG deposited onto glass coverslips. By varying the thickness of these films, they controlled conductivity of SWNT-PEG films; 10-, 30-, and 60-nm-thick SWNT-PEG films had conductivities of 0.3, 28, and 42 S/cm, respectively. Although these films had various thicknesses, they displayed similar surface roughness unlike the significantly smoother standard substrate for neuronal growth, PEI, which is a nonconductive and positively charged (at physiological extracellular pH) compound. Consequently, effects on neuronal growth achieved on the various SWNT-PEG films result from their differences in conductivity, not surface roughness. However, when the neuronal growth on various SWNT films is compared to that of neurons grown on PEI, any difference observed could be the outcome of roughness, charge and/or conductance. The total number of neurites for each neuron remained the same regardless of the conductivity of the substrate. The total outgrowth, the summed length of all processes and their branches, was significantly greater in neurons grown on the 10-nm-thick SWNT-PEG films than neurons grown on coverslips coated with the PEI standard (Fig. 3d). However, on the thicker films (30 and 60 nm) with higher conductivity, but comparable roughness to the 10-nm SWNT-PEG film, there was no difference in neurite outgrowth from the standard PEI coating. This indicates that a certain SWNT-PEG conductance (0.3 S/cm) could promote neurite outgrowth as compared to other SWNT films. Since the total outgrowth of the neurites had increased but the number of processes remained the same, the mean process length was significantly longer in neurons grown on the 10-nm films compared with other substrates. The higher conductance substrates (30-and 60-nm SWNT films) showed no difference in process length from the PEI standard, regardless that the surface of SWNT films shows significantly different roughness from PEI. There was an increase in the average area of the neuron cell body showing a trend at 0.3 S/cm substrate with a statistical significance at 28 S/cm when compared to standard and to 42 S/cm (Fig. 3d). Neurons grown on the smoother PEI had a significantly higher number of growth cones than those grown on 10- and 30-nm SWNT films, but not higher than those grown on 60-nm SWNT films (Fig. 3d). Since positively charged PEI has a much smoother surface than SWNT films, it appears that the charge, roughness, and/or conductivity could cause this effect. However, there was a significantly higher number of growth cones on neurons grown on 60-nm SWNT films when compared to measurements on 30-nm SWNT films. Since the roughness of these two conductive films is similar, it appears that the higher conductance caused an increase in the number of growth cones. These results indicate that a SWNT-PEG substrate in a narrow range of conductivity can promote neuronal growth and neurite outgrowth. As conductivity increases beyond this range, these effects on (out)growth are diminished.

Mechanical properties of substrates have been implicated as controlling factors for cell adhesion and growth (Fan et al., 2002; Nguyen-Vu et al., 2007; Xie, Chen, Aatre, Srivatsan, & Varadan, 2006; Zhang et al., 2005). DRG neurons isolated from neonatal rats grown on MWNT mats functionalized with carobxyl groups diplay neurite entanglements with CNTs (Xie et al., 2006). This effect of CNTs on neuronal adhesion was further explored in a recent study (Sorkin et al., 2009). Sorkin et al. (2009) cultured dissociated neurons, that originated from embryonic rat cortices or from locust frontal ganglion, on micropatterned as-prepared CNT islands made by CVD; the reported distribution of CNTs diameters in this study (Fig. 4 of Sorkin et al., 2009) is consistent with MWNTs. The authors have chosen to culture neurons from rats and locust due to a difference in size of cell bodies and diameter of neurites. Both cell types displayed a preference to grow on CNT islands as opposed to the surrounding quartz. The authors attribute this to surface roughness of CNT islands, although this issue was not systematically addressed here. A close examination of the rat neuronal processes revealed extensive curling and intertwining of neurites around CNTs. This was also the case with thin, but not thick, neurites outgrowing from locust neuronal somata; thick neurites show entanglement between themselves rather than with CNTs. The authors rationed that diameters of CNT within island allow entanglement of thin neurites with relatively similar diameters to CNTs and that such entanglement may represent an anchoring mechanism allowing neurons to attach to rough surfaces. These results suggest that when designing CNTs as scaffolds for neuronal growth, the mechanical properties of CNTs should be taken into consideration. Taken together, the studies discussed above show that the various qualities of planar substrates made of CNTs, most notably charge, passive conductivity, and their roughness/size can modulate neuronal growth and neurite outgrowth.

Micropatterning of cell adhesion substrates allows for precise control on localization and patterns of cellular growth. This approach is useful in generation and investigation of neural networks *in vitro*. Micropatterning of CNTs using lithography to create self-organizing neural networks have been demonstrated (Gabay, Jakobs, Ben-Jacob, & Hanein, 2005). Gabay et al. (2005) used a polydimethylsiloxane (PDMS) mold to deposit iron nanoparticles onto a quartz surface to generate islands of CNT catalyst upon which CNTs were grown via CVD. When dissociated rat cortical tissue was dispersed onto these CNT islands, neurons form ganglion-like aggregates of cells on CNT islands within 4 days in culture. The neuronal processes were seen to bridge across the nonpermissive quartz to form connections between cells of adjacent islands, leading to establishment of neuronal networks (Fig. 4). Using similar experimental approach, Sorkin et al. (2006) generated an SWNT-patterned substrate by applying iron nitrate catalyst to coverslips with a PDMS stencil and then growing CNTs by CVD. Rat cortical and hippocampal neurons were cultured on these substrates and after 2–3 days spontaneously grew into islands with neurites connecting nearby islands. These patterned networks could be used to study functional neuronal networking with CNT islands serving as electrical connections for sensing or stimulation.

Thus far, we discussed CNTs as planar substrates/scaffolds for neuronal growth. However, if CNTs would be used in neural prosthesis it is desirable that neurons can grow and show neurite outgrowth on three-dimensional CNT scaffolds. Some of the initial research in this direction has been done using cell lines such as differentiated NG108-15 (Gheith, Sinani, Wicksted, Matts, & Kotov, 2005) and PC12 cells (Nguyen-Vu et al., 2007), a cell line derived from a pheochromocytoma of the rat adrenal

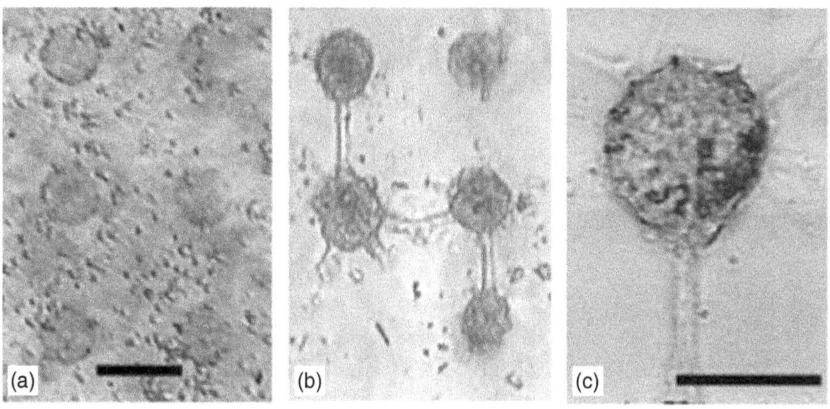

Fig. 4. Self-organizing neural networks on micropatterned arrays of CNT islands. (a) Cells were randomly dispersed 1 h after initial plating, but after 4 days in culture these cells form clusters on CNT islands with some sending neuronal processes to adjacent islands as shown in (b) and (c). Scale bar, 150 μm in (a) for (a, b), while 100 μm in (c), respectively. Reproduced from Gabay et al. (2005).

medulla; both lines were used as neuronal cell models. These studies demonstrated the permissiveness of three-dimensional CNT scaffolds for cellular adhesion and growth. Galvan-Garcia et al. (2007) reported that directionally oriented, highly purified MWNTs in the form of sheets or yarns offer an alternative presentation of CNTs to neurons. CNT sheets allowed postnatal mouse neurons from DRG explants as well as dissociated cortical and cerebellar neurons to extend processes of which the number and length were comparable to those of neurons grown on a planar permissive substrate, represented by polyornithine (PLO)-coated glass coverslips. Interestingly, the area occupied by growth cones on CNT sheets was significantly enlarged when compared to the growth cone area of neurons grown on PLO-coated coverslips. Since the growth cone appearance can be affected by the substrate qualities, the authors speculated that this enlargement could be due to complex nanotexturing of the CNT sheets. DRG neurons could also attach and grow along CNT yarns closely following the surface topography of this scaffold (Fig. 5). Thus, CNTs presented as three-dimensional substrate/scaffold can be used to grow and most likely direct the growth of primary neurons. Additionally, using a particle coagulation spinning process, carbon nanotube fibers (CNFs) in the form of threads and ribbons up to 30 cm in length and 30–100 μm in diameter can be made from SWNTs (Dubin, Callegari, Kohn, & Neimark, 2008). For testing biocompatibility of CNF, their fragments were applied to plastic tissue culture wells. CNF ribbons, when precoated with poly-D-lysine and laminin, were compatible with neural cells. After a week in culture, rat hippocampal neurons displayed a neurite outgrowth on the CNF surface; neuronal somata were not observed on the CNF surface but rather on the surface of the plastic wells. Thus, it appears that cell bodies were attached to the plastic and neurites extended onto the CNF. Similarly, collagen-coated CNF threads were permissive scaffolds for neurite-like processes and bodies of PC12 cells cultured for a week in the presence of the soluble NGF to cause their differentiation into sympathetic neuron-like cells. This demonstration of the biocompatibility of CNFs indicates that they may be a suitable material for the construction of neural prosthetic devices.

CNTs as substrate for other neural cells

Glial cells and carbon nanotubes

Formation of the glial scar remains a major obstacle in tissues repair and regeneration after injury in the CNS (reviewed in Silver & Miller, 2004). Since astrocytes are the major cellular component of the glial scar, it might be possible to use CNTs

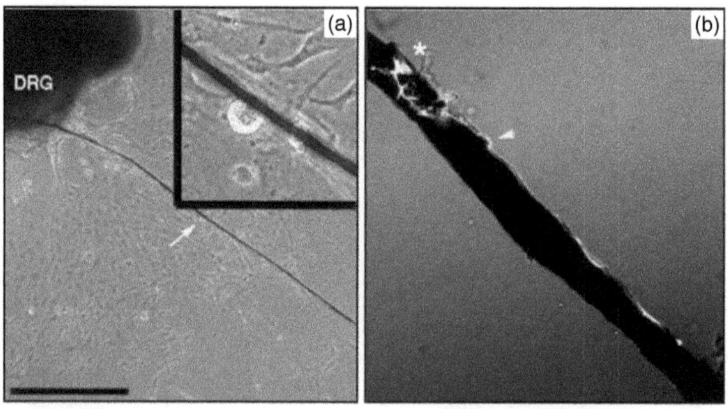

Fig. 5. Directed neuronal growth on MWNT yarns. (a) A CNT yarn is placed over a DRG explant. The arrow indicates the area shown in the inset. Shiny neuronal cell body is attached to the yarn. The cell in the (a), whose body has been detached (asterisk) after immunocytochemistry against β-tubulin, is shown in (b). Neurite extends along the yarn and closely follows the CNT surface topology (arrowhead). Scale bar, 200 μm. Modified from Galvan-Garcia et al. (2007).

to aid repair at the site of injury by inhibiting gliosis. Therefore, the understanding of CNT–astrocyte interactions is of importance. McKenzie, Waid, Shi, and Webster (2004) examined such interactions using an astrocytic cell line (ACL) established by transforming rat diencephalic astrocytes with the oncogenic early region of SV40 (ATCC #CRL-2005). MWNTs were synthesized using CVD and separated in two classes based on their diameter as conventional (diameters 125–200 nm) and nanophase (diameters 60–100 nm) MWNTs. Each type was further subclassified based on the surface energy as low (25–50 mJ/m^2) and high (125–140 mJ/m^2) surface energy MWNTs; as-prepared MWNTs displayed low surface energy. High surface energy MWNTs were obtained by pyrolytic stripping. Each of four types of MWNTs were then pressed in a single axis to obtain MWNT discs of ~1.3 cm^2. Additionally, composite discs containing a mixture of nanophase (60 nm in diameter) high surface energy MWNTs and polycarbonate urethane (PCU) were made with varying weight percents of PCU: MWNTs (100:0, 98:2, 90:10, and 75:25). Surface roughness of discs was assessed using SEM. Based on visual inspection, the surface roughness was higher when carbon discs were made of nanophase MWNTs. Composites of nanophase MWNTs and PCU also appear to have variable roughness; the 75:25 PCU:MWNT composite had higher surface roughness than the rest of them. The resistivity of composites also varied; it exponentially decreased as the weight percent of MWNTs in them increased, that is, the resistivity measured 20,500 Ωm for the 98:2 PCU:MWNT composite, while for the 75:25 PCU:MWNT composite it was 0.354 Ωm. These discs displaying various physical properties were used to grow the ACL in cultures. Glass coverslips were used as a reference substrate. As an indication of gliosis, ACL adhesion, proliferation, and activity were measured. The cells of the ACL preferentially adhered to CNTs with larger (conventional) diameter and low surface energy scaffolds as compared to glass and other three MWNT discs. The ACL adhesion was significantly greater on PCU, but it gradually decreased with the increasing content of nanophase MWNTs in PCU composites. The adhesion of ACL cells to glass coverslips was at the low end, but the ACL proliferation after 1, 3, and 5 days in culture was the highest on glass coverslips, exceeding that seen on MWNT discs. Comparison of ACL proliferation on various MWNT discs indicates that these cells prefer a conventional low surface energy substrate. Overall, ACL adhesion and proliferation were best on the conventional low surface energy scaffold, while the worst performance was seen on nanophase high surface energy discs. As a measure of ACL cell activity the authors used measurements of the enzyme alkaline phosphatase production. ACL cells grown on MWNT discs show no difference in the enzyme production after 1 week, while after 14 days in culture ALC grown on nanophase low surface energy had significantly lower production of this enzyme. Taken together, this study established some guidance in regard to qualities of materials that should be used in neural implants. This implicates that materials having nanophase features would reduce gliosis, since ACL proliferation and activity was reduced on such scaffolds. Similarly, composite materials containing large amounts of PCU should be avoided, since ACL growth was greatly enhanced as the amount of MWNTs was reduced in PCU:MWNT composites. These findings when combined with studies of CNTs as scaffold for neuronal growth (see section "CNTs as substrates/scaffolds for neuronal cell growth") warrant future testing of the use of CNTs at the site of injury in experimental models, since CNTs might retard glial scar formation while promoting the neurite outgrowth. Indeed, it would be prudent and necessary to test whether similar inhibitory effects of MWNTs on the astrocytic proliferation can be achieved with primary astrocytes isolated from variety of brain regions. Meanwhile, the studies of astrocytes grown on CNT scaffolds have been limited and only done as side projects to studies of neurons. Lovat et al. (2005) cultured primary hippocampal astrocytes on MWNT that were immobilized onto glass coverslips. Using immunocytochemistry against the astrocyte-specific marker, glial fibrillary acidic protein (GFAP), they did not find any difference in morphology and proliferation, assessed by the cell density after 8 days in culture,

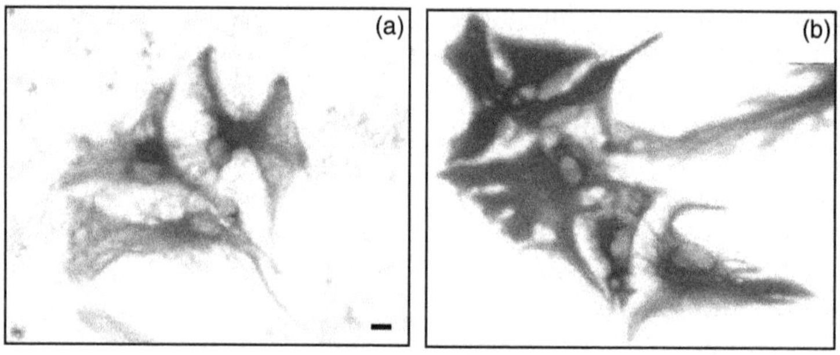

Fig. 6. Astrocytes adhere and proliferate on a CNT substrate. GFAP immunoreactivity shows no obvious morphological difference between astrocytes grown on CNT-coated coverslips (a) and bare glass coverslips (b). Scale bar, 10 μm. Modified from Lovat et al. (2005).

of astrocytes grown on MWNTs as compared to cells cultured on bare glass coverslips (Lovat et al., 2005) (Fig. 6). In a more recent study, astrocytes originating from embryonic rats, were cultured on CNT islands (Sorkin et al., 2009). These cells were able to attach to CNT islands and grow there by spreading their cell bodies over the surface without overlapping with adjacent cells, thus showing similar characteristics of growth as has been observed elsewhere when astrocytes were cultured on standard substrates, such as PEI (see, e.g., Lee, Malarkey, Reyes, & Parpura, 2008). In addition to limited examination of astrocytic adhesion properties, the function of astrocytes grown on CNT scaffolds was recently assessed (Huang et al., 2009). SWNT-PEG mats deposited onto glass overspills were used as a scaffold for astrocytic growth and also as a platform for the detection of adenosine 5'-triphosphate release from these cells. Taken together, although the above reports are an exciting indication that CNTs may find use as scaffolds for the modulation of astrocytic growth, it is hard to draw any conclusions about the effect of CNT on astrocyte growth and adhesion at present, since to date there has not been a systematic approach to study CNT–astrocyte interactions.

Besides the use of CNTs to grow astrocytes, Schwann cells have been grown on CNTs as well. Galvan-Garcia et al. (2007) grew DRG explants on MWNT sheets. Following their adhesion to MWNT sheets, DRGs grew on this scaffold with Schwann cells showing extensive migration for up to 14 days in culture. Additionally, when MWNT yarns were placed over DRG explants, Schwann cells were able to attach and grow along MWNT yarns following tubular surface topology of the scaffold (Fig. 7). These data indicate that three-dimensional CNT scaffolds presented to neural cells as MWNT sheets and yarns are good scaffolds for both neuronal (see section "CNTs as substrates/scaffolds for neuronal cell growth") and glial cell growth.

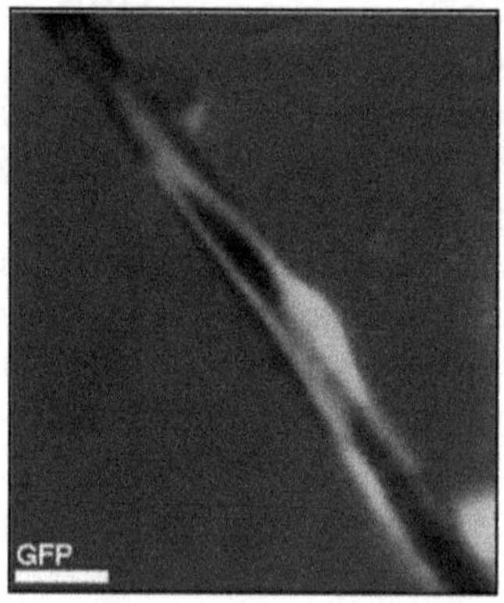

Fig. 7. Directed Schwann cell growth on MWNT yarns. For visualization of live cells, Schwann cells were transfected to express the green fluorescent protein (GFP). These glial cells spread along the CNT yarn. Scale bar, 20 μm. Modified from Galvan-Garcia et al. (2007).

Neural stem cells and carbon nanotubes

The growth and differentiation of neural stem cells on CNT substrates was recently demonstrated (Jan & Kotov, 2007). CNT substrates were assembled layer-by-layer (LBL). The alternating layer deposition of PEI and polystyrene sulfonate (PSS)-wrapped SWNTs (O'Connell et al., 2001) onto glass coverslips resulted in a bilayer; six such bilayers were stacked to form a $(PEI/SWNT)_6$ film that coated glass coverslips. Mouse embryonic cortical neurospheres containing neural stem cells (NSCs) were seeded at low density onto $(PEI/SWNT)_6$-coated, as well as onto PLO-coated glass coverslips, a standard substrate commonly used for NSC cultures. The low seeding of neurospheres was used to assess the interaction between neural cells and the substrate. Neurospheres attached to both of the substrates and after 1 day in culture developed neural processes that emanated from them to the surrounding area; the length of processes and their complexity increased during 1 week in culture. The viability of neurospheres and the length of processes were similar for neurospheres grown on $(PEI/SWNT)_6$ or PLO. Furthermore, both substrates supported a comparable differentiation of NSCs into neurons, astrocytes, and oligodendrocytes as shown by immunoreactivity of nestin, microtubule-associated protein 2 (MAP2), GFAP, and oligodendrocyte marker O4, respectively (Fig. 8). Similar results were obtained when neurospheres dissociated into single cells were plated onto $(PEI/SWNT)_5$-, PLO-, or composite $(PEI/SWNT)_5$/PLO-coated glass coverslips. These data demonstrated that environment-sensitive NSCs can be grown and differentiated on CNT substrates with similar success as on standard PLO substrate.

In a subsequent study, Kam, Jan, and Kotov (2009) fabricated LBL SWNT composites with laminin, an endogenous protein component of extracellular matrix, rather than the exogenous organic polymer PEI. Various LBL configurations were tested for the cell adhesion of dispersed NSC. The best results were obtained using glass coverslips that were covered with the heat annealed laminin/SWNT film containing multiple (up to 30) bilayers with PSS-wrapped SWNTs as the top layer. NSC plated on a surface of this scaffold containing 10 bilayers displayed more vigorous outgrowth than cells grown on control laminin-coated coverslips. After a week in culture there was a substantial differentiation of NSCs into neurons and astrocytes based on immunoreactivity for nestin, MAP2, and GFAP, respectively. Neuronal differentiation was further confirmed by the presence of synaptic protein synapsin. Since heat-annealed $(laminin/SWNT)_{10}$ has conductivity of ~ 21 S/cm, the neuronal excitability was confirmed by a direct electrical stimulation of cells via SWNT substrate which caused intracellular Ca^{2+} elevations in cells grown on top of this substrate. These results demonstrate that protein–CNT composite scaffold can serve as a permissive substrate for neural stem cells growth and differentiation. This approach might allow for the possibility to generate CNTs scaffolds that are better suitable for long-term implantation as they contain native protein. Also, this approach could generate composite materials that could contain chemoattractants, which could then guide patient's own neural stems cell population to migrate to the site of injury in the CNS in attempt to repair and replace lost cells in the region. This may prove a viable alternative to donor-based neural stem cell transplantation, in which one of the complications can be development of a donor-derived brain tumor (Amariglio et al., 2009).

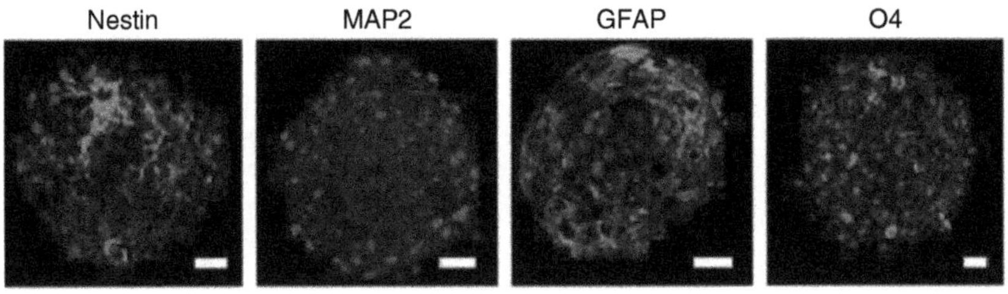

Fig. 8. Differentiated neurospheres grown on $(PEI/SWNT)_6$ substrate. Immunostaining (red) indicates the presence of neural stem cells (nestin), neurons (MAP2), astrocytes (GFAP), and oligodendrocytes (O4). Nuclei are counterstained using DAPI (blue). Scale bars, 20 μm. Modified from Jan and Kotov (2007).

Concluding remarks

The purpose of this chapter is to summarize current findings related to CNTs as a substrate for neural cells adhesion and growth. The current literature supports CNTs as a biocompatible and permissive substrate for neural cells. Due to the unique physical properties of CNTs and the ability to functionalize them with biomolecules, cellular growth can be modulated to achieve the desired effect with a relatively high degree of specificity. Consequently, CNTs hold great promise for applications as neural prostheses. Indeed, CNTs has been used to coat various metal electrodes to lower their impedance and to increase sensitivity when they are being used to record signals, and to increase the delivery of electrical charge when they are being used to stimulate neurons (Keefer, Botterman, Romero, Rossi, & Gross, 2008). This should find use in brain–machine interfaces, which can be used to help understand brain function and restore the movement in tertraplegia (Hochberg et al., 2006; Schwartz, Cui, Weber, & Moran, 2006). However, the exposure of the general populace to this material must not occur without adequate testing. At this point there is no indication that CNTs will be any more hazardous than other forms of carbon as the toxicity does not appear to be a concern in systemic applications (Liu et al., 2008). Their enormous potential in nanomedicine mandates the continued investigation of this unique nanomaterial from the biological standpoint.

Acknowledgments

We thank Randy F. Stout, Jr. for comments on previous versions of this manuscript. The authors' work is supported by a grant from the National Institute of Mental Health (MH 069791). We dedicate this chapter to the late Glenn I. Hatton, whose energy and creativity inspired new views of astrocyte–neuronal interactions.

References

Amariglio, N., Hirshberg, A., Scheithauer, B. W., Cohen, Y., Loewenthal, R., Trakhtenbrot, L., et al. (2009). Donor-derived brain tumor following neural stem cell transplantation in an ataxia telangiectasia patient. *PLoS Medicine, 6*, e29.

Bekyarova, E., Haddon, R. C., & Parpura, V. (2005). Biofunctionalization of carbon nanotubes. In C. S.S.R. Kumar (Ed.), *Biofunctionalization of nanomaterials (nanotechnologies for the life sciences)* (pp. 41–71). Wienhein-Berlin, Germany: Wiley-VCH.

Bekyarova, E., Ni, Y., Malarkey, E. B., Montana, V., McWilliams, J. L., Haddon, R. C., et al. (2005). Applications of carbon nanotubes in biotechnology and biomedicine. *Journal of Biomedical Nanotechnology, 1*, 3–17.

Dubin, R. A., Callegari, G., Kohn, J., & Neimark, A. (2008). Carbon nanotube fibers are compatible with mammalian cells and neurons. *IEEE Transactions on Nanobioscience, 7*, 11–14.

Fan, Y. W., Cui, F. Z., Hou, S. P., Xu, Q. Y., Chen, L. N., & Lee, I. S. (2002). Culture of neural cells on silicon wafers with nano-scale surface topograph. *Journal of Neuroscience Methods, 120*, 17–23.

Gabay, T., Jakobs, E., Ben-Jacob, E., & Hanein, Y. (2005). Engineered self-organization of neural networks using carbon nanotube clusters. *Physica A: Statistical Mechanics and Its Applications, 350*, 611–621.

Galvan-Garcia, P., Keefer, E. W., Yang, F., Zhang, M., Fang, S., Zakhidov, A. A., et al. (2007). Robust cell migration and neuronal growth on pristine carbon nanotube sheets and yarns. *Journal of Biomaterials Science, Polymer Edition, 18*, 1245–1261.

Gheith, M. K., Sinani, V. A., Wicksted, J. P., Matts, R. L., & Kotov, N. A. (2005). Single-walled carbon nanotube polyelectrolyte multilayers and freestanding films as a biocompatible platform for neuroprosthetic implants. *Advanced Materials, 17*, 2663–2670.

Hochberg, L. R., Serruya, M. D., Friehs, G. M., Mukand, J. A., Saleh, M., Caplan, A. H., et al. (2006). Neuronal ensemble control of prosthetic devices by a human with tetraplegia. *Nature, 442*, 164–171.

Hu, H., Ni, Y., Mandal, S. K., Montana, V., Zhao, B., Haddon, R. C., et al. (2005). Polyethyleneimine functionalized single-walled carbon nanotubes as a substrate for neuronal growth. *The Journal of Physical Chemistry B, 109*, 4285–4289.

Hu, H., Ni, Y. C., Montana, V., Haddon, R. C., & Parpura, V. (2004). Chemically functionalized carbon nanotubes as substrates for neuronal growth. *Nano Letters, 4*, 507– 511.

Huang, Y., Sudibya, H. G., Fu, D., Xue, R., Dong, X., Li, L. J., et al. (2009). Label-free detection of ATP release from living astrocytes with high temporal resolution using carbon nanotube network. *Biosensors & Bioelectronics, 24*, 2716–2720.

Ijima, S. (1991). Helical microtubules of graphite carbon. *Nature, 354*, 56–59.
Ijima, S., & Ichihashi, T. (1993). Single-shell carbon nanotubes of 1-nm diameter. *Nature, 363*, 603–605.
Jan, E., & Kotov, N. A. (2007). Successful differentiation of mouse neural stem cells on layer-by-layer assembled single-walled carbon nanotube composite. *Nano Letters, 7*, 1123–1128.
Journet, C., & Bernier, P. (1998). Production of carbon nanotubes. *Applied Physics A: Materials Science & Processing, 67*, 1–9.
Kam, N. W.S., Jan, E., & Kotov, N. A. (2009). Electrical stimulation of neural stem cells mediated by humanized carbon nanotube composite made with extracellular matrix protein. *Nano Letters, 9*, 273–278.
Keefer, E. W., Botterman, B. R., Romero, M. I., Rossi, A. F., & Gross, G. W. (2008). Carbon nanotube coating improves neuronal recordings. *Nature Nanotechnology, 3*, 434–439.
Lee, W., Malarkey, E. B., Reyes, R. C., & Parpura, V. (2008). Micropit, a new cell culturing approach for characterization of solitary astrocytes and small networks of these glial cells. *Frontiers in Neuroengineering, 1*, 2.
Liopo, A. V., Stewart, M. P., Hudson, J., Tour, J. M., & Pappas, T. C. (2006). Biocompatibility of native and functionalized single-walled carbon nanotubes for neuronal interface. *Journal of Nanoscience and Nanotechnology, 6*, 1365–1374.
Liu, Z., Davis, C., Cai, W., He, L., Chen, X., & Dai, H. (2008). Circulation and long-term fate of functionalized, biocompatible single-walled carbon nanotubes in mice probed by Raman spectroscopy. *Proceedings of the National Academy of Sciences, USA, 105*, 1410–1415.
Lovat, V., Pantarotto, D., Lagostena, L., Cacciari, B., Grandolfo, M., Righi, M., et al. (2005). Carbon nanotube substrates boost neuronal electrical signaling. *Nano Letters, 5*, 1107– 1110.
Malarkey, E. B., Fisher, K. A., Bekyarova, E., Liu, W., Haddon, R. C., & Parpura, V. (2009). Conductive single-walled carbon nanotube substrates modulate neuronal growth. *Nano Letters, 9*, 264–268.
Malarkey, E. B., & Parpura, V. (2007). Applications of carbon nanotubes in neurobiology. *Neurodegenerative Diseases, 4*, 292–299.
Malarkey, E. B., Reyes, R. C., Zhao, B., Haddon, R. C., & Parpura, V. (2008). Water soluble single-walled carbon nanotubes inhibit stimulated endocytosis in neurons. *Nano Letters, 8*, 3538–3542.
Matsumoto, K., Sato, C., Naka, Y., Kitazawa, A., Whitby, R. L., & Shimizu, N. (2007). Neurite outgrowths of neurons with neurotrophin-coated carbon nanotubes. *Journal of Bioscience and Bioengineering, 103*, 216–220.
Mattson, M. P., Haddon, R. C., & Rao, A. M. (2000). Molecular functionalization of carbon nanotubes and use as substrates for neuronal growth. *Journal of Molecular Neuroscience, 14*, 175–182.
McKenzie, J. L., Waid, M. C., Shi, R., & Webster, T. J. (2004). Decreased functions of astrocytes on carbon nanofiber materials. *Biomaterials, 25*, 1309–1317.
Monthioux, M., & Kuznetsov, V. (2006). Who should be given the credit for the discovery of carbon nanotubes? *Carbon, 44* (9), 1621–1623.
Nguyen-Vu, T. D., Chen, H., Cassell, A. M., Andrews, R. J., Meyyappan, M., & Li, J. (2007). Vertically aligned carbon nanofiber architecture as a multifunctional 3-D neural electrical interface. *IEEE Transactions on Biomedical Engineering, 54*, 1121–1128.
Ni, Y., Hu, H., Malarkey, E. B., Zhao, B., Montana, V., Haddon, R. C., et al. (2005). Chemically functionalized water soluble single-walled carbon nanotubes modulate neurite outgrowth. *Journal of Nanoscience and Nanotechnology, 5*, 1707–1712.
O'Connell, M. J., Boul, P., Ericson, L. M., Huffman, C., Wang, Y. H., Haroz, E., et al. (2001). Reversible water-solubilization of single-walled carbon nanotubes by polymer wrapping. *Chemical Physics Letters, 342*, 265–271.
Qin, L.-C., Zhao, X., Hirahara, K., Miyamoto, Y., Ando, Y., & Iijima, S. (2000). The smallest carbon nanotube. *Nature, 408*, 50.
Radushkevich, L. V., & Lukyanovich, V. M. (1952). O strukture ugleroda, obrazujucegosja pri termiceskom razlozenii okisi ugleroda na zeleznom kontakte. *Zurn Fisic Chim, 26*, 88–95.
Schwartz, A. B., Cui, X. T., Weber, D. J., & Moran, D. W. (2006). Brain-controlled interfaces, movement restoration with neural prosthetics. *Neuron, 52*, 205–220.
Silva, G. A. (2006) Neuroscience nanotechnology, progress, opportunities and challenges. *Nature Reviews. Neuroscience, 7*, 65–74.
Silver, J., & Miller, J. H. (2004). Regeneration beyond the glial scar. *Nature Reviews. Neuroscience, 5*, 146–156.
Sorkin, R., Gabay, T., Blinder, P., Baranes, D., Ben-Jacob, E., & Hanein, Y. (2006). Compact self-wiring in cultured neural networks. *Journal of Neural Engineering, 3*, 95– 101.
Sorkin, R., Greenbaum, A., David-Pur, M., Anava, S., Ayali, A., Ben-Jacob, E., et al. (2009). Process entanglement as a neuronal anchorage mechanism to rough surfaces. *Nanotechnology, 20*, 15101.
Sucapane, A., Cellot, G., Prato, M., Giugliano, M., Parpura, V., & Ballerini, L. (2009). Interactions between cultured neurons and carbon nanotubes. A nanoneuroscience vignette. *Journal of Nanoneuroscience, 1*, 10–16.
Wang, N., Tang, Z. K., Li, G. D., & Chen, J. S. (2000). Single-walled 4 A carbon nanotube arrays. *Nature, 408*, 50–51.
Xie, J. N., Chen, L. F., Aatre, K. R., Srivatsan, M., & Varadan, V. K. (2006). Somatosensory neurons grown on functionalized carbon nanotube mats. *Smart Materials and Structures, 15*, N85–N88.
Zhang, X., Prasad, S., Niyogi, S., Morgan, A., Ozkan, M., & Ozkan, C. S. (2005). Guided neurite growth on patterned carbon nanotubes. *Sensors and Actuators B, Chemical, 106*, 843–850.
Zheng, L. X., O'Connell, M. J., Doorn, S. K., Liao, X. Z., Zhao, Y. H., Akhadov, E. A., et al. (2004). Ultralong single-wall carbon nanotubes. *Nature Materials, 3*, 673–676.

H.S. Sharma (Ed.)
Progress in Brain Research, Vol. 180
ISSN: 0079-6123

CHAPTER 7

Neuromodulation: deep brain stimulation, sensory neuroprostheses, and the neural–electrical interface

Russell J. Andrews*

Smart Systems and Nanotechnology, NASA Ames Research Center, Moffett Field, CA, USA

Abstract: Although neuromodulation with implanted brain electrodes (deep brain stimulation, DBS) has been increasingly effective in treating many patients with movement disorders (e.g., advanced Parkinson's disease) over the past 20 years, the techniques have changed little for more than 50 years. After summarizing the current state of DBS, this chapter considers (1) the advances being offered by computational analysis techniques as well as (2) the benefits of monitoring and modulating brain chemical activity in addition to brain electrical activity. A review of the current state of sensory neuroprostheses follows, with consideration of emerging data on the optimal configuration of micron-sized retinal prostheses as well as on the optimal site for stimulation of cells in the retina. Very recent findings on nanotechniques to enhance charge transfer from prosthesis to cell (neuronal or glial), that is, enhancement of the neural–electrical interface, are then reviewed. The final section summarizes areas of potential cross-fertilization between those centers developing sensory neuroprostheses and those centers developing nanotechniques for DBS.

Keywords: brainstem implants; carbon nanotubes; cochlear implants; computational analysis; deep brain stimulation; nanoelectrode arrays; neural–electrical interface; neuromodulation; retinal implants; sensory neuroprostheses

Introduction

Neuromodulation for disorders of the brain — most commonly deep brain stimulation (DBS) — has shown little technological advance in the past 50 years in the devices currently used in clinical practice. Sensory neuroprostheses, however, — notably retinal and cochlear implants — have undergone rapid technological progress over the same period. This chapter reviews the current and future status of neuromodulation of both the brain and sensory neuroprostheses, the salient points being the convergence of the technologies and the intellectual cross-fertilization which is beginning to occur in the optimization of the neural–electrical interface (NEI).

Deep brain stimulation — current status

DBS at present uses electrodes 1.27 mm diameter to stimulate a small volume (roughly an ellipse several mm in diameter) of brain tissue (Fig. 1). The clinical effect is similar to ablation, but reversible. DBS has been used with varying success for the past 50 years to treat chronic

* Corresponding author.
Tel.: 1-408-829-1700; Fax: 1-408-353-0275;
E-mail: rja@russelljandrews.org

DOI: 10.1016/S0079-6123(08)80007-6

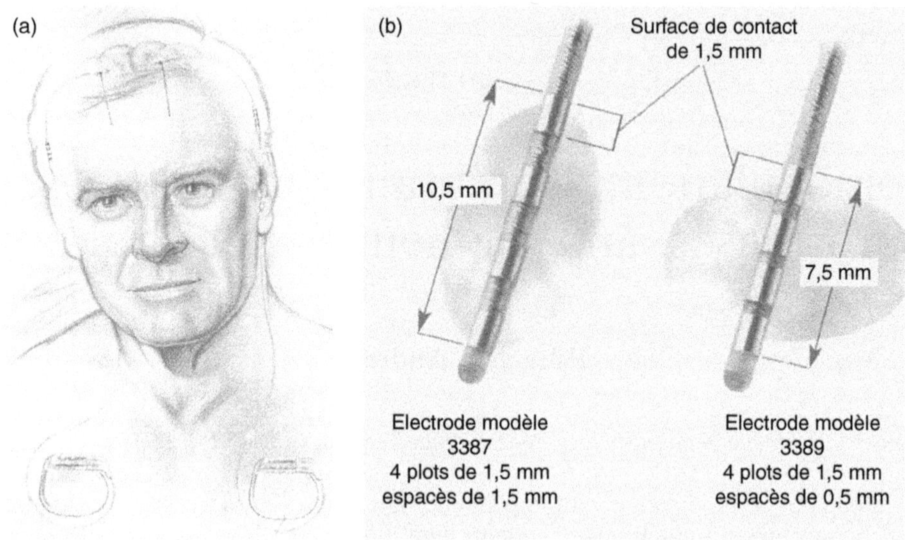

Fig. 1. (a and b) Illustrations of a typical deep brain stimulation (DBS) system (a), and the electrical contacts of typical electrodes inserted into the brain (b). Courtesy of Medtronic.

pain by stimulating regions such as the periaqueductal and periventricular gray matter (Hosobuchi, Adams, Bloom, & Guillemin, 1979). However, it first received CE mark approval in Europe over 15 years ago and FDA approval in the US over 10 years ago, for thalamic stimulation to control the refractory tremor of Parkinson's disease and essential tremor. Since then, DBS has proven to be the most significant advance in decades in the treatment of movement disorders such as advanced Parkinson's disease and dystonia (Deuschl et al., 2006; Krack et al., 2003).

Promising clinical trials are currently underway to use DBS to treat brain disorders ranging from refractory epilepsy to severe depression and obsessive-compulsive disorder (OCD) to disabling cluster headache to morbid obesity. However, clinical success with DBS for such disorders is inconsistent, with the exact sites in the brain for optimal clinical effect remaining elusive (Mallet et al., 2008; Mayberg, 2009; Schlaepfer & Bewernick, 2009). Given the increasing evidence that functional localization in the brain is often distributed in various regions (rather than confined to discrete sites), it is not surprising that the effect of DBS has been virtually identical to focal brain ablation. The advantage of DBS is that if the result is suboptimal (i.e., undesirable side effects occur), the stimulation can be turned off and – if detected intraoperatively – the electrode repositioned for better clinical efficacy.

DBS – emerging trends

Computational analysis

A major emerging trend in DBS has been the use of computational analysis in both the planning of electrode placement and the optimization of stimulation techniques. Preoperative interactive planning models allow the surgical team to predict the effects of various electrode placements prior to surgery (Chaturvedi, Butson, Cooper, & McIntyre, 2006). More significantly in the long run, computational modeling of brain electrical firing patterns in disorders such as Parkinson's disease is suggesting how the method of stimulation can improve the efficacy of DBS (Feng, Greenwald, Rabitz, Shea-Brown, & Kosut, 2007; Hauptmann, Popovych, & Tass, 2007; McIntyre, Miocinovic, & Butson, 2007). These findings include:

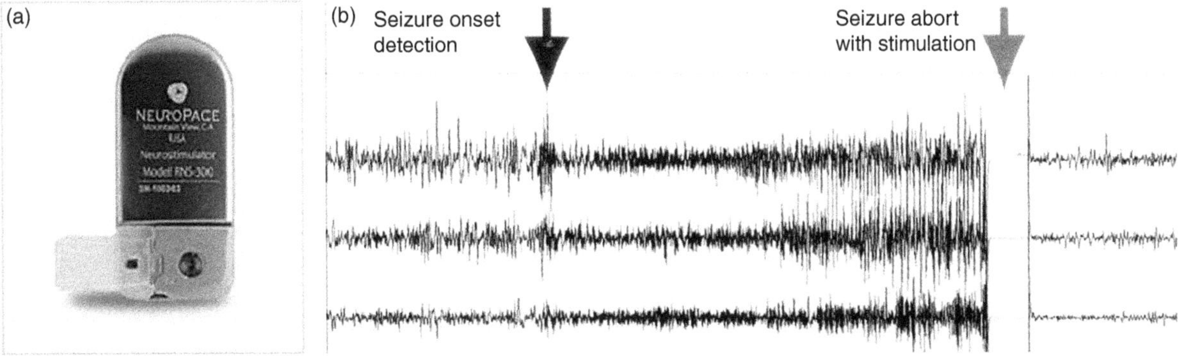

Fig. 2. (a and b) (a) The NeuroPace device for closed loop (i.e. feedback-guided) stimulation of the brain in patients with refractory epilepsy; the device is implanted under the scalp into the skull and connected to electrodes similar to those in Fig. 1 (b). (b) Electroencephalogram (EEG) demonstrating seizure arrest with brain stimulation. Courtesy of NeuroPace.

- Feedback, that is, the brain's electrical firing patterns in the disorder are monitored to guide the stimulation. The use of monitoring electrical activity to guide stimulation is in clinical trials by NeuroPace (Mountain View, CA, USA) for refractory epilepsy (Fig. 2);
- Low-frequency stimulation (<30 Hz), rather than the high-frequency stimulation (>100 Hz) used at present (Fig. 3);
- Multiple recording and stimulation electrodes. Although the large size of the present DBS electrodes (1.27 mm diameter) precludes placement of more than a few in the brain, even the use of two to four sites, recording and stimulating in a concerted fashion, should greatly improve DBS efficacy.
- Much smaller implanted batteries and microprocessors (implanted pulse generator, IPG) due to the greatly reduced power needs of the more efficient DBS. The connector leads from the brain, and the relatively large IPGs implanted under the skin of the upper chest are a major source of morbidity (infection and connector lead breakage) in the present DBS scenario (Fig. 1).

Electrical and chemical neuromodulation: neurons and glia

In Parkinson's disease the underlying disorder is a loss of dopaminergic neurons; in many mood disorders, an alteration in neurotransmitter levels, for example, dopamine, serotonin, is a major pathophysiological mechanism. To date, DBS has considered only brain electrical activity — both on the recording and the stimulating aspects. DBS efficacy will likely improve dramatically when alterations in neurotransmitter levels as well as alterations in electrical activity are considered (Wightman et al., 2007). As will be seen later, nanoelectrode arrays can monitor the level of electrochemically active neurotransmitters such as dopamine continuously.

Neurons comprise less than 10% of the human brain. Essential to the interaction of brain neurochemistry with brain electrical activity is a consideration of the cells making up the majority of brain tissue — glia. It is becoming increasingly obvious that glial cells such as astrocytes play a major role in controlling the neurotransmitter environment of the neurons, and that the glia have a substantial effect on neuronal firing patterns (Ni, Malarkey, & Parpura, 2007; Silchenko & Tass, 2008) (Fig. 4).

Sensory neuroprostheses — current status

Sensory neuroprostheses, from a technological standpoint, have developed dramatically over the past 50 years, in contrast to DBS, as noted above. Cochlear implants have incorporated very sophisticated microprocessors plus microelectrode arrays to convert sounds of different frequencies into stimulation of the cochlea at different points — resulting in stimulation of the auditory nerve

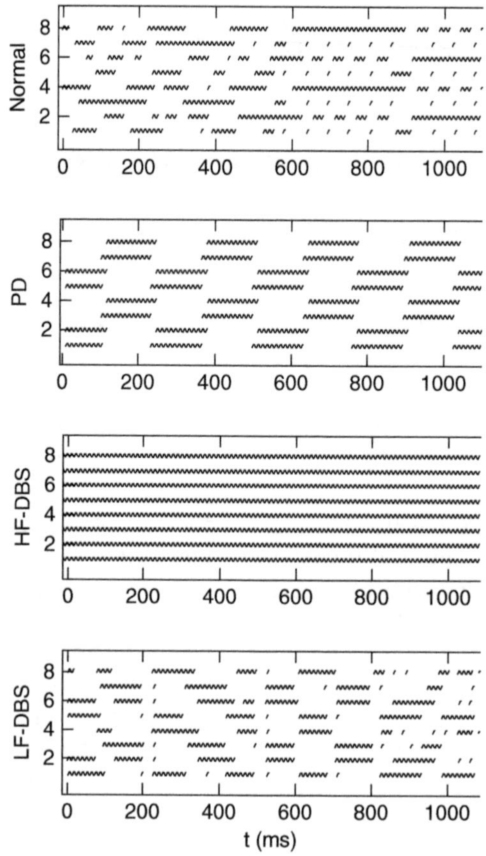

Fig. 3. Computational model of Parkinson's disease (PD): raster plots of the spike times for 8 globus pallidus internus cells. Firing patterns are much more synchronized in PD than in the normal (non-PD) situation. Standard high-frequency DBS (130 Hz; HF-DBS) results in a raster plot entrained to the HF-DBS, whereas low-frequency DBS (10 Hz; LF-DBS) results in a raster plot much more similar to the relatively unsynchronized normal situation. (Reproduced from Feng et al., *Journal of Neural Engineering* (2007) with permission.)

fibers selectively in order to reproduce the sound distinctions necessary for understanding human speech (at a minimum) (Clark, 2006).

Cochlear implants require a functioning auditory nerve and implantable cochlea. For patients lacking these (notably people with neurofibromatosis type II), an auditory brainstem implant (which stimulates the surface of the nucleus) or an auditory midbrain implant (which stimulates the inferior colliculu) are likely to be more effective sites for auditory stimulation than the cochlear nucleus and are available in situations where the cochlear nucleus has been damaged. The auditory midbrain implant consists of a 0.4 mm diameter array with 20 electrodes spaced 0.2 mm apart (Lim et al., 2007) (Fig. 5 – compare with the DBS electrode array in Fig. 1).

Much more elegant still than the auditory prostheses are the evolving retinal prostheses – in large part due to the much greater specificity required for detailed perception in the visual system than in the auditory system.

Retinal prostheses

Retinal prostheses require some function of the ganglion cells in the retina, that is, they can be successful for conditions such as age-related macular degeneration (AMD) and retinitis

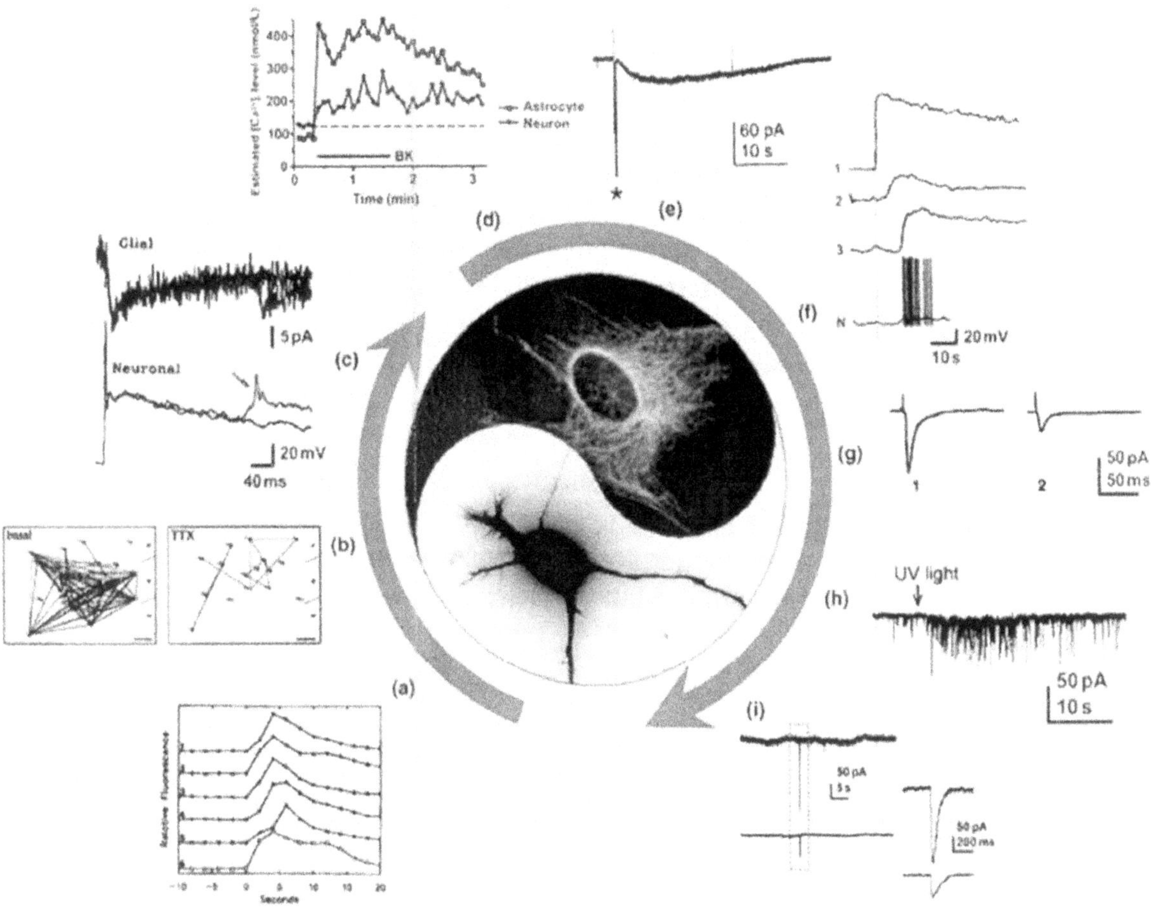

Fig. 4. Summary of glutamate-mediated astrocyte-neuron bidirectional signaling. Details are given in Ni et al., *Journal of Neurochemistry* (2007) Fig. 3. (Reproduced from Ni et al., *Journal of Neurochemistry* (2007) with permission.)

pigmentosa (RP) — which primarily involved the photoreceptors (rods and cones) — but not for conditions where all retinal function is lost. Generally speaking, there are two main categories of retinal prostheses, both of which require an array implanted onto or within the retina (Dowling, 2009).

- Optoelectronic or multiphotodiode prostheses: A multiphotodiode array is implanted, which directly transfers the incoming light to an electrical signal which is communicated by the array to the ganglion cells. Multiphotodiode arrays have not had great success to date because the energy from the incoming light is insufficient to drive the multiphotodiodes presently available.
- Multielectrode array (MEA) prostheses: These devices incorporate an external image capture (camera) and a microsystem that converts the visual signal into electrical stimulation of the retina via the implanted microelectrode array, as illustrated in Fig. 6.
- Hybrid prostheses: This technique uses a camera system which transforms the visual light into the near infrared spectrum which is projected onto the implanted multiphotodiode array. This allows residual vision to be utilized by the retina not involved with the multiphotodiode array (Loudin et al., 2007).

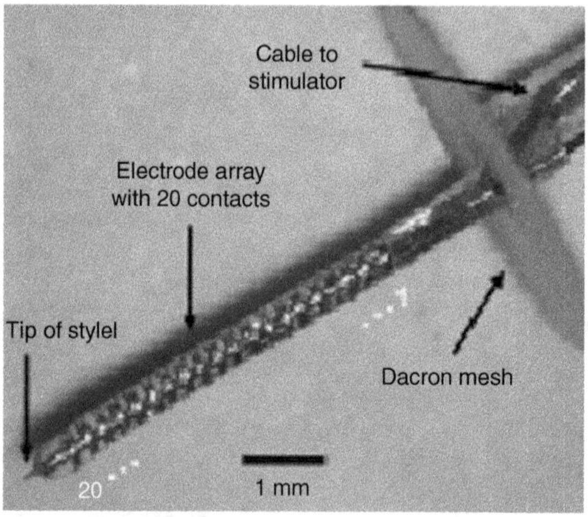

Fig. 5. Auditory midbrain implant which has been placed into the inferior colliculus of patients to improve hearing. (Reproduced from Lim et al., *Journal of Neuroscience* (2007) with permission.)

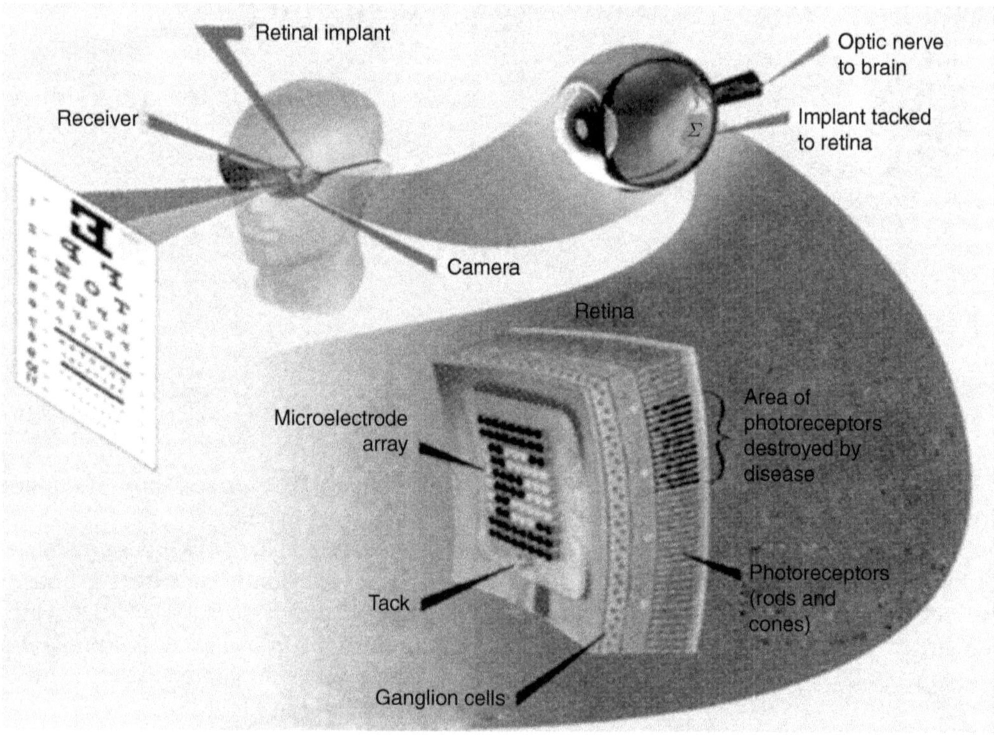

Fig. 6. Illustration of MEA retinal prosthesis. (Reproduced from Medical Design Briefs (2009) with permission.)

MEA prostheses involve extraocular and intraocular components (Fig. 6), the extraocular components being a miniature camera (typically glasses-mounted), plus an image-processing unit and an encoder. The intraocular components are a decoder and a signal generator to drive the MEAs, which in turn stimulate the ganglion cells. Radiofrequency coils outside and inside the body

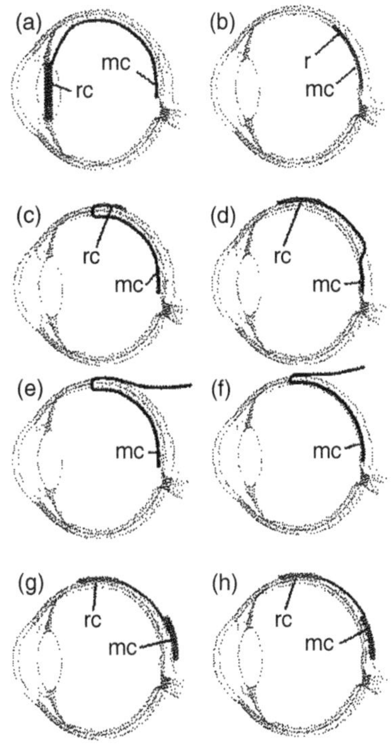

Fig. 7. Different retinal prostheses types by location: (a), (c), (e) epiretinal; (b), (d), (f) subretinal; (g), (h) extraocular. mc: microcontacts; rc: receiver coil; r: receiver. (Reproduced from Gerding, *Journal of Neural Engineering* (2007) with permission.)

transmit data and power, a recent advance being the use of dual band power (1 MHz) and data (20 MHz) telemetry to permit simultaneous power and data transmission (Wang et al., 2006). A major issue in retinal prostheses is where to place the microelectrode array and receiver coil (Fig. 7): an extraocular array is the easiest to implant, followed by an epiretinal array, with the subretinal location being the most challenging surgery. However, effective stimulation of the ganglion cells may prove to be the determining factor with regard to which location is most successful when retinal prostheses reach widespread clinical use.

Quite elegant analytical work has been done to assess the optimal characteristics of retinal prostheses (De Balthasar et al., 2008; Sekirnjak et al., 2008). Sophisticated techniques have been used to measure impedance of the array as well as retinal thickness and distance between the array and the retina (in epiretinal arrays), the thickness and distance measurements were made using optical coherence tomography (De Balthasar et al., 2008).

Sensory neuroprostheses — emerging trends

Retinal prostheses — optimal stimulation site

Micron-level research using electrodes ~10 μm in diameter and spatial analysis have recently shown that electrical activation of cells in the ganglion layer likely occurs in the region of the axon hillock (or summates in that region) (Fig. 8) (Sekirnjak et al., 2008). This has been taken a step further in very recent work which has shown that the region of lowest threshold for initiating an action potential in ganglion layer cells is on the proximal axon, and that this region differs somewhat among the differing cell types within the ganglion cell layer (directionally selective cells, local edge detector cells, etc.). Most important, however, is the finding that the region of lowest threshold for electrical stimulation corresponds quite precisely with the region of high-density sodium-channel bands on the axon (the location of the bands varying somewhat among the various cell types

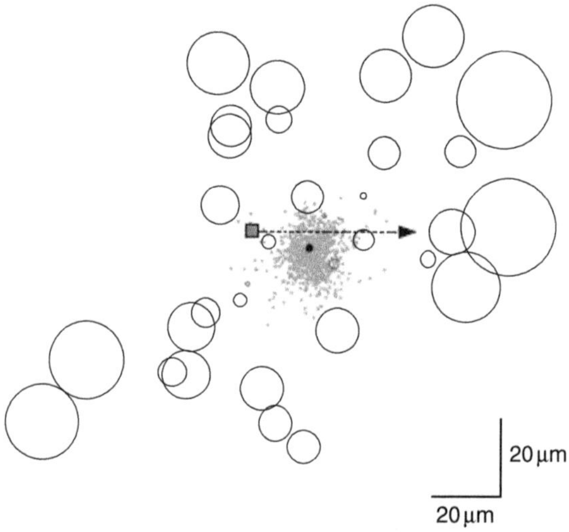

Fig. 8. Primate retinal ganglion cell stimulation with MEA – sensitivity to spatial location. Stimulation sites (circles) plotted relative to the soma center (filled square). Data from all cells aligned are at the soma center and rotated so that the direction of the axon (dotted arrow) points to the right. Circle diameter is proportional to the threshold charge (minus an offset). Lowest thresholds are near the soma and proximal portion of the axon; filled circle is center of Gaussian fit to the data; gray dots denote expected error around this center. Location of maximal sensitivity to electrical stimulation is ~13 μm from the soma center in the direction of the axon. Details are given in Sekirnjak et al., *Journal of Neuroscience* (2008) Fig. 4. (Reproduced from Sekirnjak et al., *Journal of Neuroscience* (2008) with permission.)

within the ganglion cell layer) (Fried, Lasker, Desai, Eddington, & Rizzo, 2009) (Fig. 9).

Retinal prostheses – optimal array

Although traditional "noble metal" microelectrode arrays have typically been used in retinal prostheses, recent work has focused on enhancing the NEI to improve charge transfer from the array to the retina and minimize the risk of injury to the retina. Both the configuration of the array and its coating have been considered (Butterwick et al., 2009). Using the Royal College of Surgeon's (RCS) rat model of retinal degeneration, the effects of various configurations and coatings of subretinal arrays were assessed in terms of (1) amount of fibrosis between the microelectrode array and the retina and (2) the cellular-level integration of the microelectrode array with the retina. The coatings were silicon oxide, iridium oxide, and parylene. The configurations included flat, pillars, and chambers (Fig. 10). Their findings were that (1) silicon oxide evoked a greater fibrotic response than iridium oxide or parylene and (2) apertures <10 μm in diameter precluded interdigitation with the retinal cells (only cell processes entering the chambers). Likely the pillar configuration will be optimal for maximizing the NEI in retinal prostheses (Fig. 11).

Nanotechniques to optimize the NEI

The NASA Ames Nanotechnology group has reported on nanolevel techniques to optimize the NEI (De Asis et al., 2009; Nguyen-Vu et al., 2006, 2007). In brief, it has been found that carbon nanotube (CNT) arrays, coated with the conducting polymer polypyrrole, can markedly decrease impedance and increase capacitance (i.e., improve charge transfer) under *in vitro* settings. These findings have been confirmed recently *in vivo* with rats and monkeys using standard "noble metal" electrodes coated with CNTs (Fig. 12) (Keefer, Botterman, Romero, Rossi, & Gross, 2008).

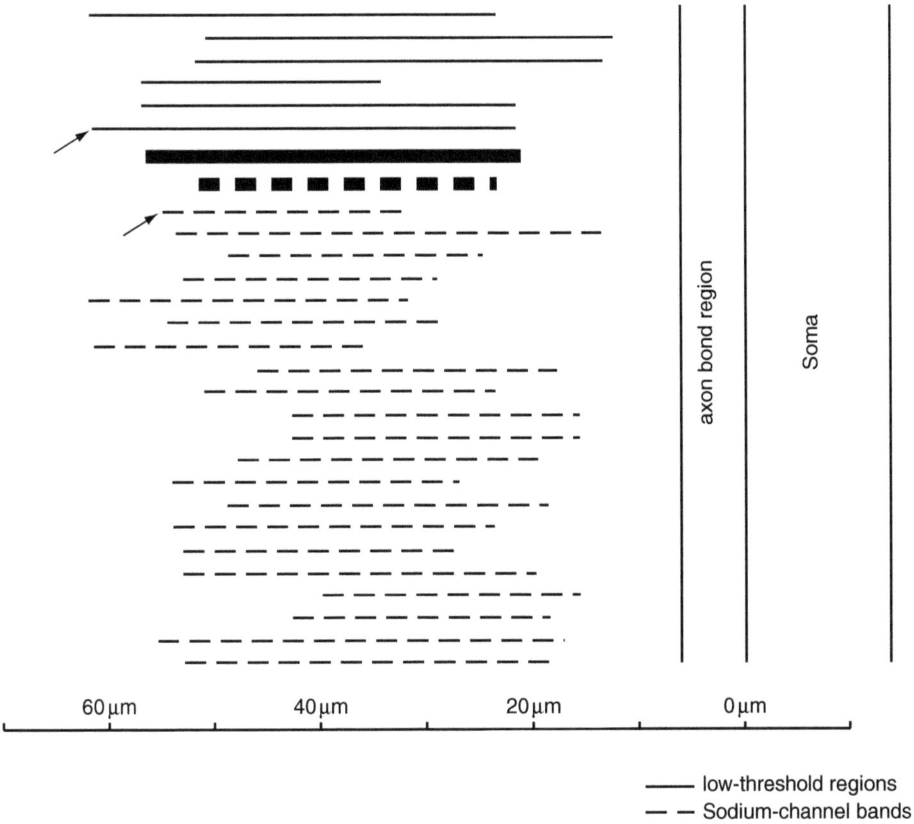

Fig. 9. Retinal low-threshold electrical stimulation bands are coextensive with sodium-channel bands. Thin dashed lines: sodium-channel bands. Thin solid lines: low-threshold stimulation regions. Thick dashed and solid lines: mean location (i.e., ~20 to 50 μm along axon from soma). (Reproduced from Fried et al., *Journal of Neurophysiology* (2009) with permission.)

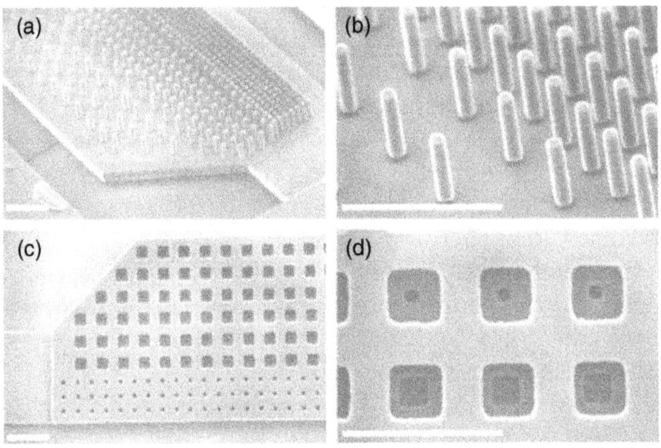

Fig. 10. (a–d) SEM of 3-D retinal implant arrays. (a) Pillar array, with center-to-center distances of 60, 40, and 20 μm. (c) Chamber array, with chamber sizes of 40 and 20 μm, aperture sizes of 20 and 10 μm. (b and d) Magnified views of (a) and (c), respectively. All scale bars are 100 μm. (Reproduced from Butterwick et al., *Experimental Eye Research* (2009) with permission.)

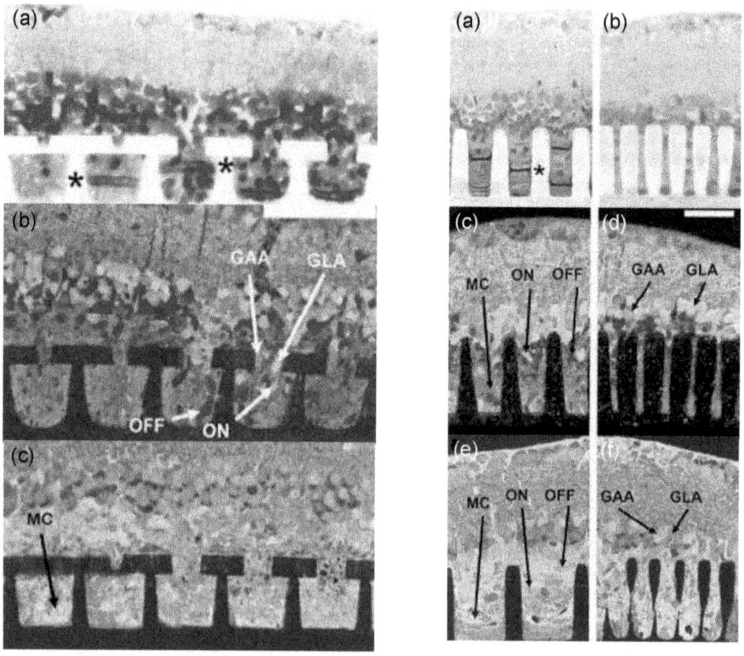

Fig. 11. Chamber arrays (left) and pillar arrays (right) implanted subretinally for 6 weeks in RCS rat. Left: cell bodies migrate through the wider (20 μm) apertures but only processes through narrower (10 μm) apertures. Right: greater penetration of cells with pillar array than chamber array. Cells identified by computational molecular phenotyping (CMP). GAA: GABAergic amacrine cell; GLA: glycinergic amacrine cell; MC: Muller cell; ON: ON-cone bipolar cell; OFF: OFF-cone/rod bipolar cell. Scale bars are 50 μm. Details are given in Butterwick et al., *Experimental Eye Research* (2009) Fig. 6 and 8. (Reproduced from Butterwick et al., *Experimental Eye Research* (2009) with permission.)

Why CNTs improve charge transfer so dramatically is beginning to be understood. CNTs form intimate contact with neuronal cell membranes (Fig. 13) (Cellot et al., 2009). In a hippocampal cell model, it has been found that CNTs are likely to provide a "short circuit" between the dendrite and the soma of the hippocampal neuron, thus enhancing after-depolarization (Fig. 14).

How to capitalize on this improvement in charge transfer in the development of neural prostheses is a major research issue. There be a relationship between the enhanced after-depolarization found in hippocampal neurons on CNT arrays (involving the dendrites and the soma) and the region of low threshold for stimulation of the neuron found in the proximal axon region (coincident with the sodium-channel bands) (Fried et al., 2009).

An additional benefit of CNT arrays is the potential to measure electrochemically active neurotransmitters, notably dopamine, with much greater sensitivity and much faster response time than present carbon fiber microelectrodes using cyclic voltammetry (typically ~500 nM detection level and ~100 msec response time) (Wightman et al., 2007). CNT arrays have been shown to measure dopamine with a detection level of ~50 nM and a response time of ~10 sec (unpublished observations, Jun Li, NASA Ames Nanotechnology Group, 2007).

Sensory neuroprostheses and DBS — cross-fertilization

Improvement in the NEI, crucial to the improvement of neuroprostheses of all types, is dependent on optimization of charge transfer between the neural tissue and the prosthesis. Some of the techniques emerging from both sensory

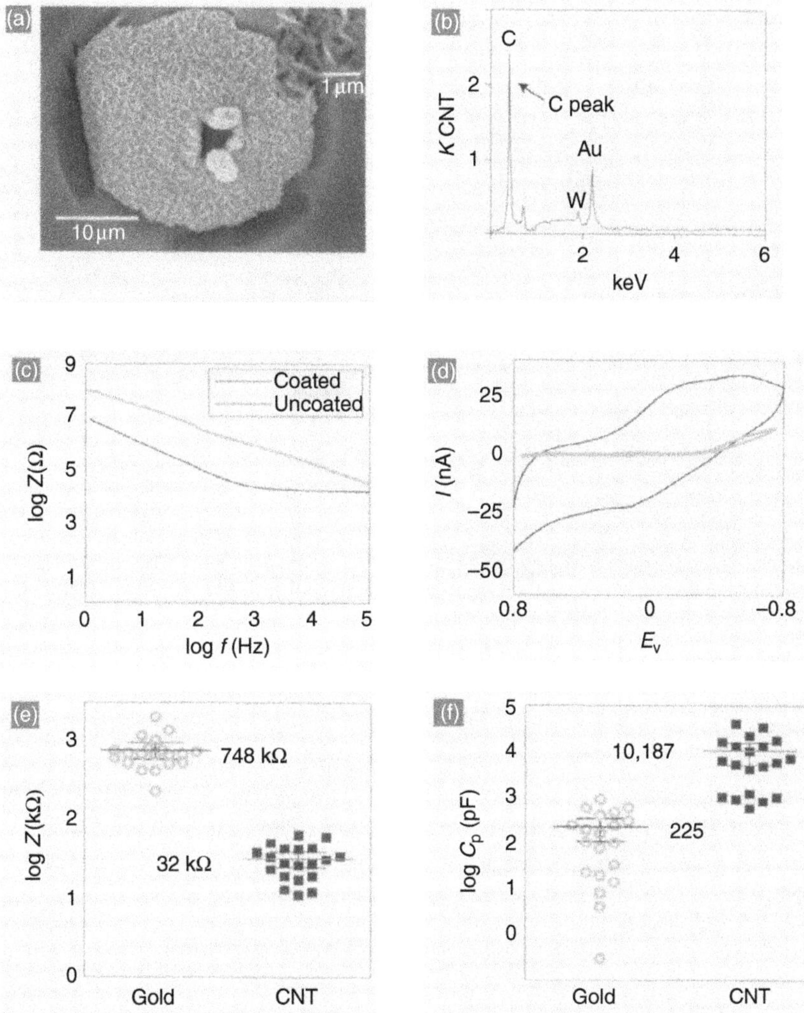

Fig. 12. CNT-coated MEA electrodes. (a) SEM of CNT-coated MEA electrode (magnified inset show porous nature of CNT coating). (b) Energy-dispersive X-ray analysis confirms carbon in the MEA coating. (c) Impedance spectroscopic scan – CNT coating results in decreased impedance at all frequencies. (d) Cyclic voltammetry scan – CNT coating results in increased charge transfer. (e) CNT coating – 23-fold decrease in impedance. (f) CNT coating – 45-fold increase in charge transfer. (Reproduced from Keefer et al., *Nature Nanotechnology* (2008) with permission.)

neuroprostheses research and nanolevel DBS (neuromodulation) research are the following:

- CNTs (either as arrays or as coatings on traditional electrodes) can decrease impedance, increase capacitance, and thus greatly improve charge transfer.
- Conducting polymer coatings (e.g., polypyrrole) contribute to this improved charge transfer.
- 3-D pillar arrays, both at the nanolevel (~10 to 100 nm – which can penetrate the cell if desired) and at the micron level (~20 to 100 µm – which allows the cells to interdigitate with the pillars), are likely to be a useful configuration for neuroprostheses.
- Dual band simultaneous data and power transfer may prove useful in applications where continuous information exchange is

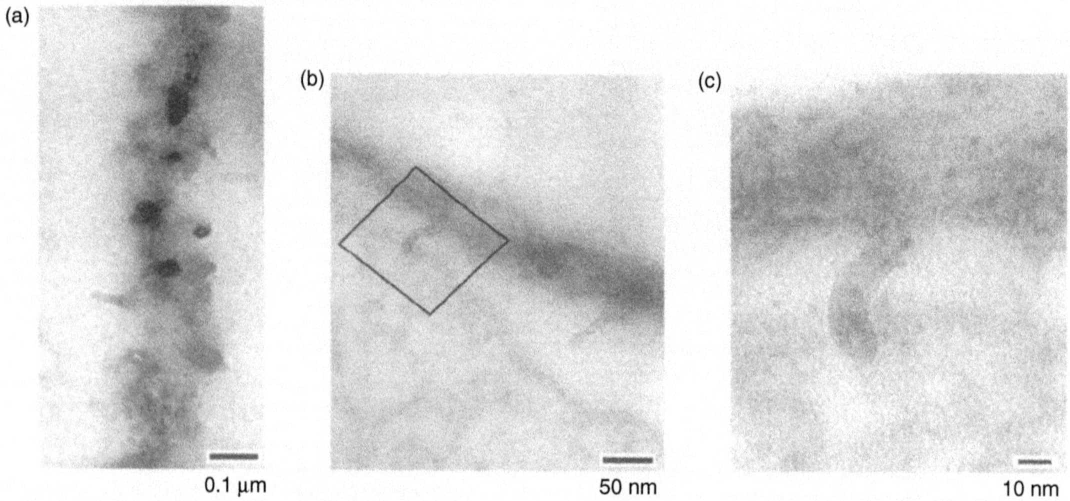

Fig. 13. Ultrastructural interaction between CNTs and neurons (TEM). (a) Morphology of CNTs. (b) CNT – membrane contact. (c) Magnified view of rectangular are in (b). Details are given in Cellot et al., *Nature Nanotechnology* (2009) Fig. 5. (Reproduced from Cellot et al., *Nature Nanotechnology* (2009) with permission.)

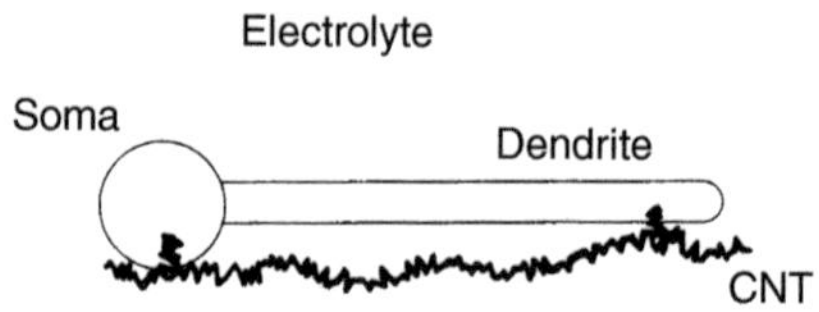

Fig. 14. Electrotonic hypothesis of CNTs and single cell activity. Intracellular compartments are electrically exposed to bundles of substrate CNTs and such contacts might result in an electrical shortcut. Details are given in Cellot et al., *Nature Nanotechnology* (2009) Fig. 4. (Reproduced from Cellot et al., *Nature Nanotechnology* (2009) with permission.)

essential (e.g., retinal prostheses, artificial limbs).

- Basic research on the optimal stimulation techniques and sites (e.g., CNTs and sodium-channel bands, respectively) will benefit from communication amongst researchers in the various neuroprostheses disciplines.

References

Butterwick, A., Huie, P., Jones, B. W., Marmor, M. F., Marc, R. E., & Palanker, D. (2009). Effect of shape and coating of a subretinal prosthesis on its integration with the retina. *Experimental Eye Research, 88*, 22–29.

Cellot, G., Cilia, E., Cipollone, S., Rancic, V., Sucapane, A., Giordani, S., et al. (2009). Carbon nanotubes might improve neuronal performance by favouring electrical shortcuts. *Nature Nanotechnology, 4*, 126–133.

Chaturvedi, A., Butson, C. R., Cooper, S. E., & McIntyre, C. C. (2006). Subthalamic nucleus deep brain stimulation: Accurate axonal threshold prediction with diffusion tensor based electric field models. Proceedings of 28th IEEE. IEEE Engineering in Medicine and Biology Society, 1240–1243.

Clark, G. M. (2006). The multiple-channel cochlear implant: The interface between sound and the central nervous system for hearing, speech, and language in deaf people – a personal perspective. *Philosophical Transactions of the Royal Society of London. Series B, Biological Sciences, 361*, 791–910.

De Asis, E. D., Nguyen-Vu, T. D.B., Arumugam, P. U., Chen, H., Cassell, A. M., Andrews, R. J., et al. (2009). High efficient electrical stimulation of hippocampal slices with vertically aligned carbon nanofiber microbrush array. *Biomedical Microdevices, 11*, 801–808, 2009.

De Balthasar, C., Patel, S., Roy, A., Freda, S., Greenwald, A., Horsager, M., et al. (2008). Factors affecting perceptual thresholds in epiretinal prostheses. *Investigative Ophthalmology and Visual Science, 49,* 2303–2314.

Deuschl, G., Schade-Brittinger, C., Krack, P., Volkmann, J., Schafer, H., Botzel, K., et al. (2006). A randomized trial of deep brain stimulation for Parkinson's disease. *New England Journal of Medicine, 355,* 896–908.

Dowling J. (2009). Current and future prospects for optoelectronic retinal prostheses. Eye, *23,* 1999–2005.

Feng, X. J., Greenwald, B., Rabitz, H., Shea-Brown, E., & Kosut, R. (2007). Toward closed-loop optimization of deep brain stimulation for Parkinson's disease: Concepts and lessons from a computational model. *Journal of Neural Engineering, 4,* L14–L21.

Fried, S. I., Lasker, A. C.W., Desai, N. J., Eddington, D. K., & Rizzo, J. F., 3rd (2009). Axonal sodium-channel bands shape the response to electrical stimulation in retinal ganglion cells. *Journal of Neurophysiology, 101,* 1972–1987.

Gerding, H. (2007). A new approach toward a minimal invasive retina implant. *Journal of Neural Engineering, 4,* S30–S37.

Hauptmann, C., Popovych, O., & Tass, P. A. (2007). Desynchronizing the abnormally synchronized neural activity in the subthalamic nucleus: A modeling study. *Expert Review of Medical Devices, 4,* 633–650.

Hosobuchi, Y., Adams, J. E., Bloom, F. E., & Guillemin, R. (1979). Stimulation of human periaqueductal grey for pain relief increases immunoreactive B-endorphin in ventricular fluid. *Science, 203,* 279–281.

Keefer, E., Botterman, B. R., Romero, M. I., Rossi, A. F., & Gross, G. W. (2008). Carbon nanotube coating improves neuronal recordings. *Nature Nanotechnology, 3,* 434–439.

Krack, P., Batir, A., Van Blercom, N., Chabardes, S., Fraix, V., Ardouin, C., et al. (2003). Five-year follow-up of bilateral stimulation of the subthalamic nucleus in advanced Parkinson's disease. *New England Journal of Medicine, 349,* 1925–1934.

Lim, H., Lenarz, T., Jospeh, G., Battmer, R. D., Anderson, D. J., Samii, A., et al. (2007).Electrical stimulation of the midbrain for hearing restoration: Insight into the functional organization of the human central auditory system. *Journal of Neuroscience, 27,* 13541–13551.

Loudin, J. D., Simanovskii, D. M., Vijayraghavan, K., Sramek, C. K., Butterwick, A. F., Huie, P., et al. (2007).Optoelectronic retinal prosthesis: System design and performance. *Journal of Neural Engineering, 4,* S72–S84.

Mallet, L., Polosan, M., Jaafari, N., Baup, N., Welter, M. L., Fontaine, D., et al. (2008). Subthalamic nucleus stimulation in severe obsessive-compulsive disorder. *New England Journal of Medicine, 359,* 2121–2134.

Mayberg, H. S. (2009). Targeted electrode-based modulation of neural circuits for depression. *Journal of Clinical Investigation, 119,* 717–725.

McIntyre, C. C., Miocinovic, S., & Butson, C. R. (2007). Computational analysis of deep brain stimulation. *Expert Review of Medical Devices, 4,* 615–622.

Nguyen-Vu, T. D.B., Chen, H., Cassell, A. M., Andrews, R., Meyyappan, M., & Li, J. (2006). Vertically aligned carbon nanofiber arrays: An advance toward electrical-neural interfaces. *Small, 2,* 89–94.

Nguyen-Vu, T. D.B., Chen, H., Cassell, A. M., Andrews, R., Meyyappan, M., & Li, J. (2007). Vertically aligned carbon nanofiber architecture as a multifunctional 3-D neural electrical interface. *IEEE Transactions on Biomedical Engineering, 54,* 1121–1128.

Ni, Y., Malarkey, E. B., & Parpura, V. (2007).Vesicular release of glutamate mediates bidirectional signaling between astrocytes and neurons. *Journal of Neurochemistry, 103,* 1273–1284.

Schlaepfer, T. E., & Bewernick, R. H. (2009). Deep brain stimulation for psychiatric disorders – state of the art. *Advances and Technical Standards in Neurosurgery, 34,* 37–57.

Sekirnjak, C., Hottowy, P., Sher, A., Dabrowski, W., Litke, A. M., & Chichilnisky, E. J. (2008). High-resolution electrical stimulation of primate retina for epiretinal implant design. *Journal of Neuroscience, 28,* 4446–4456.

Silchenko, A. N., & Tass, P. A. (2008). Computational modeling of paroxysmal depolarization shifts in neurons induced by glutamate release form astrocytes. *Biological Cybernetics, 98,* 61–74.

Wang, G., Liu, W., Sivaprakasam, M., Zhou, M., Weiland, J. D., & Humayun, M. S. (2006). A dual band wireless power and data telemetry for retinal prosthesis. Proceedings of the 28th IEEE EMBS Annual International Conference, New York, NY.

Wightman, R. M., Heien, M. L.A.V., Wassum, K. M., Sombers, L. A., Aragona, B. J., Khan, A. S., et al. (2007). Dopamine release is heterogeneous within microenvironments of the rat nucleus accumbens. *European Journal of Neuroscience, 26,* 2046–2054.

H.S. Sharma (Ed.)
Progress in Brain Research, Vol. 180
ISSN: 0079-6123

CHAPTER 8

Nanoparticles: transport across the olfactory epithelium and application to the assessment of brain function in health and disease

Michael Aschner*

Department of Pediatrics, Pharmacology and The Kennedy Center for Research on Human Development, Vanderbilt University School of Medicine, Nashville, TN, USA

Abstract: The exciting advances within nanotechnology are beginning to be harnessed by the medical field. Nanoparticles have been used for drug delivery into the brain and have been explored for imaging, sensing, and analytical purposes. The science of nanoparticles encompasses a vast array of biological, chemical, physical, and engineering research, different aspects of which are specifically addressed in each of the chapters of this volume. Nanomaterials such as nanospheres, nanotubes, nanowires, fullerene derivatives (buckyballs), and quantum dots (Qdots) are at the forefront of scientific attention, as they provide new consumer products and advance the scientific development of novel analytical tools in medicine and in the physical sciences. This chapter will briefly survey some aspects of nanoparticle biology focusing on the following: (1) the role of olfactory nanoparticle transport into the central nervous system (CNS), both as a potential route for effective drug delivery and as a route for the passage of noxious substances into the brain proper; (2) nanoparticles as sensors of cell function and toxicity; and (3) some adverse effects of nanoparticles on the dysregulation of brain redox status.

Keywords: nanoparticles; olfactory nanoparticle transport; glia; neurotoxicity

Introduction

Nanoparticles are spherical, polymeric particles composed of natural or artificial polymers. They range in size between 10 and 500 nm. As a consequence of their spherical shape and high surface area to volume ratio, these particles have a wide range of potential applications (Berry & Curtis, 2003). Nanoparticle technology is rapidly advancing, providing novel and effective treatments for various diseases (Emerich & Thanos, 2003), including neurodegenerative diseases, such as Alzheimer's and Parkinson's diseases. Nevertheless, effectively and regionally targeting drugs to the brain remain a challenge due to the restrictive properties of the blood–brain barrier (BBB). This barrier, predominantly formed by endothelial cells that are physically joined by tight junctions in their external membranes, limits the molecular exchange to transcellular transport,

* Corresponding author.
Tel: 615-322-8024;
E-mail: michael.aschner@vanderbilt.edu

DOI: 10.1016/S0079-6123(08)80008-8

thus restricting the passage of molecules across the barrier. The healthy BBB also largely protects the brain from blood-borne nanoparticle exposure; however, a number of pathologies, including hypertension and allergic encephalomyelitis, have been shown to increase BBB permeability to nanoparticles. The likely widespread future applications and impending commercialization of nanoparticles of different composition also pose risks both to humans and to environmental systems. Thus, early evaluations of the health and environmental effects of nanoparticles necessitate careful consideration. These topics are addressed in a number of other chapters of this volume and therefore will not be discussed herein. The focus of this chapter will be complimentary, namely: (1) to consider the role of olfactory nanoparticle transport into the central nervous system (CNS), both as a potential route for effective drug delivery and as a route for the passage of noxious substances into the brain proper; (2) to survey the application of nanoparticles as sensors of cell function and toxicity; and (3) some adverse effects of nanoparticles on the dysregulation of brain redox status.

Nanoparticle transport into the CNS

The olfactory epithelium is a ciliated, pseudostratified columnar type, which contains few or no goblet cells. In the human nasal passage, the surface area of the olfactory epithelium is ~2.5 cm^2. It is composed of three different cell types: the olfactory cell, support cell, and basal cell, of which only the olfactory cell is chemoreceptive. Axons of these quasi-neurons form the olfactory nerve and project to the olfactory lobe. Neurons of the olfactory epithelium have been labeled with horseradish peroxidase (HRP), a retrograde transport marker, providing evidence that these cells, as expected, can transport molecules in a retrograde fashion. Translocation of solid nanosized particles (30-nm poliovirus; 50-nm silver-coated gold colloids) along the olfactory nerve axons was documented decades ago (Bodian & Howe, 1941a,b; De Lorenzo, 1970). Inhaled elemental carbon particles (^{13}C; 35 nm, count median diameter) have also been shown to accumulate in the rat olfactory bulb after whole-body inhalation (Elder et al., 2006; Oberdorster et al., 2004), and intranasal instillation of nanosized carbon black (CB) has been shown to affect brain cytokine and chemokine mRNA expressions and increase cytokine (IL-1β, TNF-α, CCL2, and CCL3) mRNA expression and neurotransmitter levels in the mouse olfactory bulb (Tin Tin Win et al., 2008), suggesting efficient transport of CB into deep brain areas via the olfactory pathway.

Studies in rats have also established the translocation of soluble manganese (Mn) compounds from the nose along olfactory neuronal pathways to the olfactory bulb (Dorman et al., 2004; Henriksson & Tjalve, 2000; Tjalve & Henriksson, 1999; Tjalve, Henriksson, Tallkvist, Larsson, & Lindquist, 1996) upon inhalation or intranasal instillation exposure. Whether these particles can penetrate into deeper brain regions remains controversial. In an early study, Tjalve, Mejare, & Borg-Neczak (1995) demonstrated that soluble ionic Mn instilled into the olfactory chamber of pike was transynaptically transported from the olfactory tract to deeper brain areas, such as the hypothalamus. Elder et al. (2006) have also documented that the olfactory neuronal pathway efficiently translocates inhaled Mn oxide as solid ultrafine particles (UFPs) to deep nuclei of the CNS. Once there, the authors postulated, the particles, including attached contaminants, may cause oxidative stress, ultimately leading to inflammation and cellular dysfunction and damage. However, Dorman et al. (2004) failed to note the transynaptic movement of inhaled Mn from the olfactory bulb to the striatum and cerebellum of rats. Interestingly, welding produces high amounts of fumes containing Mn UFP (Zimmer, Baron, & Biswas, 2002), and evidence exists that welders express high particle count per m^2 of the nasal mucosa. Despite the differences between human and rodent olfactory systems, conflicting results regarding the role of the olfactory pathway in Mn brain deposition and neurotoxicity and the limited information that exists on nanoparticle uptake and translocation necessitate further evaluation to address the relevance of this transport system in humans.

Instilled titanium dioxide (TiO_2) nanoparticles have also been shown to exert biological effects in rats, likely reflecting their passage into the brain proper via the olfactory pathway. In a recent study, Wang et al. (2007) measured monoaminergic neurotransmitters after exposure to TiO_2. Ten days after intranasal instillation of nanoparticles in mice, brain titanium concentrations were increased, concomitant with increased norepinephrine (NE) and 5-hydroxytryptamine (5-HT) brain concentrations. In contrast, dopamine (DA), 5-hydroxyindole acetic acid (5-HIAA), 3, 4-dihydroxyphenylacetic acid (DOPAC), and homovanillic (HVA) concentrations were decreased in TiO_2-treated mice.

The olfactory pathway has also been effectively used for the delivery of drugs. For example, wheat germ agglutinin (WGA) modification on the surface of poly(ethylene glycol)-poly(lactic acid) (PEG-PLA) nanoparticles facilitated the uptake of an encapsulated fluorescence tracer in the CNS following intranasal administration, as well as the uptake of vasoactive intestinal peptide, a neuroprotective peptide, by the brain (Gao et al., 2007a). Similarly, encapsulated nimodipine (NM) within methoxy poly(ethylene glycol)-poly(lactic acid) (MPEG-PLA) nanoparticles that was intranasally administered to rats improved the efficacy of the direct nose–brain transport for drugs (Zhang et al., 2006). Furthermore, ulex europeus agglutinin I (UEA I) conjugated to PEG-PLA nanoparticles was found to increase the absorption of a fluorescent marker into the brain following intranasal administration (Gao et al., 2007b). Previous studies have also documented that PLA-PEG nanoparticles enhanced the transport of the encapsulated model protein, tetanus toxoid (TT), across the rat nasal mucosa. The transport of TT, encapsulated by the particles across the nasal mucosa of rats, was significantly enhanced, and its efficacy was related to the particle size (Vila et al., 2005). Taken together, these examples establish the nasal pathway as a tissue of considerable interest, as it is effective — at least under certain conditions — in the delivery of nanoparticles into the brain, with potential for improved drug delivery, yet concern over translocation of toxic xenobiotics as specifically exemplified with CB TiO_2 and Mn.

Nanoparticles as sensors of brain cell function and toxicity

Nanoparticles have also recently been used to monitor brain cell function. Optical nanosensors, or PEBBLEs (probes encapsulated by biologically localized embedding), have been produced for intracellular measurements of pH and calcium. PEBBLEs are fluorescent dyes that are encased in a molecular shell, protecting cells from the dye and protecting the dye from cellular degradation or manipulation. Five varieties of pH-sensitive sensors and three different calcium-selective sensors were discussed by Clark, Kopelman, Tjalkens, & Philbert (1999), with each sensor combining an ion-selective fluorescent indicator and an ion-insensitive internal standard entrapped within an acrylamide polymeric matrix. These PEBBLE sensors were effectively used for measurements of calcium transient in the cytoplasm of neural cells during the mitochondrial permeability transition, as well as for quantitative discrimination of subtle differences between the ability of human SY5Y neuroblastoma and C6 glioma to respond to a challenge from 3-dinitrobenzene (DNB), a toxin that selectively induces symmetrical gliovascular brainstem lesions reminiscent of the pathologies associated with idiopathic and chemically induced acute energy deprivation syndromes. These studies have confirmed changes in intracellular calcium transient, the precursor to cell death, in response to DNB.

In addition, optical PEBBLE nanosensors (120 nm in size) have been developed for measurements of dissolved oxygen using organically modified silicate (ormosil) nanoparticles as a matrix (Koo et al., 2004). These sensors incorporate the oxygen-sensitive platinum porphyrin dye as an indicator and an oxygen-insensitive dye as a reference for ratiometric intensity measurement. Gene gun injected PEBBLEs have been used in rat C6 glioma cells to monitor intracellular changes of dissolved oxygen (Koo et al., 2004) and have been found to be highly sensitive, responding linearly to changes in the intracellular dissolved oxygen concentration. Future applications of PEBBLEs may include measurements of oxygen or nitrogen radicals in the extracellular space and at the subcellular level, protein movement, oxidation or reduction reactions, and the production of superoxide ions within the cell. There are also design prototypes for measuring metals, such as lead and mercury (Hg) (see below), as well as phosphorylation reactions. Light emitted from PEBBLEs introduced into the volume of a cell can be viewed using confocal microscopy. The intensity of the light emitted by the dot indicates the concentration of oxygen at that particular point in the cell, thus allowing the tracking of oxygen movement within the cells. Nanoparticles can also be used to track the movement of particular cellular compartments, such as endosomes.

Since calcium almost instantly flows into neurons when they fire, and the faster the rate of firing, the higher the level of calcium, measurements of calcium transient provide a direct measure of neuronal activity. Thus, tracking calcium levels in the brain actually tracks information flow through the brain's circuitry. In order to be detected by MRI, a calcium contrast agent must include a magnetically active "paramagnetic" component. The combination of an iron oxide nanoparticle-based contrast mechanism with the versatile calcium-sensing protein, calmodulin and its targets, has allowed for calcium imaging by means of magnetic resonance imaging (MRI) with calcium-dependent responses being assayed by dynamic light scattering and MRI (Atanasijevic & Jasanoff, 2007; Atanasijevic, Shusteff, Fam, & Jasanoff, 2006). These MRI images reveal a significant contrast enhancement due to the presence of the nanoparticles. The probe is based on calcium-dependent protein–protein interactions that drive particle clustering to produce changes in T2 relaxivity. When delivered to live tissues, these particles effectively provide functional molecular imaging of biological signaling networks. Other contrast agents sensitive to aspects of brain activity have also recently been introduced, and they are reviewed by Jasanoff (2007). In addition to intracellular calcium measurements, nanoparticle-based imaging agents have also been developed for high-resolution pH determination in living animals. Both relaxation and chemical exchange saturation transfer (CEST)-based MRI contrast agents work by mechanisms that involve water or proton exchange, which are inherently pH dependent and therefore easily compatible with pH sensing (Jasanoff, 2007). These have been tested; however, it has yet to be documented whether they are able to detect changes in neuronal activity. The same group is currently developing noninvasive methods to deliver the calcium sensor to brain cells *in vivo* and are also modifying the nanoparticles so that they can target specific genetic characteristics or different populations of neurons, such as inhibitor neurons or those that produce neurotransmitters like DA or serotonin, likely permitting, in the near future, fine tuned analysis that is based on information flow involving cells and circuits, with numerous potential applications for studying learning, memory, and behavior.

Gold nanoparticles (orange and purple spheres), functionalized with one of two complementary DNA sequences (black and green curved lines, bottom) that do not form stable nanoparticle-based technologies, are also available for the detection of Hg concentrations. When gold nanoparticles are functionalized with one of two complementary DNA sequences, they fail to form stable complexes in the absence of Hg ions due to a single thymidine group mismatch (Lee, Han, & Mirkin, 2007). Hg ions bind the two types of DNA sequences at the point of the T-T mismatch and cause the otherwise red nanoparticle solutions to appear purple. The detection limit of the technique for Hg^{2+} ions in aqueous samples is at concentrations as low as 100 nM (20 ppb). In principle, this new method could also be used to detect other metal ions by substituting thymidine with artificial nucleobases that selectively bind other metal ions. While yet to be tested *in vivo*, in the future, this methodology will likely allow for the monitoring of the heavy metals in various settings (fish), including intracellular brain Hg distribution.

Electric fields (E fields) surround living cells and are critical for the optimal functioning of biological processes. These E fields are associated with neural signals, and they are also found in volumetrically small components of cells, such as the mitochondria. Generally, changes in E fields correlate with perturbations in biological function, and they are integral to neurodegenerative disorders such as Alzheimer's disease. Reduction of the E field is associated with changes in neuronal membrane potential (Blanchard, Thomas, & Ingram, 2002), and loss of the E field is associated with the reduction in mitochondrial membrane potential (Kim, He, & Lemasters, 2003). Tyner, Kopelman, and Philbert (2007) describe a 30-nm "photonic voltmeter," 1000-fold smaller than existing voltmeters, which can generate three-dimensional E field profiling throughout the entire volume of living cells. These new nanosensors greatly enhance the potential for integrating real-time measurements of intracellular/extracellular E fields with investigations of voltage-dependent cellular processes that are not immediately proximal to polarized membranes. Furthermore, better understanding of the E fields will permit integration with chemical and physical imaging of the live cell, resulting in real-time, high-dimensional images of cells, including multiple physical (space, fields, temperature, and viscosity) and chemical (ions, molecules, and radicals) dimensions (Tyner et al., 2007). Improved understanding of the role of cellular E fields in influencing and/or regulating biological processes (cellular biology, biophysics, and biochemistry) will also allow for pertinent studies on subtle changes in E fields as biomarkers of toxicity.

Neurotrauma, brain ischemia, and neurodegenerative diseases are characterized by reactive astrocytes (commonly referred to as reactive gliosis). Cellular hallmarks of reactive gliosis include the hypertrophy of astrocytic processes and the upregulation of cytoskeletal proteins (intermediate filaments), which are composed of nestin, vimentin, and glial fibrillary acidic protein (GFAP). Accordingly, GFAP expression can be effectively used to probe for gliosis and astrocytic activation in pathological conditions. Liu et al. (2008) have recently utilized a magnetic resonance (MR) probe that contains superparamagnetic iron oxide nanoparticles (SPIONs, a T2 susceptibility contrast agent) linked to a short DNA sequence complementary to the cerebral mRNA of GFAP found in glia and astrocytes. SPION-gfap was delivered to the CNS after BBB leakage introduced by a puncture wound, global cerebral ischemia, and cortical spreading depression in C57BL6 mice. $T2^*$ MR images and $R2^*$ ($R2^*=1/T2^*$) maps using a transcription MRI technique in live mice elegantly established that the SPION-gfap probe reported foci with an elevated signal in subtraction $R2^*$ maps and that these foci matched areas identified as having extensive glial network (gliosis) in postmortem immunohistochemistry (Liu et al., 2008). Magnetic sensor design with magnetic particle and assay development, specifically tuned toward the detection of the predominantly astrocytic protein S100β, a marker for brain damage, is also currently underway (http://www.imec.be/wwwinter/mediacenter/en/SR2006/681420.html). The resolution of 9 L gliosarcoma in the rat brain has also been shown to be greatly enhanced by nanoparticles (The National Academics, 2005). When ruthenium (Sol Gel Ru-DNPs) is included within the nanoparticle, Sol Gel Ru-DNPs can produce a large amount of singlet oxygen upon excitation by laser with exquisite resolution of the tumor's periphery due to enhanced contrast. The ability of these probes to reflect gliosis and to delineate tumors in living animals represents a seminal development in providing a real-time technique for surveying CNS injury (Kopelman et al., 2005). The utility of these methods for therapeutically targeting tissues is beyond the scope of this chapter.

Similarly, brain inflammation is dominated by cells of the innate immune system, with resident microglia/brain macrophages and blood-derived monocytes/macrophages being the most important cell types involved. MRI with iron oxide nanoparticles has been used to track macrophages in multiple sclerosis, ischemic stroke lesions, and tumors (Petry, Boiziau, Dousset, & Brochet, 2007). Iron oxide nanoparticles, such as ultrasmall superparamagnetic iron oxide (USPIO), are taken up by circulating phagocytic cells, providing characteristic signal changes in MRI of the infarcted brain parenchyma (Jander, Schroeter, & Saleh, 2007). Thus, MRI of USPIO provides a tool for probing the role of macrophages in the etiology of brain disorders associated with inflammation. Characterization of these events will assist in monitoring disease progression, defining precise time intervals for therapeutic interventions and facilitating the development of new therapeutic strategies.

Nanoparticles have also been used in the assessment of remyelination. Demyelination is a common pathological finding in many neurological disorders and is often irreversible due to absence or failed endogenous repair. Bulte et al. (1999) and Frank et al. (2003) have exploited the transferrin receptor as an efficient intracellular delivery device for magnetic nanoparticles and have transplanted tagged oligodendrocyte progenitor cells into the spinal cord of myelin-deficient rats. Both groups elegantly demonstrated that cell migration could be detected by using three-dimensional MR microscopy. Furthermore, a close correlation between the areas of contrast enhancement and the achieved extent of myelination were noted. Accordingly, MR tracking of transplanted magnetic nanoparticle-tagged oligodendrocyte progenitors offers a method that could be extended to other transplantation studies where desired outcomes may be followed with conventional imaging technologies.

Another method involving metal nano-optical sensors is based on localized surface plasmon resonance (LSPR) spectroscopy, which relies on the LSPR spectral shifts caused by the surrounding dielectric environmental change in a binding event (Zhao, Zhang, Yonzon, Haes, & Van Duyne, 2006). Specifically, LSPR detects proteins and other biomolecules based on the excitation of a surface plasmon at the interface between a dielectric and a thin layer of metal, typically gold. The applicability of the system for the detection of cellular metabolites was recently reported (Endo, Yamamura, Kerman, & Tamiya, 2008; Nazem & Mansoori, 2008). Specifically, LSPR detection has been utilized as a biomarker of Alzheimer's disease and amyloid-β-derived diffusible ligands (ADDL) in human brain extract and cerebrospinal fluid samples. LSPR spectroscopy was highly informative in establishing the interaction between the antigen, ADDL, and specific anti-ADDL antibodies. The nanosensor also provided quantitative binding information for both the antigen and the second antibody detection which permitted the determination of the ADDL concentration (Haes, Chang, Klein, & Van Duyne, 2005). Accordingly, LSPR will prove useful for enhancing the understanding of the aggregation mechanisms of amyloid-β. Finally, the ability of LSPR to perform analysis of human brain extract and cerebrospinal fluid samples from control and Alzheimer's disease patients holds great promise in its potential for the diagnosis of Alzheimer's disease.

Oxidative stress appears to play a major etiological role in initiating and promoting the neurodegeneration of Alzheimer's disease. Concomitantly, this disease is characterized by abnormally high concentrations of redox-active metals, particularly iron, in affected brain areas. While the specific role of iron in the etiology of this disease has yet to be fully clarified, the conjugation of iron chelators with nanoparticles has been posited to create limitations in delivering iron chelators in the traditional way, thereby necessitating transport across the BBB. In recent *in vitro* studies, Liu et al. (2006) have shown that a chelator-nanoparticle system and the chelator-nanoparticle system complexed with iron, when incubated with human plasma, preferentially adsorb apolipoprotein E and apolipoprotein A-I, indicative of the potential of the system to facilitate transport out of the brain via mechanisms that transport low-density lipoprotein.

Glutamate is the main excitatory neurotransmitter in the mammalian brain; however, excessive intrasynaptic levels can trigger neurotoxicity. Overstimulation of glutamate receptors by high extracellular levels of glutamate leads to cell death via a mechanism referred to as excitotoxicity. Extracellular levels of glutamate are strictly controlled by a high-affinity sodium-dependent uptake system (system X_{AG}^{-}), which is predominantly localized on astrocytes. Alterations in glutamate transport have been implicated in many neurodegenerative insults, both acute, like ischemia, and chronic, like amyotrophic lateral sclerosis. Accordingly, methods that can survey glutamate CNS levels produce considerable benefits. Genetically encoded nanosensors have recently been used to monitor glutamate levels both inside and at the surface of living cells

(Okumoto et al., 2005). The fluorescent indicator protein for glutamate (FLIPE) consisted of the glutamate/aspartate binding protein, ybeJ, from *Escherichia coli* fused to two variants of the green fluorescent protein (GFP). In the presence of the ligand, FLIPE displayed a concentration-dependent decrease in fluorescence resonance energy transfer (FRET) efficiency. When expressed on the surface of rat hippocampal neurons or PC12 cells, the sensors responded to extracellular glutamate with a reversible concentration-dependent decrease in FRET efficiency. Depolarization of neurons was associated with a reduction in FRET efficiency, and no change was evident when cells expressing sensors in the cytosol were superfused with up to 20 mM of glutamate, consistent with a minimal contribution of glutamate uptake to cytosolic glutamate levels (Okumoto et al., 2005). These results establish that FLIPE sensors can be effectively used for real-time monitoring of glutamate metabolism in living cells and in tissues, providing the necessary tools for studying the metabolism of glutamate in drug discovery, as well as consequences of toxic injury to brain tissue.

γ-aminobutyric acid (GABA) is the main inhibitory neurotransmitter, and it plays a major role in neural function and neurodegenerative disorders, such as Huntington's and Parkinson's diseases, seizures, myoclonic discharges, and alcohol addiction. Existing methods for extracellular GABA determination are based on microdialysis followed by liquid chromatography combined with pre/postcolumn derivatization. Liquid chromatography-based measurements are dyscontinuous, and microdialysis itself is an invasive procedure, causing neuronal death and reactive gliosis. Furthermore, microdialysis has very poor spatiotemporal resolution. In addition, direct electrochemical monitoring of extracellular GABA in the brain is very difficult because a specific active enzyme capable of generating a redox species does not exist for GABA. Finally, GABA is insensitive to ultraviolet (UV)-visible spectroscopic determination methods. Accurate, frequent on-line assessment of changes in GABA's extracellular content offers a unique opportunity to better understand its functions. Antibody-linked biocompatible nanomaterials immobilized onto transducers to produce high-sensitivity biosensors for real-time monitoring of GABA concentrations have recently been designed and tested (http://www.inframat.com/biosens2.htm). The development of this selective, sensitive, and rapidly responding GABA electrochemical immunomicroprobe will make it possible to measure dynamic events associated with GABA neurotransmission and will offer new insights into the role of this neurotransmitter in health and disease.

Voltammetric sensors that are based on gold nanoparticles have also been developed for the analysis of biological molecules such as cysteine (Li, Duan, Liu, & Du, 2006) and NE (Lu, Wang, & Lin, 2004). Gold nanoparticle-distributed, poly-(4-aminothiophenol)-modified glassy carbon electrodes have also been successfully utilized for the determination of DA (Gopalan et al., 2006), and a nano-Au self-assembly gold electrode has been used for determination of epinephrine (Hu, Zhang, Wu, & Yang, 2008; Wang, Bai, Huang, & Wang, 2006). A nano-encapsulated, self-referenced fluorescent nitric oxide sensor for wide-field optical imaging has also been designed and evaluated (Zhang, Shu, & Robinson, 2007).

Nanometer-scale crystals composed of semiconductor materials, such as cadmium selenide (CdSe) or cadmium telluride (CdTe), commonly called quantum dots (Qdots), have been used to monitor mitochondrial function. The interested reader is referred to an elegant review by Foster, Galeffi, Gerich, Turner, and Muller (2006), which details Qdots' advantages and pitfalls. The applicability of Qdots will be briefly reviewed herein. In general, Qdots are characterized by large extinction coefficients, very broad excitation spectra, narrow emission peaks, intense fluorescence emission, and exceptional photostability, giving them an edge over commonly used dye molecules (Foster et al., 2006). Of utmost importance is the fact that the high quantum yield of Qdots is not attenuated upon conjugation to biomolecules, which often occurs with common fluorescence dyes. A critical limitation of Qdots is their poor permeability across membranes. Furthermore, Qdots are often taken up by endocytosis, resulting in vesicular or lysosomal sequestration rather than cytoplasmic distribution (Foster et al., 2006). Long-term studies in intact tissue or *in vivo* also suggest that Qdots may have potential neurotoxicity, further limiting their applicability. The neurotoxicity is characterized by release of Cd^{2+} under oxidative conditions (Derfus, Chan, & Bhatia, 2004a,b). Another concern has been the potential of Qdots to generate reactive oxygen species (ROS) (Lovric et al., 2005a; Lovric, Cho, Winnik, & Maysinger, 2005b).

Finally, a critical feature of recovery from toxic injury is associated with neurogenesis, a highly dynamic process that is also inherent to the adult brain and is characterized by the existence of neural stem cells with the potential to produce new neurons. Accordingly, tools that allow the assessment of neurogenesis are deemed critical in evaluating recovery from brain injury. Couillard-Despres et al. (2008) have recently described a new tool permitting *in vivo* imaging of neurogenesis. It is based on a doublecortin promoter driving the expression of the luciferase reporter gene (DCX-promo-luciferase). *In vivo* optical imaging in mice proved profitable for detecting the onset of and increases in neurogenesis both in developing fetal brains and in intact adult mice. The DCX-promo-luciferase transgenic tool was also effective in detecting neuronal migration after transplantation, thus offering advantages over common methods, such as bromodeoxyuridine incorporation or labeling with iron oxide nanoparticles. Sykova and Jendelova (2007) have utilized T2-weighted MR images of superparamagnetic iron oxide (SPIO) nanoparticles. Specifically, these authors have followed, in rats, the fate of embryonic stem cells (ESCs) and bone marrow mesenchymal stem cells (MSCs) labeled with iron oxide nanoparticles (Endorem). Additional studies followed CD34+ cells labeled with magnetic MicroBeads (Miltenyi) with a cortical or spinal cord lesion, models of stroke and spinal cord injury (SCI), respectively. Grafted MSC and ESC migrated during the first week to the lesion site in the cortex and in the spinal cord and were visible in the lesion on MR images as a hypointensive signal, persisting for more than 30 days. In rats with an SCI, the authors reported an increase in functional recovery after the implantation of MSC, concomitant with an increase in white matter volume in lesioned brain areas in cell-treated animals. These studies establish the utility of MRI in surveying grafted adult cells as well as ESC labeled with iron oxide nanoparticles for cellular migration in a brain injury model. Combined, the studies by Couillard-Despres et al. (2008), Sykova & Jendelova (2006, 2007), and others (Dousset, Tourdias, Brochet, Boiziau, & Petry, 2008) establish the means for real-time monitoring of neurogenesis/remyelination in intact animals without the requirement of cellular prelabeling. These methods hold tremendous promise for significantly enhancing the understanding of neurogenesis and neurodevelopment in relationship to candidate compounds for regenerative potential and for considerably improving the understanding of disease pathophysiology and the specific biological processes inherent to neuroregeneration. As new developments are optimized for choices of appropriate cellular markers to achieve the required imaging while minimizing the side effects, novel contrast agent design for rational approaches in imaging will improve (Dousset et al., 2008).

Some adverse effects of nanoparticles on the dysregulation of brain redox status

Due to their small size, reactive properties, and high surface area to material amount/weight, nanoparticles can adversely affect human health. Their properties, namely small size and large surface area, favor their entry into cells and tissues, where they may persist for long duration. Several studies have addressed the toxicity of nanoparticles in tissue culture models. For example, Hussain et al. (2006) studied the effects of Mn-40 nm and Ag-15 nm particles in PC12 cells, demonstrating that submicromolar concentrations of these compounds efficiently induced changes in DA homeostasis and generated significant increase in intracellular ROS. Increase in brain iron levels has also been also reported, possibly due to the release from peripheral degradation of nanoparticles and subsequent delivery to the CNS by transferrin. Iron can precipitate oxidative stress via the Fenton reaction. Carbon nanotubules have also been shown to exert neurotoxicity. The toxicity of fullerenes is mediated by lipid peroxidation. A recent study of largemouth bass exposure to pure unmodified fullerenes showed that C60 fullerenes localized to the brain and led to an increase in lipid peroxidation, as well as a decrease of glutathione in the gill (Oberdorster, 2004). Work from our laboratory (Au et al., 2007) assessed the effects of iron oxide nanoparticles on plating patterns and the cytotoxic effects of nanoparticles on cultured neonatal rat astrocytes. Membrane integrity and mitochondrial function were measured using colorimetric analysis lactate dehydrogenase (LDH) and 3-(4, 5-dimethylthiazol-2-yl)-2,5-diphenyltetrazolium bromide (MTS), respectively. Treatment with nanoparticles did not significantly alter astrocytic LDH release ($p > 0.05$) in the control group ($100\% \pm 1.56$) *versus* the group receiving treatment ($97.18\% \pm 2.03$), yet, a significant increase in MTS activity ($p < 0.05$) between the control ($100\% \pm 3.65$) and treated groups ($112.8\% \pm 3.23$) was observed. These studies are consistent with astrocytic mitochondrial uncoupling by nanoparticles. Furthermore, the nanoparticles were shown to impede the attachment of astrocytes to the substratum (culture plates), also consistent with a cytotoxic effect.

Allergic encephalomyelitis has been associated with increased permeability of the BBB to nanoparticles in experimental set ups and nanoparticle surface charges have also been noted to alter BBB integrity (Lockman, Koziara, Mumper, & Allen, 2004). Inhalation exposure of BALB/c mice to particulate matter showed activation of proinflammatory cytokines in the brain of exposed mice (Campbell et al., 2005); however, in general there is a paucity of data on the toxic effects of these particles in the CNS, despite the fact that have already found applications into MRI methods, where imaging of different cell types has proved useful experimentally (Jendelova et al., 2004), and it has been suggested that it might be useful to track the development of stem cell grafts used to treat neurodegenerative diseases. The reader is referred to an elegant article by Borm et al. (2006) on health risks associated with exposures to nanomaterials.

Summary

As described in this chapter, nanoparticles may gain access into the CNS via the nasal pathway. This route may offer advantages over common techniques associated with drug delivery that necessitate breaching the restrictive properties of the BBB. However, as described in this chapter, the consequences of efficient olfactory pathway transport of inhaled environmental metals (described for Mn and TiO_2) and other xenobiotics as solid UFPs (CB) will also require further experimental scrutiny to determine the contribution of this pathway to neurodegenerative diseases.

The implementation of successful applications for biological and chemical analysis of nanoscale optical biosensors when combined with specific and novel biological materials lend promise for the development of novel and successful nanotechnology-based devices, which will offer highly sensitive and selective (give low false positives, low false negatives) opportunities for sampling the CNS microenvironment and will facilitate enhanced understanding of the biology of the CNS processes associated with neurodegeneration and neuroregeneration. Imaging techniques to visualize stem cells for monitoring, control, and treatment of biological systems, in particular, neurodegenerative disorders, are at the forefront of current investigations and promise to generate new therapeutic modalities. In addition, novel and innovative nanoparticle technologies will likely lead to the development of high throughput systems for biomarkers associated with disease diagnosis, as already exemplified by amyloid-β measurements in Alzheimer's disease. Undoubtedly, many of these diagnostic agents will also multitask as carriers of therapeutic molecules. While the potential utility of these probes in biology is limitless, it is incumbent upon the scientific community to assess their potential risks to optimal nervous system function, as well as their neurotoxic characteristics.

Acknowledgments

This review was partially supported by grants from NIEHS 10563 and DoD W81XWH-05-1-0239 (MA).

References

Atanasijevic, T., & Jasanoff, A. (2007). Preparation of iron oxide-based calcium sensors for MRI. *Nat Protocols*, *2*, 2582–2589.

Atanasijevic, T., Shusteff, M., Fam, P., & Jasanoff, A. (2006). Calcium-sensitive MRI contrast agents based on superparamagnetic iron oxide nanoparticles and calmodulin. *Proceedings of the National Academy of Sciences of the United States of America*, *103*, 14707–14712.

Au, C., Mutkus, L., Dobson, A., Riffle, J., Lalli, J., & Aschner, M. (2007). Effects of nanoparticles on the adhesion and cell viability on astrocytes. *Biological Trace Element Research*, *120*, 248–256.

Berry, C. C., & Curtis, A. S.G. (2003). Functionalisation of magnetic nanoparticles for applications in biomedicine. *Journal of Physics D: Applied Physics*, *36*, R198–R206.

Blanchard, B. J., Thomas, V. L., & Ingram, V. M. (2002). Mechanism of membrane depolarization caused by the Alzheimer Abeta1-42 peptide. *Biochemical and Biophysical Research Communications*, *293*, 1197–1203.

Bodian, D., & Howe, H. A. (1941a). Experimental studies on intraneural spread of poliomyelitis virus. *Bulletin of the Johns Hopkins Hospital*, *68*, 248–267.

Bodian, D., & Howe, H. A. (1941b). The rate of progression of poliomyelitis virus in nerves. *Bulletin of the Johns Hopkins Hospital*, *69*, 79–85.

Borm, P. J., Robbins, D., Haubold, S., Kuhlbusch, T., Fissan, H., Donaldson, K., et al. (2006). The potential risks of nanomaterials: A review carried out for ECETOC. *Particle and Fibre Toxicology*, *3*, 11.

Bulte, J. W., Zhang, S., van Gelderen, P., Herynek, V., Jordan, E. K., Duncan, I. D., et al. (1999). Neurotransplantation of magnetically labeled oligodendrocyte progenitors: Magnetic resonance tracking of cell migration and myelination. *Proceedings of the National Academy of Sciences of the United States of America*, *96*, 15256–15261.

Campbell, A., Oldham, M., Becaria, A., Bondy, S. C., Meacher, D., Sioutas, C., et al. (2005). Particulate matter in polluted air may increase biomarkers of inflammation in mouse brain. *Neurotoxicology*, *26*, 133–140.

Clark, H. A., Kopelman, R., Tjalkens, R., & Philbert, M. A. (1999). Optical nanosensors for chemical analysis inside single living cells. 2. Sensors for pH and calcium and the intracellular application of PEBBLE sensors. *Analytical Chemistry*, *71*, 4837–4843.

Couillard-Despres, S., Finkl, R., Winner, B., Ploetz, S., Wiedermann, D., Aigner, R., et al. (2008). In vivo optical imaging of neurogenesis: Watching new neurons in the intact brain. *Molecular Imaging*, *7*, 28–34.

De Lorenzo, A. J. D. (1970). The olfactory neuron and the blood-brain barrier. *Taste and Smell in Vertebrates*. In G. E.W. Wolstenholme & J. Knight (Eds.), (pp. 151–176). London: J &A Churchill.

Derfus, A. M., Chan, W. C.W., & Bhatia, S. N. (2004a) Intracellular delivery of quantum dots for live cell labeling and organelle tracking. *Advanced Materials*, *16*, 961–966.

Derfus, A. M., Chan, W. C.W., & Bhatia, S. N. (2004b). Probing the cytotoxicity of semiconductor quantum dots. *Nano Letters*, *4*, 11–18.

Dorman, D. C., McManus, B. E., Parkinson, C. U., Manuel, C. A., McElveen, A. M., & Everitt, J. I. (2004). Nasal toxicity of manganese sulfate and manganese phosphate in young male rats following subchronic (13-week) inhalation exposure. *Inhalation Toxicology*, *16*, 481–488.

Dousset, V., Tourdias, T., Brochet, B., Boiziau, C., & Petry, K. G. (2008). How to trace stem cells for MRI evaluation? *Journal of Neurological Science*, *265*, 122–126.

Elder, A., Gelein, R., Silva, V., Feikert, T., Opanashuk, L., Carter, J., et al. (2006). Translocation of inhaled ultrafine manganese oxide particles to the central nervous system. *Environmental Health Perspectives*, *114*, 1172–1178.

Emerich, D. F., & Thanos, C. G. (2003). Nanotechnology and medicine. *Expert Opinion on Biological Therapy*, *3*, 655–663.

Endo, T., Yamamura, S., Kerman, K., & Tamiya, E. (2008). Label-free cell-based assay using localized surface plasmon resonance biosensor. *Analytica Chimica Acta*, *614*, 182–189.

Foster, K. A., Galeffi, F., Gerich, F. J., Turner, D. A., & Muller, M. (2006). Optical and pharmacological tools to investigate the role of mitochondria during oxidative stress and neurodegeneration. *Progress in Neurobiology*, *79*, 136–171.

Frank, J. A., Miller, B. R., Arbab, A. S., Zywicke, H. A., Jordan, E. K., Lewis, B. K., et al. (2003). Clinically applicable labeling of mammalian and stem cells by combining superparamagnetic iron oxides and transfection agents. *Radiology*, *228*, 480–487.

Gao, X., Chen, J., Tao, W., Zhu, J., Zhang, Q., Chen, H., et al. (2007b). UEA I-bearing nanoparticles for brain delivery following intranasal administration. *International Journal of Phamaceutics*, *340*, 207–215.

Gao, X., Wu, B., Zhang, Q., Chen, J., Zhu, J., Zhang, W., et al. (2007a). Brain delivery of vasoactive intestinal peptide enhanced with the nanoparticles conjugated with wheat germ agglutinin following intranasal administration. *Journal of Controlled Release*, *121*, 156–167.

Gopalan, A. I., Lee, K. P., Manesh, K. M., Santhosh, P., Kim, J. H., & Kang, J. S. (2006). Electrochemical determination of dopamine and ascorbic acid at a novel gold nanoparticle distributed poly-(4-aminothiophenol) modified electrode, (published online) doi:10.1016/j.2006.08.026 Talanta.

Haes, A. J., Chang, L., Klein, W. L., & Van Duyne, R. P. (2005). Detection of a biomarker for Alzheimer's disease from synthetic and clinical samples using a nanoscale optical biosensor. *Journal of the Amercian Chemical Society*, *127*, 2264–2271.

Henriksson, J., & Tjalve, H. (2000). Manganese taken up into the CNS via the olfactory pathway in rats affects astrocytes. *Toxicological Sciences*, *55*, 392–398.

Functional Nanoparticles. IMEC, Scientific Report 2006. http://www.imec.be/wwwinter/mediacenter/en/SR2006/681420.html, h.w.i.b.w.m.e.S.h., Imec, Functional nanoparticles (2008).

Functional Nanoparticles. IMEC, Scientific Report 2006. http://www.inframat.com/biosens2.htm, h.w.i.c.b.h., Inframat, Functionalized nanoparticle based biosensor description (2008).

Hu, G. Z., Zhang, D. P., Wu, W. L., & Yang, Z. S. (2008). Selective determination of dopamine in the presence of high concentration of ascorbic acid using nano-Au self-assembly glassy carbon electrode. *Colloids and Surfaces, B: Biointerfaces, 62*, 199–205.

Hussain, S. M., Javorina, A. K., Schrand, A. M., Duhart, H. M., Ali, S. F., & Schlager, J. J. (2006). The interaction of manganese nanoparticles with PC-12 cells induces dopamine depletion. *Toxicological Sciences, 92*, 456–463.

Jander, S., Schroeter, M., & Saleh, A. (2007). Imaging inflammation in acute brain ischemia. *Stroke, 38*, 642–645.

Jasanoff, A. (2007). MRI contrast agents for functional molecular imaging of brain activity. *Current Opinion in Neurobiology, 17*, 593–600.

Jendelova, P., Herynek, V., Urdzikova, L., Glogarova, K., Kroupova, J., Andersson, B., et al. (2004). Magnetic resonance tracking of transplanted bone marrow and embryonic stem cells labeled by iron oxide nanoparticles in rat brain and spinal cord. *Journal of Neuroscience Research, 76*, 232–243.

Kim, J. S., He, L., & Lemasters, J. J. (2003). Mitochondrial permeability transition: A common pathway to necrosis and apoptosis. *Biochemical and Biophysical Research Communications, 304*, 463–470.

Koo, Y. E., Cao, Y., Kopelman, R., Koo, S. M., Brasuel, M., & Philbert, M. A. (2004). Real-time measurements of dissolved oxygen inside live cells by organically modified silicate fluorescent nanosensors. *Analytical Chemistry, 76*, 2498–2505.

Kopelman, R., Lee Koo, Y. E., Philbert, M., Moffat, B. A., Reddy, G. R., & McConville, P. (2005). Multifunctional nanoparticle platforms for in vivo MRI enhancement and photodynamic therapy of a rat brain cancer. *Journal of Magnetism and Magnetic Materials, 1*, 404–410.

Lee, J. S., Han, M. S., & Mirkin, C. A. (2007). Colorimetric detection of mercuric ion (Hg2+) in aqueous media using DNA-functionalized gold nanoparticles. *Angewandte Chemie International Edition in English, 46*, 4093–4096.

Li, Z. P., Duan, X. R., Liu, C. H., & Du, B. A. (2006). Selective determination of cysteine by resonance light scattering technique based on self-assembly of gold nanoparticles. *Analytical Biochemistry, 351*, 18–25.

Liu, C. H., You, Z., Ren, J., Kim, Y. R., Eikermann-Haerter, K., & Liu, P. K. (2008). Noninvasive delivery of gene targeting probes to live brains for transcription MRI. *FASEB Journal, 22*, 1193–1203.

Liu, G., Men, P., Harris, P. L., Rolston, R. K., Perry, G., & Smith, M. A. (2006). Nanoparticle iron chelators: A new therapeutic approach in Alzheimer disease and other neurologic disorders associated with trace metal imbalance. *Neuroscience Letters, 406*, 189–193.

Lockman, P. R., Koziara, J. M., Mumper, R. J., & Allen, D. D. (2004). Nanoparticle surface charges alter blood-brain barrier integrity and permeability. *Journal of Drug Targeting, 12*, 635–641.

Lovric, J., Bazzi, H. S., Cuie, Y., Fortin, G. R., Winnik, F. M., & Maysinger, D. (2005a). Differences in subcellular distribution and toxicity of green and red emitting CdTe quantum dots. *Journal of Molecular Medicine, 83*, 377–385.

Lovric, J., Cho, S. J., Winnik, F. M., & Maysinger, D. (2005b). Unmodified cadmium telluride quantum dots induce reactive oxygen species formation leading to multiple organelle damage and cell death. *Chemistry & Biology, 12*, 1227–1234.

Lu, L. P., Wang, S. Q., & Lin, X. Q. (2004). Fabrication of layer-by-layer deposited multilayer films containing DNA and gold nanoparticle for norepinephrine biosensor. *Analytica Chimica Acta, 519*, 161–166.

Nazem, A., & Mansoori, G. A. (2008). Nanotechnology solutions for Alzheimer's disease: Advances in research tools, diagnostic methods and therapeutic agents. *Journal of Alzheimers Disease, 13*, 199–223.

Oberdorster, E. (2004). Manufactured nanomaterials (fullerenes, C60) induce oxidative stress in the brain of juvenile largemouth bass. *Environmental Health Perspectives, 112*, 1058–1062.

Oberdorster, G., Sharp, Z., Atudorei, V., Elder, A., Gelein, R., Kreyling, W., et al. (2004). Translocation of inhaled ultrafine particles to the brain. *Inhalation Toxicology, 16*, 437–445.

Okumoto, S., Looger, L. L., Micheva, K. D., Reimer, R. J., Smith, S. J., & Frommer, W. B. (2005). Detection of glutamate release from neurons by genetically encoded surface-displayed FRET nanosensors. *Proceeding of the National Academy of Science. United States of America., 102*, 8740–8745.

Petry, K. G., Boiziau, C., Dousset, V., & Brochet, B. (2007). Magnetic resonance imaging of human brain macrophage infiltration. *Neurotherapeutics, 4*, 434–442.

Sykova, E., & Jendelova, P. (2006). Magnetic resonance tracking of transplanted stem cells in rat brain and spinal cord. *Neuro-degenerative Disease, 3*, 62–67.

Sykova, E., & Jendelova, P. (2007). In vivo tracking of stem cells in brain and spinal cord injury. *Progress in Brain Research, 161*, 367–383.

The National Academics (2005). The promise of nanotechnology. *Implications of Nanotechnology for Environmental Health Research.* In L. Goldman & C. Coussens (Eds.). Washington, DC: National Academic Press.

Tin Tin Win, S., Mitsushima, D., Yamamoto, S., Fukushima, A., Funabashi, T., Kobayashi, T., et al. (2008). Changes in neurotransmitter levels and proinflammatory cytokine mRNA expressions in the mice olfactory bulb following nanoparticle exposure. *Toxicology and Applied Pharmacology, 226*, 192–198.

Tjalve, H., & Henriksson, J. (1999). Uptake of metals in the brain via olfactory pathways. *Neurotoxicology, 20*, 181–195.

Tjalve, H., Henriksson, J., Tallkvist, J., Larsson, B. S., & Lindquist, N. G. (1996). Uptake of manganese and cadmium from the nasal mucosa into the central nervous system via olfactory pathways in rats. *Pharmacology & Toxicology, 79*, 347–356.

Tjalve, H., Mejare, C., & Borg-Neczak, K. (1995). Uptake and transport of manganese in primary and secondary olfactory neurones in pike. *Pharmacology & Toxicology, 77*, 23–31.

Tyner, K. M., Kopelman, R., & Philbert, M. A. (2007). "Nanosized voltmeter" enables cellular-wide electric field mapping. *Biophysical Journal, 93*, 1163–1174.

Vila, A., Sanchez, A., Evora, C., Soriano, I., McCallion, O., & Alonso, M. J. (2005). PLA-PEG particles as nasal protein carriers: The influence of the particle size. *International Journal of Pharmaceutics, 292*, 43–52.

Wang, J. X., Li, Y. F., Zhou, G. Q., Li, B., Jiao, F., Chen, C. Y., et al. (2007). Influence of intranasal instilled titanium dioxide nanoparticles on monoaminergic neurotransmitters of female mice at different exposure time. *Zhonghua Yu Fang Yi Xue Za Zhi, 41*, 91–95.

Wang, L., Bai, J., Huang, P., & Wang, H. (2006). Self-assembly of gold nanoparticles for the voltammetric sensing of epinephrine. *Electrochemistry Communications, 8*, 1035.

Zhang, G., Shu, F. P., & Robinson, C. J. (2007). Design and characterization of a nano-encapsulated self-referenced fluorescent nitric oxide sensor for wide-field optical imaging. *Conference Proceeding: Annual International Conference of the IEEE Engineering in Medicine and Biology Society, 2007*, 103–106.

Zhang, Q. Z., Zha, L. S., Zhang, Y., Jiang, W. M., Lu, W., Shi, Z. Q., et al. (2006). The brain targeting efficiency following nasally applied MPEG-PLA nanoparticles in rats. *Journal of Drug Targeting, 14*, 281–290.

Zhao, J., Zhang, X., Yonzon, C. R., Haes, A. J., & Van Duyne, R. P. (2006). Localized surface plasmon resonance biosensors. *Nanomed, 1*, 219–228.

Zimmer, A. T., Baron, P. A., & Biswas, P. (2002). The influence of operating parameters on number-weighted aerosol size distribution generated from a gas metal arc welding process. *Journal of Aerosol Science, 33*, 519–531.

SECTION IV

Nanoneurotoxicity and Nanoneuroprotection

H.S. Sharma (Ed.)
Progress in Brain Research, Vol. 180
ISSN: 0079-6123

CHAPTER 9

Nanoparticles influence pathophysiology of spinal cord injury and repair

Hari Shanker Sharma[1,*], Dafin F Muresanu[2], Aruna Sharma[1], Ranjana Patnaik[3], and José Vicente Lafuente[4]

[1]*Laboratory of Cerebrovascular and Pain Research, Department of Surgical Sciences, Anesthesiology and Intensive Care Medicine, University Hospital, Uppsala University, SE-75185 Uppsala, Sweden*
[2]*Department of Neurology, University of Medicine and Pharmacy "Iuliu Hatieganu," Cluj-Napoca, Romania*
[3]*Department of Biomaterials, School of Biomedical Engineering, Institute of Technology, Banaras Hindu University, Varanasi, India*
[4]*Lab Neurociencias Clínicas y Experimentales (LaNCE), Dpt. de Neurociencias, Universidad del País Vasco – EuskalHerriko Unibertsitatea, Bilbao, España*

Abstract: Spinal cord injury (SCI) is a serious clinical problem for which no suitable therapeutic strategies have been worked out so far. Recent studies suggest that the SCI and its pathophysiological responses could be altered by systemic exposure to nanoparticles. Thus, SCI when made in animals intoxicated with engineered nanoparticles from metals or silica dust worsened the outcome. On the other hand, drugs tagged with titanium (TiO_2) nanoparticles or encapsulated in liposomes could enhance their neuroprotective efficacy following SCI. Thus, to expand our knowledge on nanoparticle-induced alterations in the spinal cord pathophysiology further research is needed. These investigations will help to develop new strategies to achieve neuroprotection in SCI, for example, using nanodrug delivery. New results from our laboratory showed that nanoparticle-induced exacerbation of cord pathology following trauma can be reduced when the suitable drugs tagged with TiO_2 nanowires were administered into the spinal cord as compared to those drugs given alone. This indicates that nanoparticles depending on the exposure and its usage could induce both neurotoxicity and neuroprotection. This review discusses the potential adverse or therapeutic utilities of nanoparticles in SCI largely based on our own investigations. In addition, possible mechanisms of nanoparticle-induced exacerbation of cord pathology or enhanced neuroprotection following nanodrug delivery is described in light of recently available data in this rapidly emerging field of nanoneurosciences.

Keywords: nanoparticles; spinal cord injury; nanodrug delivery; blood–spinal cord barrier; spinal cord edema; cord pathology

* Corresponding author.
Tel.: +46-18-243899; Fax: +46-18-243899;
E-mail: Sharma@surgsci.uu.se

DOI: 10.1016/S0079-6123(08)80009-X

Introduction

Spinal cord injury (SCI) is a serious clinical problem for which no suitable therapeutic strategies have been developed so far (Sharma, 2007a, 2008; Sharma et al., 2007a). Victims of SCI are permanently disabled and often show paraplegia or quadriplegia that needs lifelong support from the society causing huge expenditure for their maintenance (Babb, Brasser, Hosack, Severe, & Stone, 2009; Leung, Wong, Wong, & Hutchins, 2009). Thus, the need of the hour is to explore improved therapeutic strategies to treat SCI patients and also to expand our knowledge on the cellular and molecular aspects of the pathophysiology of spinal cord trauma (see Sharma, 2004; Stålberg et al., 1988).

One of the earliest events in SCI is alteration in the microfluid environment of the spinal cord that makes it vulnerable to various harmful factors coming from the microvasculature (Sharma, 2004, 2009a; Sharma & Westman, 2004a; Sharma, Westman, Cervós-Navarro, Dey, & Nyberg, 1998a). Since normal spinal cord is strictly regulated by the blood–spinal cord barrier (BSCB), several vasoactive agents, toxins, or other cytotoxic agents could not get access to the spinal cord microenvironment (see Sharma, 2004, 2007a, 2008, 2009; Sharma & Westman, 2004a). However, SCI leads to breakdown of the BSCB allowing several harmful or toxic agents to enter into the spinal cord microenvironment leading to cord pathology (Sharma, 2004, 2007a, 2008). This breakdown of the BSCB may be caused either by the physical trauma or by the release of various cytotoxic agents and harmful substances within the cord and in the peripheral circulation. These agents in turn could induce causing disruption of the cell membrane of the spinal cord endothelial cells resulting in breakdown of the BSCB (see Sharma, 2004; Sharma & Westman, 2004a, 2004b). This aspect of SCI pathophysiology is reported quite extensively in both experimental and clinical situations (see Sharma, 2007, 2008).

Breakdown of the BSCB will also lead to penetration of serum proteins into the spinal cord microenvironment through leaky microvessels (Sharma, 2004, 2007; Sharma & Olsson, 1990; Sharma, Olsson, & Westman, 1995; Olsson, Sharma, & Pettersson, 1990; Pettersson, Sharma, & Olsson, 1990). An increased accumulation of proteins into the cord could lead to vasogenic edema formation (Kiyatkin, Brown, & Sharma, 2007; Sharma & Kiyatkin, 2009; Sharma, Westman, & Nyberg, 1998; Sharma, Sjöquist, & Westman, 2001; Sharma et al., 1995). Swelling of the spinal cord in the vertebral column and spread of edema fluid across the longitudinal and transverse direction in the nontraumatized cord may result in paralysis and loss of several vital functions that are normally regulated by the intact spinal cord ascending and descending fiber tracts (Sharma, 2008; Stålberg, Sharma, & Olsson, 1998; Winkler, Sharma, Stålberg, Westman, 1998). Accumulation of edema fluid and its spread may also induce neuronal and other nonneuronal cells, for example, myelin, endothelial, and glial cell damages (Olsson, Sharma, Nyberg, & Westman, 1995; Olsson, Sharma, Pettersson, & Cervós-Navarro, 1992; Sharma, Olsson, & Cervós-Navarro, 1993a, 1993b, 1995; see Stålberg et al., 1988; Winkler et al., 1988). Loss of nerve cells, degeneration of myelin, and altered functions of astrocytes could all lead to spinal cord dysfunction resulting in severe disability within the short duration after SCI (see Sharma, 2004, 2007a, 2008; Sharma & Westman, 2004b).

Experiments carried out in our laboratory previously demonstrated breakdown of the BSCB in rats within minutes after SCI resulting in concomitant development of spinal cord edema formation (Sharma & Olsson, 1990; see Sharma, 2004). This BSCB breakdown to protein tracers and edema formation increased progressively up to 12–24 h after the initial insult (Sharma, 2004, 2005, 2007). Interestingly, an injury into the spinal cord T10–11 level results in BSCB disruption and edema formation in far remote segments, for example, C5 and T4 to L5 and S2 segments within 3–5 h (Sharma, 2005; Sharma et al., 1995). This indicates that spread of edema fluid in both rostral and caudal segments following SCI could occur very rapidly (Sharma, 2004). Microscopical examination revealed a close correlation between edema formation and neuronal, glial, and myelin damage in the cord (Sharma, 2005, 2007, 2008; Sharma, Winkler, Stålberg, Olsson, & Dey, 1991;

Winkler et al., 1998). Thus, most of the damaged neurons, activated astrocytes, and myelin vesiculation could be seen at the light and or electron microscopy in the regions associated with BSCB disruption (see Sharma, 2004; Sharma & Westman, 2004b; Sharma et al., 1995, Sharma, Sjöquist, & Westman, 2001). This suggests that breakdown of the BSCB is crucial in edema formation and cell injuries.

Recent experiments carried out in several laboratories show that the extent of neural injury could be exacerbated in an environment where microfine particles (now often referred to as "nanoparticles") are high in numbers as compared to other areas where the concentration of nanoparticles are much limited (Baun, Hartmann, Grieger, & Hansen, 2009; Sharma & Sharma, 2007; Shimizu et al., 2009). This suggests that neural injury could be affected by the presence of nanoparticles in the surrounding environments. However, whether spinal cord pathophysiology is also altered by the presence of nanoparticle-rich environment is still not well known. Thus, it is still unclear whether soldiers who receive traumatic injuries in the brain or spinal cord in an environment where lots of gun powder explosion releasing sulfur, copper, or carbon nanoparticles, or in desert environments that is rich in silica dust (SiO_2 nanoparticles) have different pathophysiological outcome as compared to persons receiving injury in a relatively clean environment (see Sharma, 2007b; Sharma et al., 2009a).

Our laboratory was the first to show that under laboratory conditions, rats exposed to fine silica dust (SiO_2) for 1 week resulted in exacerbation of the SCI-induced cord pathology as compared to normal animals kept at controlled laboratory environment (Sharma et al., 2009a). This suggests that nanoparticles could influence the pathophysiology of SCI (Sharma et al., 2007a, 2009a). However, further studies are definitely needed to explore possible mechanisms behind such phenomena.

However, nanoparticles are also used to enhance therapeutic efficacy of drugs to treat diseases (Sharma et al., 2009b). Thus, drugs tagged with nanoparticles or encapsulated within the liposomes (soft nanoparticles) that could be released at the target site is shown to be much more effective than the same drugs administered alone (Mistry et al., 2009; Modi et al., 2009; Ren et al., 2009). This is because of the fact that nanodrug combination could resist degradation within the biological system leading to its long-term effects as compared to drug alone administered *in vivo* situations (Yin et al., 2009). Furthermore, with the help of nanoparticles, nanodrug combination can effectively penetrate into the CNS and may reach to deeper areas to induce direct effects (see Sharma et al., 2007a). Thus, nanodrug delivery could be used to effectively treat diseases of the nervous system and to enhance or prolong the therapeutic effects of the compounds (see Sharma et al., 2009a). However, use of nanodrug delivery in SCI is still not examined in details to enhance recovery or to reduce cord pathology.

Previous experiments in our laboratory for the first time showed that drugs if delivered to the spinal cord topically tagged with titanium (TiO_2) nanowires are much more effective in reducing trauma-induced BSCB disruption, edema formation, and cell injury as compared to the parent compounds given without nanowires (Sharma et al., 2007a, 2009a). Interestingly, tagging of drugs with nanowire did not alter their properties to act as neuroprotective drugs, but enhanced their therapeutic efficacy only (Sharma et al., 2007a). Thus, drugs that were not effective in reducing spinal cord edema formation or BSCB breakdown in SCI were still not effective in reducing these parameters after tagging them with nanowires (Sharma et al., 2007a, 2009a). However, the efficacy of those drugs that is quite effective in reducing SCI-induced pathophysiology of cord injury was considerably enhanced when administered with nanowires (Sharma, 2007; Sharma & Sharma, 2007; Sharma et al., 2007, 2009a). These observations suggest that nanowired drugs could be used to achieve greater therapeutic advantages in SCI. However, this is a new area that required additional investigation to achieve good therapeutic strategies in SCI.

This review describes the potential usefulness and disadvantages of nanoparticles influencing the pathophysiology of SCI that is largely based on our own investigations. In addition, the possible mechanisms of nanoparticle-induced exacerbation of cord pathology or nanodrug delivery-induced enhanced neuroprotection in SCI are discussed.

Nanoparticles influence brain function

"Nanoparticles" derived from metals, semiconductor materials, or from semisolid or soft substances in the size range of 1–200 nm have great potential for their use in biomedical applications, for example, diagnosis or treatment (Euliss, Dupont, Gratton, & Desimone, 2006; Farokhzad, & Langer, 2006; Lanone & Boczkowski, 2006; Wagner, Dullaart, Bock, & Zweck, 2006). Liposomes are semisolid or soft nanoparticles that are often used for drug delivery within the biological system that could reach the target organs more effectively to treat diseases (Fang, 2006). However, it is also likely that these nanoparticles when reaching the target organs may have some toxic effects of their own. Thus, further research is needed to understand the nanoneurotoxicities in the biological system before they are routinely used as nanoneuropharmaceuticals in clinical practices.

Any material when reaches the size of nanoscale, its basic properties are dramatically altered. Thus, copper (Cu) when reaching below 50 nm in size becomes extremely hard material and its malleability and ductility are entirely different than the bulk copper (Zhang et al., 2003). This suggests that nanoparticles from metals could behave entirely in different manners when reaching the biological systems. However, the influence of nanoparticles *in vivo* on brain function in normal or in pathological conditions is still not well known. It is quite likely that the basic physiopathology following any insult to the CNS may be altered by the presence of nanoparticles in the biological system (Sharma & Sharma, 2007). However, it remains to be investigated whether changes in biological system caused by nanoparticles could depend on their sizes or their basic constituent properties from which they are derived.

Nanoparticles enter into the body fluid system through inhalation from the environment (Borm et al., 2006; Lam, James, McCluskey, Arepalli, & Hunter, 2006; Mills et al., 2006). It is believed that the inhaled nanoparticles can enter into a variety of nonneural cells through endocytosis (Kim et al., 2006) and may reside there for few weeks to several months (Dunning et al., 2004). There are indications that absorption, distribution, metabolism, excretion, and toxicity of nanoparticles will depend on their physicochemical properties and the surrounding environmental conditions (Lam et al., 2006; Teeguarden, Hinderliter, Orr, Thrall, & Pounds, 2006). However, the potential risks of nanoparticle-induced cellular toxicity are still unknown (see Xia et al., 2006). Whether external or internal stressful situations, for example, disease processes and/or environmental conditions, may further modify the effects of intracellular nanoparticles on cellular toxicity is not known and requires additional investigation.

Nanoparticles and neurotoxicity

Effect of nanoparticles on neurotoxicity *in vivo* situations is still not well known. However, a possibility exists that nanoparticles when entering into the body fluid system may have some adverse effects on the CNS (see Sharma, 2007b, 2009; Sharma & Sharma, 2007). This idea is supported by the fact that nanoparticles exert higher inflammatory potential because of their small sizes (10–100 nm) as compared to the large particles of the same materials to the exposed cells or tissues (Oberdörster, Elder, & Rinderknecht, 2009; Oberdörster et al., 2005). Thus, an intense inflammatory neutrophil response was observed in the lungs of rats or mice when titanium dioxide (TiO_2) particles were applied intratracheally into rats or mice in the range of 20 nm, as compared to the TiO_2 in the range of 250-nm size in identical doses (Oberdörster et al., 2009; see Brown, Wilson, MacNee, Stone, & Donaldson, 2001). This suggests that size of nanoparticles is important factor in inducing cellular toxicity. Furthermore, few evidences point out that the size and particulate chemistry also plays important roles in determining nanoparticle-induced toxicity. Thus, exposure of rats and humans to polytetrafluoroethylene (PTFE) fume (~18 nm) induces high acute toxicity and mortality (Cavagna, Finulli, & Vigliani, 1961;

Coleman, Scheel, Kupel, & Larkin, 1968; Griffith, Stephens, & Tayfun, 1973), whereas, the particle size >100 nm results in a considerable loss of toxicity (Johnston et al., 2000).

Apart from size of the nanoparticles, their toxicity effects are also dependent on the species used. Thus, administration of carbon nanotubes (2–50 nm, 0.3–1.3 mg/kg) within the trachea induces acute inflammatory effects in the mice lung (Shvedova et al., 2005) but failed to elicit any toxic response in rats even when administered in high doses (1–5 mg/kg; Warheit et al., 2004).

It appears that nanoparticles are able to induce profound cellular stress leading to cell toxicity or cell death (Sharma, 2009; Sharma & Sharma, 2007; Sharma et al., 2009c). This is evident from the *in vitro* findings that the carbon nanotubes could induce profound oxidative stress (Sharma et al., 2009c). Thus, increased formation of free radicals, accumulation of peroxidative products, and depletion of cellular antioxidants are seen in several *in vitro* studies using carbon nanotube experiments (Fenoglio et al., 2006; Manna et al., 2005; Shvedova et al., 2005). This indicates that nanoparticle-induced cellular and molecular reactions in the biological system are important factors in inducing cell and tissue injuries.

However, further *in vivo* studies are needed to characterize the physicochemical characteristics, aggregation states, and concentration (number, mass, surface area) of nanoparticles in inducing neurotoxicity.

Exposure of nanoparticles induces breakdown of the blood–brain barrier function

Experiments carried out in our laboratory revealed that engineered nanoparticles from metals, for example, Ag, Cu, or Al (50–60 nm) when administered systemically are able to induce breakdown of the blood–brain barrier (BBB) to protein tracers, for example, Evans blue albumin and radioiodine (Sharma et al., 2009d). These effects of nanoparticles were most pronounced when administered either intravenously or intracerebroventricularly. The Cu and Ag nanoparticles exert most pronounced effects on the BBB to protein tracers as compared to Al nanoparticles (Sharma, 2007; Sharma et al., 2009d). Acute exposure of nanoparticles (4–24 h) has mild effects on the BBB disruption whereas chronic administration of these nanoparticles (1 week) exerted moderate to high degree of BBB disruption depending on the type of nanoparticles used (Sharma HS unpublished observations). On the other hand, intraperitoneal administration of nanoparticles exerted very mild or almost no changes in the BBB function following their administration either for 4 h or for 1 week duration (Sharma et al., 2009d). These observations suggest that nanoparticles could induce BBB dysfunction depending on their route and dose of administration.

Opening of the BBB with nanoparticles in our experiments was associated with neurotoxicity. These neurotoxic effects were most pronounced following intravenous and intracerebral administration of Ag nanoparticles followed by Cu and Al nanoparticles. This indicates that breakdown of the BBB by nanoparticles is instrumental in neurotoxicity. A lack of BBB leakage and neurotoxicity in animal following intraperitoneal administration of nanoparticles is in line with this idea (Sharma et al., 2009d). However, further studies using dose and size-related effects of nanoparticles are needed to understand their effects on neurotoxicity *in vivo*.

In addition, whether nanoparticles may alter the pathophysiological response following CNS injuries is a new feature and requires further investigation in great details. Thus, it is also not known whether nanoparticle exposure will alter the biological effects of drugs on the CNS. Keeping these views in consideration, our laboratory has initiated a series of investigations on nanoparticle-induced alteration in CNS injuries and their possible effects on neuroprotective efficacy of various drugs in animal models of brain and spinal cord injuries.

Nanoparticles aggravate pathophysiology of spinal cord injury

Nanoparticles from the ambient air could get access into the respiratory system and then enter into the brain probably through lymphatic channels (Oberdörster, 1996; Sharma & Sharma, 2007). Once these nanoparticles enter into the body fluid microenvironments they can either stay there or gain access into various cells and organelles quickly and stay there for long time (Bennett, 2002).

Thus, it is quite likely that military personnel engaged in peacekeeping or combat operations in Middle East or elsewhere are vulnerable to various nanoparticles present in the ambient environment (see Sharma & Sharma, 2007; Sharma et al., 2009d). These military operations especially in the desert areas make them vulnerable to silica dust (SiO_2 nanoparticles) and thus could affect their physical or mental health conditions (Bahrami & Mahjub, 2003; Vierikko et al., 2009). In addition, gunpowder explosion and other related activities would also expose them to Cu, S, and C nanoparticles. Since SiO_2 is a known toxic agent in few *in vivo* and in several *in vitro* studies involving liver, lung, and kidney cells, it is quite likely that CNS is also affected by SiO_2 nanoparticles (Chen et al., 2008; Fubini, Giamello, Volante, & Bolis, 1990; Ichinose et al., 2008; Nicolini et al., 1995; Wang et al., 1994; Yang, Liu, Yang, Zhang, & Xi, 2009). Thus, if our military personnel receive head trauma or SCI under these circumstances, it is unclear whether the pathophysiology of CNS injuries in such cases will be comparable to those who got injured in normal environment (see Sharma & Sharma, 2007). Furthermore, whether drug treatments having neuroprotective effects in normal cases of CNS injuries can also induce beneficial effects in humans who are inflicted with CNS injuries in a nanoparticle-rich environment.

To further expand our knowledge on the influence of nanoparticles on CNS injuries and its repair mechanisms, we have undertaken a series of investigations on chronic exposure of SiO_2, S, C, Cu, Ag, and Al nanoparticles (50–60 nm) in a rat model of SCI and have examined their potential adverse effects on spinal cord pathophysiology. In addition, we have also interest to see whether the effects of potential neuroprotective agents in the pathophysiology of SCI are altered in animals that are previously exposed to nanoparticles. Thus, a comparison is made between nanodrug delivery of neuroprotective compounds following SCI in nanoparticles exposed and unexposed groups. A brief summary of our current investigation using SiO_2 nanoparticles in SCI is summarized below.

Previous exposure of SiO_2 exacerbates pathophysiology of SCI

We investigated the effects of chronic exposure of SiO_2 (50–60 nm) nanoparticles in a well-established rat model of SCI (Sharma & Olsson, 1990; Sharma, et al., 1993a, 1993b, 1995). Our aim is to find out whether SCI in nanoparticle-exposed animals could show different pathophysiological symptoms as compared to saline-treated animals. Since nanoparticles could induce profound oxidative stress (Oberdörster et al., 2009; Sharma et al., 2009c), we also examined the effects of a well-known antioxidant compound H-290/51 (Sharma et al., 2001) on pathophysiology of SiO_2-treated animals after SCI.

Animal model of SCI

We developed a new model of rats SCI that essentially consists of a longitudinal incision to the right dorsal horn of the T10–11 segments (Fig. 1). The deepest part of the lesion is mainly limited to the Rexed's lamina VIII (Fig. 1) and comprises only about 4% of the total volume of the rat spinal cord making it quite comparable to that often seen in human cases (Sharma, 2004, 2007, 2009; Sharma & Olsson, 1990). The model offers unique possibilities to understand precise development of cord pathology in relation to functional paralysis because the physical injury, that is, the knife wound, is exclusively limited to the gray matter leaving white matter largely intact (Sharma, 2005, 2007; Stålberg et al., 1988; Winkler et al., 1988).

In this model, cell and tissue injury or spread of edema fluid and disruption of the BSCB could easily be examined in the nontraumatized proximal (T9) or distal (T12) segments of the cord after

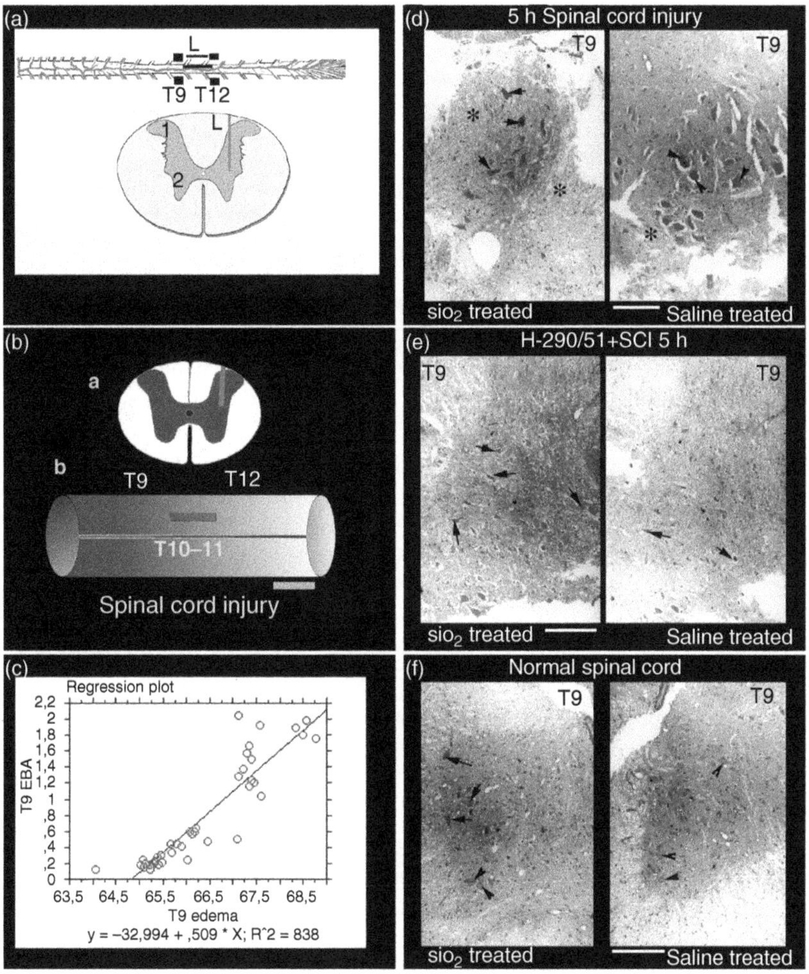

Fig. 1. Pathophysiology of spinal cord injury (SCI) in SiO_2-treated rats and their modification with H-290/51 treatment. In a new model of SCI a unilateral incision was made on the right dorsal horn (a) and the tissue samples from the T9 and T12 (b) were used to analyze spinal cord pathology, water content, and edema formation. Leakage of Evans blue albumin and edema formation after SCI showed a very strong correlation in both saline and SiO_2-treated injured rats (c). However, SiO_2-treated injured rats show much more exacerbation of cell injury, tissue loss, and spinal cord damage as compared to saline-treated rats after SCI (d). Pretreatment with H-290/51 showed slightly attenuated pathological changes in spinal cord in SiO_2-treated rats; however, this effect was most pronounced in saline-treated injured animals (e). Normal animals also showed mild to moderate neuronal damage by SiO_2 treatment in the spinal cord as compared to saline-treated rats (f). Bar = 25 μm. Data modified after Sharma et al. (2009a).

trauma (Olsson et al., 1990, 1992, 1995; Sharma, 2005; Sharma & Olsson, 1990). In addition, a comparison between ipsilateral (right side) and contralateral (left side) spinal cord can also be made while assessing neuroprotective effects of the drugs and compounds in this model, a feature that is nonexistent in any other current animal model of SCI (Sharma, 2004, 2005, 2008, 2009).

Nanoparticles administration

In separate groups of animals, SiO_2 (40–50 nm) nanopowders (Denzlingen, Germany) was suspended in 0.9% saline and administered intraperitoneally once daily (50 mg/kg) for 7 days (see Sharma & Sharma, 2007; Sharma et al., 2007a, 2009c) On the 8th day, the animals were anesthetized and one group was subjected to SCI, whereas the other group was left intact. Control animals received only saline daily in identical manner (Sharma & Sharma, 2007; Sharma et al., 2007b, 2009a, 2009d).

SiO_2 nanoparticles exacerbated BSCB breakdown

We examined the BSCB using two protein tracers, for example, Evans blue albumin and radioiodine ([131]Iodine) as described earlier (see Sharma, 2007b, 2007c; Sharma et al., 2007b, 2009a, 2009b, 2009c, 2009d). In brief, both the tracers were administered into femoral vein after the end of the experiment and allowed to circulate for 5 min. The intravascular tracer was washed out with physiological saline followed by 4% buffered paraformaldehyde perfusion through cardiac puncture (Sharma, 2005). After perfusion, the spinal cord tissue was taken out and dissected into T9 and T12 segments. These segments were then visualized for Evans blue leakage and weighed immediately and counted in a 3-inch well type gamma counter (Packard, Ramsey, Minnesota, 55303 USA). Immediately before saline perfusion, about 1 ml of whole blood was withdrawn from the left cardiac ventricle for determination of the whole blood radioactivity. Leakage of the radiotracer in the cord was expressed as percentages over whole blood radioactivity (Sharma et al., 1993a, 1993b, 1995). After counting the radioactivity in the spinal cord, samples were processed for colorimetric determination of the Evans blue dye using standard protocol (Sharma, Olsson, & Dey, 1990, 1991, 1995, 2001).

Chronic exposure of SiO_2 induced a mild but significant increase in the BSCB permeability to Evans blue and radioiodine in the spinal cord T9 and T12 segments (Table 1) in normal animals.

When these SiO_2-treated animals were subjected to SCI, the leakage of Evans blue and radioiodine were much more pronounced at 5 h as compared to saline-treated spinal cord-traumatized rats (Table 1). This increase in the BSCB permeability to protein tracers was most pronounced in the caudal spinal cord segment (T12) as compared to the rostral T9 segment (Table 1, see Sharma et al., 2009a).

These observations clearly show that nanoparticles significantly exacerbate the BSCB leakage to proteins in SCI. This suggests that exposure of nanoparticles is associated with pronounced aggravation of pathophysiological changes in the organisms in response to additional trauma (see Sharma et al., 2009a).

SiO_2 nanoparticles exacerbated edema formation

Leakage of plasma proteins into the cord microfluid environment leads to vasogenic edema formation. Thus, we examined spinal cord edema formation in SiO_2-treated rats before and after SCI. Spinal cord edema was determined by measuring water content of the T9 and T12 segments of the spinal cord (Sharma & Olsson, 1990). For this purpose, the T9 and T12 segments are removed, weighed, and then placed in an oven at 90°C for 72 h to determine their dry weights (Sharma et al., 1991). The water content was calculated from the differences between the wet and dry weights of the samples (Sharma et al., 1993a, 1993b, 1995, 2009a, 2009c, 2009d).

We observed that normal animals treated with SiO_2 nanoparticles showed a mild but significant increase in spinal cord water content as compared to saline-treated rats (Table 1A). When these SiO_2-treated animals were subjected to 5 h SCI, edema development was much more aggravated compared to saline-treated injured rats (Sharma et al., 2009a). Thus, in SiO_2-treated animals, injured rats showed ~12% increase in volume swelling, whereas saline-treated animals showed 8% increase in volume swelling after SCI. These observations are in line with the idea that prior exposure of nanoparticles exacerbates trauma-induced edema development and thus increases the chances morbidity of the victims (Sharma, 2007a,b 2009; Sharma & Sharma, 2007; Sharma et al., 2009a, 2009b, 2009c, 2009d).

Regression analysis showed a very tight correlation between Evans blue albumin leakage and edema formation in animals treated with either saline or SiO_2 in normal and in spinal cord traumatized rats (Fig. 1). This suggests that leakage of plasma proteins within the spinal cord

Table 1. Effect of SiO_2 nanoparticles on pathophysiology of 5 h spinal cord injury (SCI) in rats

	EBA (mg %)		^{131}Iodine (%)		Spinal cord water (%)		SCBF (mL/g/min)		Neuronal damage nr	
Type of Expt.	T9	T12	T9	T12	T9	T12	T9	T12	T9	T12
A. Control										
Saline treated	0.23 ± 0.04	0.26 ± 0.08	0.34 ± 0.06	0.38 ± 0.05	65.34 ± 0.14	65.64 ± 0.12	0.98 ± 0.05	1.04 ± 0.04	Nil	Nil
SiO_2 treated	$0.42 \pm 0.08^{*}$	$0.44 \pm 0.06^{*}$	$0.56 \pm 0.04^{*}$	$0.58 \pm 0.06^{*}$	$65.89 \pm 0.21^{*}$	$65.84 \pm 0.21^{*}$	0.94 ± 0.03	$0.96 \pm 0.03^{*}$	$8 \pm 2^{*}$	$8 \pm 4^{*}$
B. Spinal cord injury										
Saline treated	$1.16 \pm 0.12^{*a}$	$1.28 \pm 0.21^{*a}$	$1.89 \pm 0.25^{*a}$	$2.24 \pm 0.32^{*a}$	$67.34 \pm 0.23^{*a}$	$67.76 \pm 0.43^{*a}$	$0.76 \pm 0.08^{*a}$	$0.70 \pm 0.04^{*a}$	$24 \pm 6^{*a}$	$32 \pm 8^{*a}$
SiO_2 treated	$1.89 \pm 0.14^{*b}$	$1.98 \pm 0.23^{*b}$	$2.38 \pm 0.32^{*b}$	$2.67 \pm 0.33^{*b}$	$68.34 \pm 0.43^{*b}$	$68.89 \pm 0.28^{*b}$	$0.68 \pm 0.04^{*b}$	$0.62 \pm 0.06^{*b}$	$36 \pm 8^{*b}$	$43 \pm 6^{*b}$
C. H-290/51 treatment+ Control										
Saline treated	$0.18 \pm 0.06^{*}$	$0.14 \pm 0.04^{*}$	0.28 ± 0.08	$0.24 \pm 0.06^{*}$	$65.04 \pm 0.09^{*}$	$65.12 \pm 0.08^{*}$	1.08 ± 0.05	1.06 ± 0.03	Nil	Nil
SiO_2 treated	$0.22 \pm 0.10^{*c}$	$0.23 \pm 0.08^{*c}$	$0.34 \pm 0.02^{*c}$	$0.38 \pm 0.06^{*c}$	65.34 ± 0.12	65.23 ± 0.34	0.96 ± 0.08	0.94 ± 0.04	$2 \pm 2^{*}$	$2 \pm 3^{*}$
D. H-290/51 treatment+SCI										
Saline treated	$0.56 \pm 0.08^{*b}$	$0.66 \pm 0.07^{*b}$	$0.78 \pm 0.12^{*b}$	$0.82 \pm 0.14^{*b}$	$66.12 \pm 0.31^{*b}$	$66.28 \pm 0.18^{*b}$	$0.84 \pm 0.04^{*b}$	$0.80 \pm 0.08^{*b}$	$8 \pm 4^{*b}$	$12 \pm 6^{*b}$
SiO_2 treated	$1.67 \pm 0.32^{*b}$	$1.74 \pm 0.44^{*b}$	$1.90 \pm 0.21^{*b}$	$1.68 \pm 0.21^{*b}$	$67.34 \pm 0.23^{*b}$	$67.65 \pm 0.32^{*b}$	$0.70 \pm 0.04^{*b}$	$0.74 \pm 0.06^{*b}$	$20 \pm 8^{*a}$	$26 \pm 10^{*b}$

Values are mean $\pm$ SD of 5–7 rats at each point.

[*] $P < 0.05$, from saline control, ANOVA followed by Dunnet's test for multiple group comparison.

[a] $P < 0.05$, from SiO_2 treated.

[b] $P < 0.05$, SCI group.

[c] $P < 0.05$, SiO_2-treated group. Student's unpaired *t*-test.

Morphological analyses, $^{*}P < 0.05$, chi-square test (Data after Sharma et al., 2009a).

microenvironment is crucial for edema formation (Sharma et al., 2009a, 2009c).

SiO_2 nanoparticles exacerbated spinal cord blood flow reduction

We examined the spinal cord blood flow (SCBF) using radiolabeled microsphere techniques in the T9 and T12 segments as described earlier (Sharma, 2005; Sharma et al., 1990, 1993c, 1995). For this purpose, about 10^6 carbonized microspheres (15 ± 0.6 μm o.d.) radiolabeled with $^{[125]}$Iodine were administered into the left common carotid artery retrogradely toward heart over 90 s (Sharma et al., 1990, 1993c). Starting from 30 s before microsphere injection, timed serial samples of arterial blood from femoral artery were withdrawn at the rate of 0.8 ml/min at every 30 s that continued up to 30 s after the completion of microsphere administration (see Sharma et al., 1993c, 1995). At the end of the experiment, the rats were decapitated and the T9 and T12 spinal cord segments were dissected out. The large superficial blood vessels along with blood clots, if any, were removed and these segments were weighed separately and counted for radioactivity in a Gamma counter (Packard, Ramsey, Minnesota, 55303 USA). Total radioactivity in all the blood samples was also determined. The SCBF was calculated from the radioactivity present in the spinal cord at the time of killing in relation to the radioactivity in the peripheral blood withdrawn multiplied by the rate of withdrawal of the blood (see Sharma et al., 1990, 1993c, 1995; Olsson et al., 1992, 1995).

Our observations showed that SiO_2-treated normal animals exhibited a slight but significant ischemia that was most pronounced in the T12 segment. When these SiO_2-treated rats were subjected to SCI, they showed much more pronounced decrease in the SCBF at 5 h as compared to saline-treated traumatized rats (Table 1B; Sharma et al., 2009a). Interestingly, the magnitude and intensity of trauma-induced reduction in the SCBF was also most marked in the T12 segment (see Table 1).

These observations suggest that SiO_2 treatment exacerbates trauma-induced ischemia in the spinal cord that is most severe in the segment below the lesion site. Damage of microvessels, swelling of spinal cord, and accumulation of various vasoactive substances in the spinal cord below the lesion site could be the main reasons for such severe ischemia in the T12 segment. However, how SiO_2 could contribute to a reduction in the SCBF in normal animals is still unclear and requires further investigations (see Sharma et al., 2009a, 2009b, 2009c, 2009d).

SiO_2 nanoparticles exacerbated cord pathology

The neuronal changes in the spinal cord were examined using histopathological investigations. For this purpose, animals were perfused transcardially with 4% paraformaldehyde preceded with a brief saline rinse (Sharma & Olsson, 1990). After perfusion, the spinal cord samples from the T9 and T12 segments were dissected out and embedded in paraffin. About 3-μm thick sections were cut and stained with Nissl or Haematoxylin or Eosin (H&E) for analyzing nerve cell damage (Sharma, Olsson, Nyberg, & Dey, 1993c; Sharma et al., 1993a, 1993b; Sharma et al., 1995).

Treatment with SiO_2 nanoparticles induces mild neuronal damage in normal spinal cord (Table 1, Fig. 1). When these SiO_2-treated animals were subjected to SCI, they showed profound increase in neuronal injuries. These neuronal damages were most marked in the caudal segment (T12) to the lesion site as compared to the rostral spinal cord segment (T9) (see Table 1). Thus, several neurons in the spinal cord showed chromatolysis, vacuolation, and loss of cell nucleus with condensed cytoplasm. In some neurons where nucleus is present, an eccentric nucleolus is clearly seen. In addition, loss of neurons or presence of dark and distorted nerve cells is most pronounced in the SiO_2-treated injured rats as compared to normal animals after SCI (Fig. 1). In addition, a general sponginess and expansion of the cord were also evident in SiO_2-treated traumatized rats as compared to saline-treated injured animals (Table 1). This observation suggests that SiO_2-treated animals are most susceptible to SCI leading to exacerbation of cord pathology and damage to motor and sensory neurons. Thus it appears that SiO_2 exposure could enhance pathology of the CNS following trauma (for details see Sharma et al., 2009a).

SiO_2 nanoparticles exacerbated motor dysfunction

We examined the functional outcome after SCI using Rota rod and inclined plane angle tests (Sharma, 2006; Sharma & Sjöquist, 2002; Sharma et al., 2006, 2007a, 2007b).

To measure the sensory motor coordination of injured animals, the animals were placed on a Rota rod apparatus fixed at 16 rpm on which normal animals can stay up to 120 s without falling (see Sharma, 2006). The spinal cord-traumatized rats could not maintain their stay on the Rota rod for that long. The duration of their stay on Rota rod was counted manually (see Sharma et al., 2007a; Table 2).

Another test for sensory motor function in traumatized animals was determined using the inclined plane angle test. For this purpose, the animals were placed on an inclined plane platform and the angle of the plane was adjusted in such a way that the animals can stay on the platform for 5 s without falling (see Sharma, 2006; Sharma et al., 2007a, 2007b). Normal animals can stay at an angle of 60° without any problems (Sharma, 2007a). However, after SCI, the angle of inclined plane requires to be lowered to ~30° or lesser (see Sharma et al., 2007a, 2007b, 2009a) so that they can stay on the platform up to 5 s (see Sharma et al., 2007a, 2009a).

SiO_2-treated normal animals did not show any apparent dysfunction in their motor or cognitive functions seen on either Rota rod or capacity angle performance tests (Table 2). However, when these SiO_2-treated animals were subjected to SCI, a pronounced decrease in their performances on Rota rod and the capacity angle tests was observed (Table 2). Mild to moderate changes in the physiological variables were observed in SiO_2-treated injured animals (Table 2). This suggests that SiO_2 treatment exacerbates sensory motor disturbances after SCI (Sharma et al., 2009a).

SiO_2 attenuated neuroprotective effects of H-290/51 in SCI

Since nanoparticles could induce profound oxidative stress, we examined the effects of a potent antioxidant compound H-290/51 on SiO_2-induced exacerbation of spinal cord pathophysiology following trauma (see Sharma et al., 2009a).

Treatment with H-290/51 In separate groups of saline or SiO_2-treated animals, H-290/51 (Astra-Zeneca, Mölndal, Sweden) was administered (50 mg/kg, p.o.) once 30 min before SCI (Sharma & Sjöquist, 2002; Sharma, Sjöquist, & Alm, 2003; Sharma et al., 2006). This treatment is well known to induce profound neuroprotection in SCI (see Sharma & Sjöquist, 2002; Sharma et al., 2006). Pretreatment with H-290/51 30 min before SCI in SiO_2-treated animals failed to reduce the leakage of Evans blue and radioiodine tracers within the spinal cord T9 or T12 segments (Table 1D) or spinal cord edema formation (Table 1D). However, in saline-treated rats H-290/51 was able to effectively reduce the leakage of protein tracers or edema formation at 5 h SCI (Table 1D). Pretreatment with H-290/51 markedly prevented the SCI-induced reductions in the SCBF in saline-treated rats — a feature that was not so pronounced in SiO_2-treated injured animals (Table 1).

The H-290/51 treatment induced pronounced neuroprotection following SCI in normal animals. However, its effects on reducing neuronal damages in SiO_2-treated spinal cord-traumatized rats were much less apparent (Fig. 1; Table 1). Interestingly, H-290/51 did not improve the behavioral outcome of SiO_2-treated spinal cord-injured rats, although, the compound was effective in improving sensory motor disturbances in normal rats after SCI (see Table 2). These observations clearly demonstrate that nanoparticle exposure not only exacerbates the spinal cord pathophysiology following trauma but also attenuates the neuroprotective effects of the drug, for example, H-290/51 (Sharma et al., 2009a, 2009b, 2009c, 2009d). The probable mechanisms of such reduction in the neuroprotective efficacy are not understood. It appears that excessive production of oxidative stress caused by trauma and nanoparticles could

Table 2. Effect of SiO_2 nanoparticles on pathophysiology of 5 h spinal cord injury (SCI) in rats

	Motor function		Physiological variables			
Type of expt.	Rota Rod	Capacity	MABP (torr)	Arterial pH	Arterial PaO_2 (torr)	Arterial $PaCO_2$ (torr)
A. Control						
Saline treated	120 ± 0	60 ± 0	118 ± 4	7.38 ± 0.04	80.56 ± 0.34	34.32 ± 0.21
SiO_2 treated	108 ± 2	56 ± 3	107 ± 2	7.36 ± 0.06	80.78 ± 0.12	33.89 ± 0.21
B. Spinal cord injury						
Saline treated	$72 \pm 4^{*}$	$36 \pm 6^{*}$	$89 \pm 8^{*}$	7.36 ± 0.08	$81.34 \pm 0.33^{*}$	$33.34 \pm 0.10^{*}$
SiO_2 treated	$68 \pm 6^{*}$	$26 \pm 4^{*a}$	$93 \pm 6^{*}$	7.35 ± 0.08	$81.67 \pm 0.21^{*a}$	$33.45 \pm 0.08^{*a}$
C. H-290/51 treatment+Control						
Saline treated	118 ± 2	58 ± 1	116 ± 4	7.37 ± 0.04	80.43 ± 0.22	34.56 ± 0.21
SiO_2 treated	$110 \pm 4^{*a}$	$58 \pm 3^{*}$	115 ± 4	7.36 ± 0.08	80.44 ± 0.21	34.10 ± 0.10
D. H-290/51 treatment+SCI						
Saline treated	$86 \pm 7^{*b}$	$48 \pm 4^{*b}$	$108 \pm 4^{*b}$	7.37 ± 0.06	81.04 ± 0.32	33.65 ± 0.18
SiO_2 treated	$71 \pm 4^{*a}$	$30 \pm 4^{*a}$	$107 \pm 8^{*a}$	7.35 ± 0.08	81.34 ± 0.21	33.21 ± 0.32

Values are mean ± SD of 5–7 rats at each point.
[*] $P < 0.05$, from saline control, ANOVA followed by Dunnet's test for multiple group comparison.
[a] $P < 0.05$, from SiO_2 treated.
[b] $P < 0.05$, SiO_2-treated injured group. Student's unpaired t-test (Data after Sharma et al., 2009a).

limit the ability of antioxidants to induce marked neuroprotection.

Treatment with nanowired-H-290/51 There are reasons to believe that nanodrug delivery enhances the neuroprotective efficacy of drugs in the CNS (see below). Thus, we examined the effects of nanowired H-290/51 in SiO_2-treated spinal cord-traumatized rats. The compound H-290/51 was tagged to TiO_2 nanowires using standard procedures as described in detail earlier (see Sharma et al., 2007a).

In brief, the TiO_2 nanowires were synthesized from 0.30 g of TiO_2 powder (Degussa P25) by mixing it into 40 ml of 10 M alkali solution in a 150 ml Teflon-lined autoclave container. After a hydrothermal reaction in an oven for 7 days above 160°C, long nanowires were collected, washed with distilled water or dilute acid, fabricated into the white and flexible membrane on a Teflon template, and then dried at room temperature (Sharma et al., 2007a). X-ray diffraction (XRD) patterns of the nanofibers confirm that the nanowire resemble the titanate in lattice structure (see Sharma et al., 2007a, 2009a). The energy dispersive X-ray (EDX) elementary analysis result of this film shows the existence of Ti, Na, and O in the nanowire membrane as reported earlier (Fig. 2; Sharma et al., 2007a; see Sharma et al., 2009b).

After purification of the TiO_2 nanowires, the compound H-290/51 (5 mg concentration) was tagged separately as described earlier (Sharma et al., 2009a, 2009c). For this purpose, the film was first sterilized in 70% ethanol and then rinsed in sterile 0.9% saline. Subsequently, the membrane (1.0 cm × 1.0 cm) was soaked in a 1.0 ml solution of 5mg/L H-290/51 at room temperature for 12 h and then washed with deionized (DI) water before use (Sharma et al., 2009a, 2009c). The H-290/51 labeling on the nanowires was confirmed by scanning electron microscopy (results not shown).

Pretreatment with nanowired H-290/51 markedly improved the sensory motor function in the SiO_2-treated spinal cord-traumatized animals at 5 h. Also the nanowired H-290/51 was able to significantly reduce the BSCB leakage of protein tracers, edema formation, and neuronal damage in nanoparticle-treated injured rats (results not shown). A marked improvement in SCBF was also seen by the nanowired H-290/51 in SiO_2-treated injured animals (Sharma HS unpublished observations). These observations suggest that nanowired drugs could be able to induce marked neuroprotection following nanoparticle-induced exacerbation of cord pathology in SCI.

Possible mechanisms of SiO_2-induced exacerbation of neurotoxicity in SCI

There are reasons to believe that nanoparticles could enter into the body fluid compartments from the environment through the respiratory tracts (Bennett, 2002; Brooking, Davis, & Illum, 2001) and depending on their sizes, they are translocated into other tissues and organs including the CNS (Bennett, 2002; Florence, Hillery, Hussain, & Jani, 1995; Oberdörster, 2009; Sharma, 2009a). Thus, it is likely that the toxic effects of nanoparticles when they enter into the CNS could contribute to exacerbation of brain or spinal cord pathology following trauma (see Gao & Jiang, 2006; Sharma et al., 2009a, 2009b, 2009c).

Nanoparticles when entering into the body fluid microenvironment induce profound oxidative stress (Fubini et al., 1990; Sharma & Sharma, 2007; Sharma et al., 2009a; Wang et al., 1994). An increased oxidative stress leads to production of free radicals damaging the plasma membranes of the neurons, glial cells, and the endothelial cells (Bao, Chen, Schneider, & Weaver, 2008; Deng-Bryant, Singh, Carrico, & Hall, 2008; Neretti et al., 2009; Xiong & Hall, 2009). An exacerbation of the BSCB breakdown to protein tracers in the spinal cord-traumatized animals treated with SiO_2 nanoparticles is in line with this hypothesis. Entry of proteins into the spinal cord microenvironment leads to edema formation and subsequently results in neuronal injury (Sharma & Olsson, 1990; Sharma & Sharma, 2007; Sharma & Sjöquist, 2002; Sharma, Gordh, et al., 2006; Sharma, Wiklund et al., 2006; Sharma et al., 2001, 2003, 2007a). Exacerbation of edema formation and neuronal damages in the spinal cord of SiO_2-treated traumatized animals further support this hypothesis (see Sharma et al., 2009a).

Exacerbation of cell and tissue damage following SCI in nanoparticle-treated rats may be due to an enhanced production of free radical formation caused by nanoparticles alone and induce by spinal cord trauma in coherence (Fubini et al., 1990; Nicolini et al., 1995; Sharma & Sharma, 2007; Sharma & Sjöquist, 2002; Sharma et al., 2003, 2006, 2009a, 2009c; Wang et al., 1994). Available evidences suggest that apart from nanoparticles exposure, SCI itself induces severe oxidative stress and results in lipid peroxidation and formation of oxygen radicals (Sharma, 2005, 2006, 2007a, 2007b; Sharma & Sjöquist, 2002; Sharma, Gordh, et al., 2006; Sharma, Wiklund et al., 2006; Sharma et al., 2001, 2003, 2007a, 2007b). Formation of free radicals could be directly or indirectly responsible for cell and tissue damage (Bao et al., 2008; Deng-Bryant et al., 2008; Xiong & Hall, 2009; see Sharma et al., 2009a). Thus, a combination of two noxious events, for example, nanoparticles and SCI, could obviously lead to an enhanced production of oxidative stress and cell injury (Sharma et al., 2009c, 2009d).

This idea is further strengthened by our results obtained with the antioxidant compound H-290/51. The H-290/51 is a powerful chain-breaking antioxidant and thus has the capacity to inhibit formation of free radicals (Sharma & Sjöquist, 2002; Sharma et al., 2001, 2003, 2006). Although, the drug was able to induce marked neuroprotection in normal animals subjected to SCI, the compound failed to protect spinal cord pathology or associated functional disturbances in SiO_2-treated traumatized rats (Sharma et al., 2009a).

It appears that in nanoparticle-treated injured animals repeated administration of H-290/51 and/or high doses of the compound are needed to block the excessive production of oxidative stress (see Sharma et al., 2007a). Keeping these views in mind when we administered nanowired H-290/51 in SiO_2-treated spinal cord-traumatized rats, profound neuroprotection was seen. This suggests that nanowired H-290/51 when administered into the spinal cord may have prolonged effects and rapid penetration within the cord to neutralize excessive production of oxidative stress (Sharma et al., 2007a, 2007b, 2009a, 2009b, 2009c, 2009d). However, further studies are needed to understand the molecular mechanisms behind nanowired H-290/51-induced neuroprotection.

Nanoparticles enhance drug delivery

Recent studies on nanodrug delivery show great enthusiasm regarding the use of engineered nanoparticles in medicine to treat several diseases besides their use in diagnostic purposes (see Costigan, 2008; de Jong & Borm, 2008; Sharma et al., 2009d). For this purpose, nanoparticles ≤100 nm are used that have several advantages for drug labeling because of their much larger surface to mass ratio. Thus, it would be easier for these nanoparticles to bind, adsorb, and carry drugs, probes, or other proteins, for example, antibodies for selective binding processes with the antigens *in vivo* situations (Costigan, 2008; ESF, 2005; SCENIHR, 2008; see Sharma et al., 2009b). In addition, nanoparticles made from semiconductor materials known as quantum dots (Mertens, Biteen, Atwater, & Polman, 2006) are often very good carriers for drug delivery and/or as imaging agents (Fang, 2006; Sharma et al., 2007a; Williams, 2006). In this regard, liposome nanoparticles are extremely used to deliver anticancer drugs and/or vaccines to the specific target sites to enhance the effectiveness of the therapeutic agents (see Fang, 2006; He et al., 2006; Moghimi, 2006).

However, the potential adverse effects of nanoparticles used in nanodrug delivery are still not well known and requires detailed investigations (see Borm et al., 2006; Koziara, Lockman, Allen, & Mumper, 2006).

For nanodrug delivery, loading of sufficient amount of drugs to nanoparticles >100 nm is needed to have the desired effects. This loading of drugs to nanocarriers, however, varies with the type of metals or particles used for nanodrug delivery (Sharma et al., 2007a). In addition, the drugs itself could be formulated at nanoscale that could function as its own carrier (Duncan, 2003; Kipp, 2004) (see Table 3).

The main advantages of nanodrug delivery are to achieve enhanced drug delivery to the target cells, and/or a reduction in the toxicity of the free drug to the nontarget cells. This allows nanodrug delivery of compounds to an increased therapeutic index in desired tissues or organs (see Sharma et al., 2009c). However, some portions of the drug could be trapped by mononuclear phagocytes in the liver and spleen (Gibaud et al., 1996; Moghimi, Hunter, & Murray, 2001). To

Table 3. Drugs or nanoparticles used for biomedical applications and/or drug delivery

Nanoparticles/materials	Type	Biomedical Application
Liposomes Gelatin Dextran Chitosan Starch Albumin	Natural material/derivative	Drug delivery
Carbon-based carriers	Fullerenes	
Polyethylenimine Polycaprolactone Poly(cyano) acrylates Polysorbate Hydrogels	Polymer carriers	
Cd/Zn-selenides	Quantum dots	Imaging/diagnostic
Silica Magnetic iron oxide	Nanoparticle	Gene/drug delivery
Solid lipid formulation	Nanoparticle	

Compiled from various sources. (Data after Sharma et al., 2009c).

encounter such situations, coating of nanoparticles with polyethylene glycol (PEG) or Polyvinylpyrrolidone (PVP) or other natural polymers, for example, dextran or chitosan (see Table 3) is helpful (Gupta & Gupta, 2005).

Size and materials of nanoparticles affecting drug delivery

The size and material of the nanoparticles appear to play important roles in nanodrug delivery. This is especially true for lipid vesicles that showed lower liver uptake for small size of 25–50 nm nanoparticles compared to large sizes of 200–300 nm liposomes (Seki et al., 2004). Small differences in size of the nanoparticles will also influence their actual distribution and bioavailability (Fang et al., 2006; Saez et al., 2000; Sharma & Sharma, 2007) and could also influence their clearance (de Jong & Borm, 2008). Thus, liposome with sizes >100 nm showed an increased clearance rate by mononuclear phagocytic system that correlates well the size of the nanoparticles (Senior et al., 1985). However, in case of nanoparticles with sizes less than 100 nm, the electrostatic charges could largely determine their clearance (Senior & Gregoriadis, 1982). Furthermore, not only the size or charge of the nanoparticles alone determines the bioavailability, absorption, or clearance of nanoparticles in the biological systems *in vivo*, but their composition is also important to influence their bioactivity. Thus, all nanoparticles below the sizes of 100 nm do not behave similarly in the biological system either *in vivo* or *in vitro* situations (de Jong & Borm, 2008). This suggests that the actual shape and size of nanomaterials are important for drug delivery (Senior et al., 1985).

However, apart from their physical properties, nanoparticles are often used for drug delivery that could be activated by heat or light (Gomes, Lunardi, Marchetti, Lunardi, & Tedesco, 2006). These thermosensitive or photosensitive nanoparticles could thus be used to release the desired amount of drugs at selective or specific sites (Na et al., 2006; Ricci Junior & Marchetti, 2006).

Furthermore, the surface chemistry of nanoparticles also play important roles in drug distribution or uptake, especially *in vivo* situations (Schins et al., 2002). Thus, quartz if coated with polyvinyl pyridine-*N*-oxide (PVNO)-polymer is nontoxic when taken up by macrophages, whereas, quartz without coating induces severe cellular toxicity (Tomazic-Jezic, Merritt, & Umbreit, 2001). Similarly, two different types of polymer particles could be redistributed differently in various tissues or organs after their *in vivo* administration (Tomazic-Jezic et al., 2001). In addition, modification of nanoparticle surface with different polymer coating can result in great variations in nanoparticle-induced blood coagulation, mitochondrial reactive oxygen species (ROS) activation, and/or cellular oxidative burst phenomena (Xia et al., 2006). Some of the nanoparticle coatings, for example, PEG are able to increase their circulation time *in vivo* situations (Duncan & Izzo, 2005).

Surface modification of nanoparticles will also allow them to be in circulation for long time due to inhibition of their recognition and consequently their phagocytosis by the mononuclear phagocytic system (Niidome, Yamagata, & Okamoto, 2006; Peracchia et al., 1999).

Thus, suitable choice of nanoparticles and their coating with specific material is important for drug delivery without the loss of drug activity to the target organs *in vivo* (Dyer et al., 2002).

Nanoparticles enhance therapeutic effects of drugs in the spinal cord

Drug delivery to the spinal cord using nanotechnology is still a very new subject as most of the nanodrug delivery studies are limited to brain diseases (see Sharma et al., 2007a, 2009c). However, there are reasons to believe that spinal cord like brain also possesses a BSCB, thus the principles of drug delivery to the spinal cord may be quite similar to that of the brain (See Sharma, 2004, 2009a, 2009b; Sharma & Westman, 2004a).

One of the main problems in treating SCI is to maintain a high concentration of the active ingredients of the drug within the spinal cord cell and tissues to achieve neuroprotection (Sharma, 2008). Using systemic administration of drugs thus poses serious problems in SCI as the adverse effects of the compounds on other organs may occur due to repeated use of the drug (Sharma, 2004). Furthermore, a high concentration of the compound is hard to achieve at the injury site. Thus, development of suitable therapeutic strategies using nanodrug delivery is needed to overcome these situations in SCI.

Clinically, methylprednisolone even when given in very high doses after SCI intravenously (35 mg/kg) did not induce neuroprotection in the spinal cord but resulted in serious adverse side effects (see Sharma, 2004, 2007a, 2007b, 2008). To overcome these problems, methylprednisolone in some gel formulations was used at the site of injury to achieve good neuroprotection (see Sharma, 2008). Chvatal, Kim, Bratt-Leal, Lee, and Bellamkonda (2008) demonstrated remarkable protective effects on regeneration and functional outcome after 7 days of the primary insult when methylprednisolone was applied topically at the injury site encapsulated in poly-lactic-co-glycolic acid (PLGA) nanoparticle formulations. This suggests that nanoparticle drug formulation could be useful to treat SCI-induced cord pathology and functional disability (see Sharma et al., 2009c).

However, it is unclear whether nanodrug delivery using other nanoparticles, for example, TiO_2 may have similar effects on the neuroprotective efficacy of methylprednisolone or other drugs in SCI. In addition, it is not known, whether vascular reactions that are important in setting the cascade of spinal cord pathophysiology are also reduced by local nanodrug delivery. Moreover, whether there is any correlation between functional outcome and cord pathology after injury and this balance is restored by nanodrug delivery is still not well-known.

Keeping these views in consideration, we formulated several drugs having potential neuroprotective efficacy in SCI using TiO_2 nanowires and compared their effects on spinal cord pathology-induced vascular reaction and functional outcome in our rat model (see below).

Nanoparticles enhance neuroprotection in SCI

Recent observations suggest that drugs attached to innocuous nanowires could provide a suitable alternate for enhanced nanodrug delivery within the CNS. Several *in vitro* models indicate that nanowired drugs have better therapeutic potentials than other methods of nanodrug delivery in enhancing the therapeutic efficacies of the compounds (Sharma & Sharma, 2007, 2008; Sharma et al., 2007a).

We used nanodrug-delivery to the spinal cord *in vivo* situations using TiO_2 nanowire to enhance therapeutic potential of drugs in SCI (Sharma et al., 2007a, 2009a, 2009b). These nanowired drugs with known neuroprotective efficacy in CNS injuries were applied topically over the traumatized spinal cord in our rat model and various functional and pathological outcome measures were examined.

Titanium nanowires for nanodrug delivery

As mentioned above, we used titanium wire-mesh to tag drugs for nanodrug delivery to achieve neuroprotection in the spinal cord trauma because titanium nanowires are largely innocuous in nature (Dong et al., 2006). The nanowire we used is a TiO_2-based (hydrogen titanate) single crystalline ceramic biomaterial with a typical diameter ranging from 50 to 60 nm with a superb chemical stability that can be used to enhance drug delivery within the CNS (Dong et al., 2006; Sharma et al., 2007a).

For labeling of drugs, 0.20 g of TiO_2 powder (Degussa P25) was introduced into 40 ml of 10 M alkali solution in a 150-ml Teflon-lined autoclave container as mentioned above. After the hydrothermal reaction in an oven for 1–15 days at temperatures above 180°C, the white paper-like flexibility product was collected from the Teflon rod template and washed with distilled water. The compounds AP173, AP713, and AP364 (10 μg concentration) were tagged separately to nanowires (Sharma et al., 2007a, 2009a-d). In our study, the film was first sterilized in 70% ethanol and then rinsed in sterile 0.9% saline. Subsequently, the membrane (1.0 cm × 1.0 cm) was soaked in a 1.0 ml solution of 10 μg /L AP173 or AP713 at room temperature for 12 h and then washed with DI water before use (Dong et al.,

2006). These nanowired compounds were administered separately over the traumatized spinal cord 5 min after injury in identical manner. These animals were also allowed to survive 5 h after injury (Sharma et al., 2007a).

Nanodrug delivery and functional outcome in SCI

The functional outcome after SCI was examined using Tarlov scale and inclined plane angle tests as described earlier (Sharma, 2006; Sharma, 2007a, 2007b; Sharma et al., 2009a, 2009c). The Tarlov scale for hind limb function was graded as total paraplegia = 0; no spontaneous movement but responds to pinch = 1; spontaneous movement = 2; able to support weight but unable to walk = 3; walk with gross deficits = 4; walk with mild deficits = 5; normal walk = 6 (Sharma et al., 2007a).

Topical application of nanowired compounds significantly improved the hind limb function from 3 h after injury up to 5 h period, a feature not seen with parent compounds alone (Fig. 2). This improved motor function after SCI was most

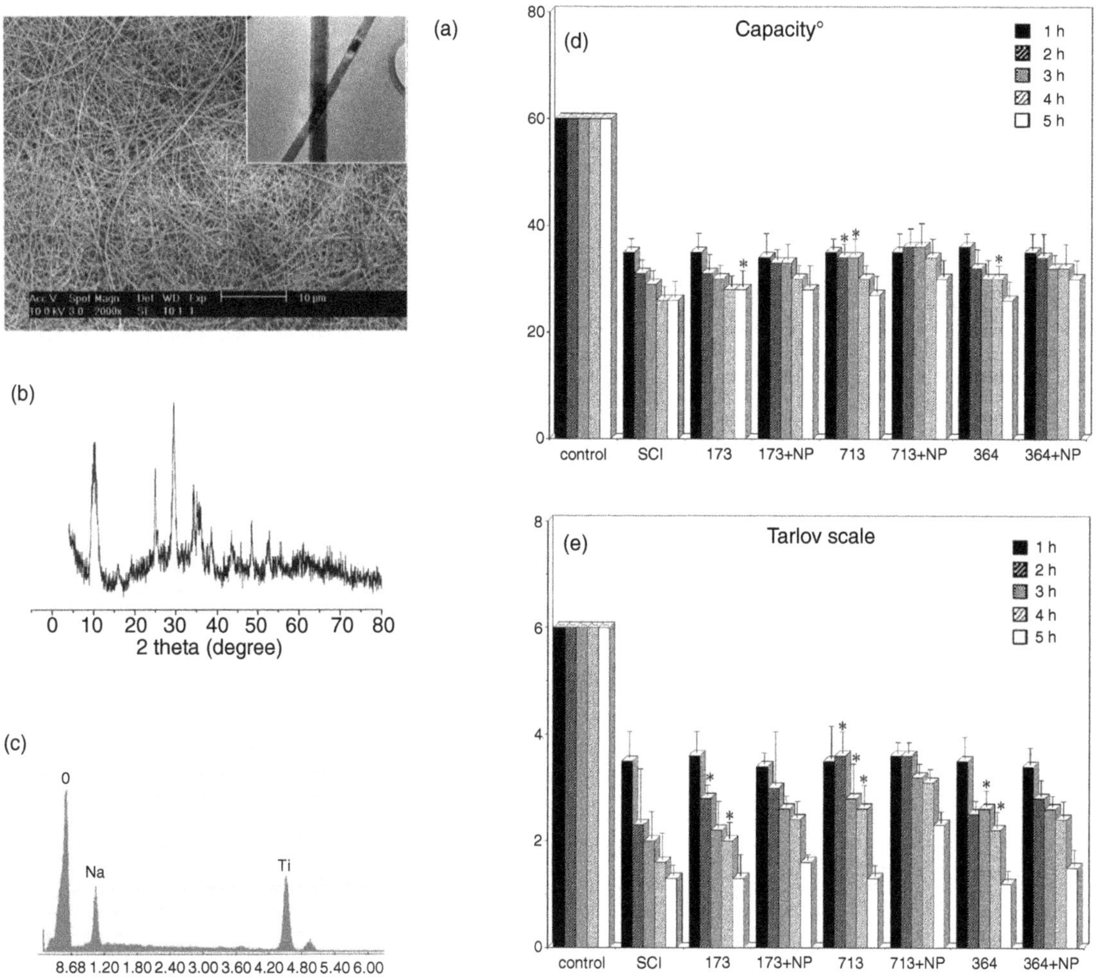

Fig. 2. Nanowiring of neuroprotective drugs using TiO_2 nanowires (a–c) and their effects on spinal cord motor functions as seen using capacity angle test (d) and Tarlov scale (e) after SCI. for details see text. Data modified after Sharma et al. (2007a, 2009c).

Table 4. Effects of nanowired compounds on blood–spinal cord barrier (BSCB) permeability and edema formation following spinal cord injury (SCI) in the rat. SCI was made on the right dorsal horn of the T10–11 segments and the animals were allowed to survive 5 h. Neuroprotective compounds AP173, AP713, or AP364 (10 μg) were applied over the traumatized cord topically in a volume of 20 μL 5 min after trauma. These compounds equivalent to 10 μg nanowired were applied in identical manner. For details, see text.

Expt. type	n	BSCB permeability [131]Iodine (%) T9	BSCB permeability [131]Iodine (%) T12	Spinal cord water content (%) T9	Spinal cord water content (%) T12	Cell injury
Control	5	0.34±0.07	0.36±0.08	65.23±0.21	65.46±0.17	Nil
Untreated SCI	6	1.96±0.21**	2.04±0.23**	68.56±0.13**	68.76±0.34**	++++
Neuroprotective compounds (10 μg in 20 μL) 5 min after SCI						
AP173+SCI	5	1.08±0.23aa	1.34±0.31aa	66.98±0.14aa	67.44±0.22aa	+++
AP713+SCI	5	0.96±0.28aa	0.98±0.54aa	66.78±0.21aa	66.96±0.42aa	++
AP364+SCI	5	1.24±0.23aa	1.67±0.43aa	67.04±0.11a	67.58±0.38a	+++?
Neuroprotective compounds tagged with nanowires (equivalent to 10 μg in 20 μL) 5 min after SCI						
AP173	5	0.98±0.21aa	0.96±0.43aa	66.56±0.12aa	67.02±0.37a	+++
AP713	5	0.76±0.23aa#	0.84±0.41aa#	65.89±0.32aa#	66.05±0.21aa#	+
AP364	5	0.94±0.33aa	0.96±0.45aa#	66.78±0.34aa	67.04±0.23aa	+++?
TiO_2-nanowires (<50 nm in 20 μL) 5 min after SCI						
TiO_2	5	1.88±0.32**	1.96±0.54**	68.88±0.54**	68.89±0.34**	++++

Values are mean±SD.
* P <0.05,** and P <0.01 from untreated control.
a P<0.05 and aa$P$ <0.01, from drug treated control
# P < 0.05, compared from same compound without nanowiring.
Nonparametric chi-square test. + = faint, ++ = mild, +++ = moderate, ++++ = extensive, ? = not clear (Data after Sharma et al., 2007a, 2009c).

marked with nanowired AP713 as compared to other nanowired compounds (Fig. 2).

Interestingly, these nanowired compounds also improved the performance of injured animals on the inclined plane angle tests (Fig. 2d,e). Rats treated with nanowired compounds can stay at a higher capacity angle from 2 to 5 h after injury and this effect was most pronounced with nanowired compound AP713 (Fig. 2). On the other hand, TiO_2 nanowire treatment alone in a comparable dose did not influence motor functions as compared to the untreated traumatized group (see Sharma et al., 2007, 2009c).

These observations indicate that nanodrug delivery could considerably enhance the therapeutic effects of the compounds in SCI probably due to a greater accessibility and lesser degradability within the cord. Obviously, nanowired drugs may have then greater effects on neural cells and receptors as compared to the normal compounds. However, nanowires alone did not induce any beneficial effects on motor function or spinal cord pathology following trauma. This suggests that nanowired drugs and not the nanowires themselves have superior neuroprotective effects in SCI.

Nanodrug delivery and BSCB breakdown in SCI

The nanowired drugs are able to reduce SCI-induced BSCB breakdown to Evans blue and [131]Iodine tracers in the cord (Sharma et al., 2007a, 2009c). Our observations show that treatment with nanowired compounds resulted in a considerably higher protection of the BSCB function following trauma as compared to the normal compounds (Table 3). This effect was most pronounced by the nanowired compound AP713 followed by AP374 and AP173. Interestingly, nanowire treatment alone did not reduce the leakage of protein tracers in the cord after injury (Table 3). These observations show that nanowired drugs are more efficient in reducing BSCB permeability in SCI.

Nanodrug delivery and edema formation in SCI

When these neuroprotective compounds were administered tagged with nanowires, their potential antiedematous efficacies were further enhanced as compared to the normal compounds (Fig. 3). This effect was most marked with nanowired AP713 (Fig. 3). The TiO_2 alone did not reduce water content of the spinal cord following injury (Table 3).

A significant reduction in water content and volume swelling occurred in the traumatized portion of the cord (T10–11 segments) (see Table 3) after treatment with the nanowired-compounds. This indicates that drugs that are able to reduce BSCB breakdown to proteins are also antiedematous and nanowired drugs are the most effective agents in reducing edema formation and volume swelling in SCI (see Sharma, 2008, 2009a; Sharma & Olsson, 1990; Sharma, Westman, Cervós-Navarro, et al., 1998; Sharma, Westman, & Nyberg, 1998; Sharma et al., 1991, 2003a.c, 1995; Winkler et al., 1988).

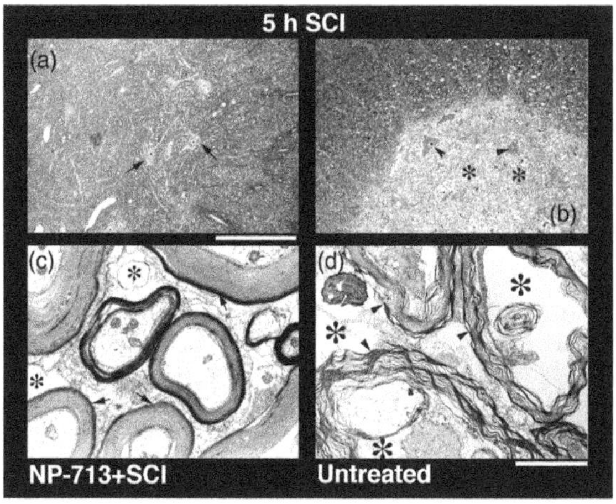

Fig. 3. Effect of nanowired compound 713 on SCI at light (a) and electron microscopy (c) as compared to saline-treated injured cord (b, d). Nanowired compound induced marked neuroprotection at light and electron microscopy as compared to untreated group following 5 h SCI. Bar = a, b = 25 μm; c, d = 1 μm. For details see text. Data modified after Sharma (2007a, 2009a).

Nanodrug delivery and cord pathology in SCI

Using light microscopy we observed that spinal cord cell and tissue destruction was minimal in nanowired AP713-treated group as compared to AP713 treatment alone at 5 h after SCI (Fig. 3). On the other hand, only mild to moderate neuroprotection was observed when AP173 and AP364 nanowired compounds were given under identical conditions (see Sharma et al., 2007a, 2009c). Electron microscopy also revealed preservation of myelin and neuropil in nanowired AP713-treated rats after injury (Fig. 3) indicating that nanowired drugs are able to attenuate cell and tissue injury at the ultrastructural level as well (Sharma et al., 2009c).

Possible mechanisms of nanowired drugs-induced neuroprotection

Our observations clearly demonstrate that nanodrug delivery enhances the therapeutic efficacy of the compounds when administered locally in SCI. This indicates that the nanowired compounds have greater accessibility within the cord cells and tissues to exert their actions on neural and nonneuronal cells more effectively than the respective parent compounds. Interestingly, the nanowires alone did not induce any neuroprotection in SCI. This suggests that suitable drugs in combination with nanowires are needed to achieve superior neuroprotective effects in CNS injuries (see Sharma et al., 2007a, 2009b, 2009c, 2009d).

It appears that increased bioavailability of nanowired drugs (Yao, Doose, Novak, & Bialer, 2006) and reduced biodegradation (Olivier et al., 1999) of compounds tagged with nanowires leading to long-term bindings with their receptors

(Assie et al., 2007) and/or potentiation of the receptor-mediated intracellular signal transduction (Zhao et al., 2006). Obviously, all these factors could be responsible for an enhanced neuroprotective effect of nanowired drugs in SCI (see Sharma & Sharma, 2007; Sharma et al., 2007a, 2009c). However, additional research is needed to clarify the detailed mechanisms of superior neuroprotective effects induced by nanowired drugs.

We used biocompatible titanate nanofiber scaffolds on the surface of titanium foil/mesh for nanowired drug delivery. This mode of nanowired drug delivery is technically superior in inducing controlled release of compounds in the cell culture or within the brain or spinal cord tissues as compared to other methods of drug delivery, for example, using nanoparticles and/or nanotubes (Kane & Stroock, 2007). The TiO_2-based nanowires are currently being used in medicine for drug delivery, imaging, and diagnostic pathology (Gommersall et al., 2007). Our new method for fabricating the nanowire bioscaffold directly on Ti (Dong et al., 2006) could also be responsible for an enhanced drug delivery through TiO_2 nanowries in the spinal cord tissues *in vivo* situation (Sharma & Sharma, 2007; Sharma et al., 2007a, 2009c). Thus, the highly scaffolding TiO_2 nanofibers could provide controlled release of drugs within the tissue and dense cellular binding sites that are needed for enhanced cellular activity and/or regeneration (Brokx, Bisland, & Gariepy, 2002; Chi, Victorio, & Jin, 2007; Silva et al., 2004).

A possibility that nanowired compounds penetrate faster, deeper, and widespread into the spinal cord tissues to exert greater beneficial effects in SCI compared to the parent compounds is also quite likely (Betbeder et al., 2000). Obviously, a rapid and widespread transport of nanowired drugs across the spinal cord will induce more effective neuroprotection in SCI. This idea is further supported by the fact that the nanowiring did not alter the pharmacological profile of the drugs. Thus, nanowired AP713 was the most effective compound to induce neuroprotection in SCI as compared to AP173 and AP364 treatment.

It remains to be seen whether these drugs if labeled to nanowires of other metals or encapsulated in liposomes, for example FeO2 nanoparticles or soft nanoparticles, and given various time intervals after trauma could also be able to induce effective neuroprotection in SCI. It is possible to incorporate drugs into nanocapsules or nanospheres so that the compound in question could be restricted within a vesicular system of nanoparticles (Brigger et al., 2002). Using this technology, drugs may be targeted to specific tissue sites, such as brain or spinal cord, to achieve their maximal effects (Kreuter, 2001). This is a subject that is currently being investigated in our laboratory in CNS injury using several animal models.

General conclusion

The research reported in this review demonstrates that nanoparticles are able to enhance the biological response of injury if given alone in animals prior to trauma. Similarly, nanoparticles when attached to the drug are administered into the system; they are capable to enhance the therapeutic efficacy of the compounds. This suggests that nanoparticles are important tools to influence biological responses of injury or drugs. Thus, further research is needed to find out how these properties of nanoparticles may be used for the advantages of patients of CNS trauma for faster recovery in future clinical strategies. Our investigations also highlighted the fact that nanoparticle exposure could aggravate the pathophysiology of trauma. This suggests that military exercises performed in the environment where nanoparticles from silica dust, or gunpowder explosion are present in high amount; CNS injury occurring to soldiers in such environments may have higher impact in victims than occurring in normal environments. Thus, additional precautionary measures may be taken in such cases during treatments with available drugs so that the magnitude and intensities of cell or tissue injury may be kept under good control. Initial investigations in our laboratories show that high concentration of drugs for long periods of time are needed to control trauma-induced pathology in nanoparticle-exposed cases. Thus, nanodrug delivery could be the need of the hour in nanoparticle-exposed subjects following injury to minimize the progression and persistence of their brain pathology. In addition, a multiple combination of different neuroprotective drugs are needed to regulate different aspects of cellular and molecular stress of the injured tissues to enhance neurorepair. The possibility that the bioavailability of drugs and their actions on the membrane receptors are altered in nanoparticle-exposed cases may also be considered. Thus, these factors may be taken into account while designing suitable therapeutic measures for CNS injury in human cases of nanoparticles exposure in the battlefield. Our laboratory is currently engaged in animal experiments to find multiple combinations of drugs with or without nanowires in order to attenuate brain and spinal cord pathology in nanoparticle-intoxicated groups.

Acknowledgments

This investigation is partially supported by the European Office of Aerospace Research and Development (EOARD)(London), UK Air Force Material Command, USAF, under grant number FA8655-05-1-3065. The U.S. Government is authorized to reproduce and distribute reprints for Government purpose notwithstanding any copyright notation thereon. The views and conclusions contained herein are those of the authors and should not be interpreted as necessarily representing the official policies or endorsements, either expressed or implied, of the Air Force Office of Scientific Research or the U.S. Government. Financial support from Swedish Medical Research Council (Grant Nr. 2710, HSS); Astra-Zeneca, Mölndal, Sweden (HSS); Alexander von Humboldt Foundation, Germany (HSS); IPSEN Medical, France (HSS); and University Grants Commission, New Delhi, India (HSS); and Indian Council of Medical Research, New Delhi, India (HSS) is gratefully acknowledged. The authors have no conflict of interest with any financial agencies mentioned above. We express our sincere gratitude to Col Robert Kang (EROARD, London, UK), Tomas Winkler, Torbjörn Lundstedt (Uppsala), Syed F Ali, and R Z Tian (USA) for providing important input to this review. Their help and contribution in part in some of the original investigations on which this review is based is acknowledged with thanks.

References

Assie, M. B., Bardin, L., Auclair, A., Consul-Denjean, N., Sautel, F., Depoortere, R., et al. (2007). F15063, a potential antipsychotic with dopamine D(2)/D(3) antagonist, 5-HT (1A) agonist and D(4) partial agonist properties: (IV) duration of brain D2-like receptor occupancy and antipsychotic-like activity versus plasma concentration in mice. *Naunyn-Schmiedeberg's Archives of Pharmacology*, *375*(4), 241–250. Epub 2007 Apr 24.

Babb, R., Brasser, B. A., Hosack, K. R., Severe, E., & Stone, K. (2009). New year, new cost constraints. *Rehab Management*, *22*(1), 26, 28–31.

Bahrami, A. R., & Mahjub, H. (2003). Comparative study of lung function in Iranian factory workers exposed to silica dust. *Eastern Mediterranean Health Journal*, *9*(3), 390–398.

Bao, F., Chen, Y., Schneider, K. A., & Weaver, L. C. (2008). An integrin inhibiting molecule decreases oxidative damage and improves neurological function after spinal cord injury. *Experimental Neurology*, *214*(2), 160–167. Epub 2008 Sep 26.

Baun, A., Hartmann, N. B., Grieger, K. D., & Hansen, S. F. (2009). Setting the limits for engineered nanoparticles in European surface waters – are current approaches appropriate? *Journal of Environmental Monitoring*, *11*(10), 1774–1781. Epub 2009 Aug 12.

Bennett, W. D. (2002) Rapid translocation of nanoparticles from the lung to the bloodstream? *American Journal of Respiratory and Critical Care Medicine*, *165*(12), 1671–1672.

Betbeder, D., Sperandio, S., Latapie, J. P., de Nadai, J., Etienne, A., Zajac, J. M., et al. (2000). Biovector nanoparticles improve antinociceptive efficacy of nasal morphine. *Pharmaceutical Research*, *17*(6), 743–748.

Borm, P. J., Robbins, D., Haubold, S., Kuhlbusch, T., Fissan, H., Donaldson, K., et al. (2006). The potential risks of nanomaterials: A review carried out for ECETOC. *Particle and Fibre Toxicology*, *3*, 11.

Brigger, I., Morizet, J., Aubert, G., Chacun, H., Terrier-Lacombe, M. J., Couvreur, P., et al. (2002). Poly(ethylene glycol)-coated hexadecylcyanoacrylate nanospheres display a combined effect for brain tumor targeting. *Journal of Pharmacology and Experimental Therapeutics*, *303*(3), 928–936.

Brokx, R. D., Bisland, S. K., & Gariepy, J. (2002). Designing peptide-based scaffolds as drug delivery vehicles. *Journal of Controlled Release*, *78*(1–3), 115–123.

Brooking, J., Davis, S. S., & Illum, L. (2001). Transport of nanoparticles across the rat nasal mucosa. *Journal of Drug and Targeting*, *9*(4), 267–279.

Brown, D. M., Wilson, M. R., MacNee, W., Stone, V., & Donaldson, K. (2001). Size-dependent proinflammatory effects of ultrafine polystyrene particles: A role for surface area and oxidative stress in the enhanced activity of ultrafines. *Toxicology and Applied Pharmacology*, *175*, 191–199. [PubMed]

Cavagna, G., Finulli, M., & Vigliani, E. C. (1961). [Experimental study on the pathogenesis of fevers caused by the inhalation of Teflon (polytetrafluoroethylene) fumes.] [Article in Italian] *Medicina del Lavoro*, *52*, 251–261.

Chen, Z., Meng, H., Xing, G., Yuan, H., Zhao, F., Liu, R., et al. (2008). Age-related differences in pulmonary and cardiovascular responses to SiO2 nanoparticle inhalation: Nanotoxicity has susceptible population. *Environmental Science and Technology*, *42*(23), 8985–8992.

Chi, B., Victorio, E. S., & Jin, T. (2007). Synthesis of TiO2-based nanotube on Ti substrate by hydrothermal treatment. *Journal of Nanoscience and Nanotechnology*, *7*(2), 668–672.

Chvatal, S. A., Kim, Y. T., Bratt-Leal, A. M., Lee, H., & Bellamkonda, R. V. (2008). Spatial distribution and acute anti-inflammatory effects of methylprednisolone after sustained local delivery to the contused spinal cord. *Biomaterials*, *29*(12), 1967–1975. Epub 2008 Feb 5

Coleman, W. E., Scheel, L. D., Kupel, R. E., & Larkin, R. L. (1968). The identification of toxic compounds in the pyrolysis products of polytetrafluoroethylene (PTFE). *American Industrial Hygiene Association Journal*, *29*(1), 33–40.

Costigan, S. (2006). The toxicology of nanoparticles used in health care products. Available at the website of the Medicines and Healthcare products Regulatory Agency, Department of Health, UK. Retrieved November 15, 2008, from http://www.mhra.gov.uk/home/idcplg?IdcService=SS_GET_PAGE&nodeId=996

De Jong, W. H., & Borm, P. J. (2008). Drug delivery and nanoparticles: Applications and hazards. *International Journal of Nanomedicine*, *3*(2), 133–149. Review.

Deng-Bryant, Y., Singh, I. N., Carrico, K. M., & Hall, E. D. (2008). Neuroprotective effects of tempol, a catalytic scavenger of peroxynitrite-derived free radicals, in a mouse traumatic brain injury model. *Journal of Cerebral Blood Flow and Metabolism*, *28*(6), 1114–1126. Epub 2008 Mar 5.

Dong, W., Zhang, T., McDonald, M., Padilla, C., Epstein, J., & Tian, Z. R. (2006). Biocompatible nanofiber scaffolds on metal for controlled release and cell colonization. *Nanomedicine*, *2*(4), 248–252.

Duncan, R. (2003). The dawning era of polymer therapeutics. *Nature Reviews Drug Discovery*, *2*, 347–360.

Duncan, R., & Izzo, L. (2005). Dendrimer biocompatibility and toxicity. *Advanced Drug Delivery Reviews*, *57*, 2215–2237.

Dunning, M. D., Lakatos, A., Loizou, L., Kettunen, M., ffrench-Constant, C., Brindle, K. M., et al. (2004). Superparamagnetic iron oxide-labeled Schwann cells and olfactory ensheathing cells can be traced in vivo by magnetic resonance imaging and retain functional properties after transplantation into the CNS. *Journal of Neuroscience*, *24*(44), 9799–9810.

Dyer, M., Hinchcliffe, M., Watts, P., Castile, J., Jabbal-Gill, I., Nankervis, R., et al. (2002). Nasal delivery of insulin using novel chitosan based formulations: A comparative study in two animal models between simple chitosan formulations and chitosan nanoparticles. *Pharmaceutical Research*, *19*, 998–1008.

European Science Foundation. (2005). Policy Briefing (ESF), ESF Scientific Forward Look on Nanomedicine IREG Strasbourg, France, ISBN 2-912049-520.

Euliss, L. E., Dupont, J. A., Gratton, S., & Desimone, J. (2006). Imparting size, shape, and composition control of materials for nanomedicine. *Chemical Society Reviews*, *35*(11), 1095–1104. Epub 2006 Sep 20.

Fang, C., Shi, B., Pei, Y. Y., Hong, M. H., Wu, J., & Chen, H. Z. (2006). In vivo tumor targeting of tumor necrosis factor-alpha-loaded stealth nanoparticles: Effect of MePEG molecular weight and particle size. *European Journal of Pharmaceutical Science*, *27*, 27–36.

Fang, J. Y. (2006). Nano- or submicron-sized liposomes as carriers for drug delivery. *Chang Gung Medical Journal*, *29* (4), 358–362. Review.

Farokhzad, O. C., & Langer, R. (2006). Nanomedicine: Developing smarter therapeutic and diagnostic modalities. *Advanced Drug Delivery Reviews*, *58*(14), 1456–1459.

Fenoglio, I., Tomatis, M., Lison, D., Muller, J., Fonseca, A., Nagy, J. B., et al. (2006). Reactivity of carbon nanotubes: Free radical generation or scavenging activity? *Free Radical Biology & Medicine*, *40*(7), 1227–1233. Epub 2005 Dec 9.

Florence, A. T., Hillery, A. M., Hussain, N., & Jani, P. U. (1995). Factors affecting the oral uptake and translocation of polystyrene nanoparticles: Histological and analytical evidence. *Journal of Drug and Targeting*, *3*(1), 65–70.

Fubini, B., Giamello, E., Volante, M., & Bolis, V. (1990). Chemical functionalities at the silica surface determining its reactivity when inhaled. Formation and reactivity of surface radicals. *Toxicology and Industrial Health*, *6*(6), 571–598. Review.

Gao, K., & Jiang, X. (2006). Influence of particle size on transport of methotrexate across blood brain barrier by polysorbate 80-coated polybutylcyanoacrylate nanoparticles. *International Journal of Pharmaceutics, 310*(1–2), 213–219. Epub 2006 Jan 19.

Gibaud, S., Demoy, M., Andreux, J. P., Weingarten, C., Gouritin, B., & Couvreur, P. (1996). Cells involved in the capture of nanoparticles in hematopoietic organs. *Journal of Pharmaceutical Sciences*, *85*, 944–950.

Gomes, A. J., Lunardi, L. O., Marchetti, J. M., Lunardi, C. N., & Tedesco, A. C. (2006). Indocyanine green nanoparticles useful for photomedicine. *Photomedicine and Laser Surgery*, *24*, 514–521.

Gommersall, L., Shergill, I. S., Ahmed, H. U., Hayne, D., Arya, M., Patel, H. R., et al. (2007). Nanotechnology and its relevance to the urologist. *European Urology*, *52*, 368–375.

Griffith, F. D., Stephens, S. S., & Tayfun, F. O. (1973). Exposure of Japanese quail and parakeets to the pyrolysis products of fry pans coated with Teflon and common cooking oils. *American Industrial Hygiene Association Journal*, *34*(4), 176–178.

Gupta, A. K., & Gupta, M. (2005). Synthesis and surface engineering of iron oxide nanoparticles for biomedical applications. *Biomaterials*, *26*, 3995–4021.

He, W., Yong, T., Ma, Z. W., Inai, R., Teo, W. E., & Ramakrishna, S. (2006). Biodegradable polymer nanofiber mesh to maintain functions of endothelial cells. *Tissue Engineering*, *12*(9), 2457–2466.

Ichinose, T., Yoshida, S., Sadakane, K., Takano, H., Yanagisawa, R., Inoue, K., et al. (2006). Effects of Asian sand dust, Arizona sand dust, amorphous silica and aluminum oxide on allergic inflammation in the murine lung. *Inhalation Toxicology*, *20*(7), 685–694.

Johnston, C. J., Finkelstein, J. N., Mercer, P., Corson, N., Gelein, R., & Oberdorster, G. (2008). Pulmonary effects induced by ultrafine PTFE particles. *Toxicology and Applied Pharmacology*, *168*(3), 208–215.

Kane, R. S., & Stroock, A. D. (2007). Nanobiotechnology: Protein-nanomaterial interactions. *Biotechnology Progress*, *23*(2), 316–319. Epub 2007 Mar 3.

Kim, J. S., Yoon, T. J., Yu, K. N., Noh, M. S., Woo, M., Kim, B. G., et al. (2006). Cellular uptake of magnetic nanoparticle is mediated through energydependent endocytosis in A549 cells. *Journal of Veterinary Science*, *7*(4), 321–326.

Kipp, J. E. (2004). The role of solid nanoparticle technology in the parental delivery of poorly water-soluble drugs. *International Journal of Pharmaceutics*, *284*, 109–122.

Kiyatkin, E. A., Brown, P. L., & Sharma, H. S. (2007). Brain edema and breakdown of the blood-brain barrier during methamphetamine intoxication: Critical role of brain hyperthermia. *European Journal of Neuroscience*, *26*(5), 1242–1253.

Koziara, J. M., Lockman, P. R., Allen, D. D., & Mumper, R. J. (2006). The blood-brain barrier and brain drug delivery. *Journal of Nanoscience and Nanotechnology*, *6*(9–10), 2712–2735.

Kreuter, J. (2001). Nanoparticulate systems for brain delivery of drugs. *Advanced Drug Delivery Reviews*, *47*(1), 65–81. Review.

Lam, C. W., James, J. T., McCluskey, R., Arepalli, S., & Hunter, R. L. (2006). A review of carbon nanotube toxicity and assessment of potential occupational and environmental health risks. *Critical Reviews in Toxicology*, *36*(3), 189–217. Review.

Lanone, S., & Boczkowski, J. (2006). Biomedical applications and potential health risks of nanomaterials: Molecular mechanisms. *Current Molecular Medicine*, *6*(6), 651–663.

Leung, A. K., Wong, A. F., Wong, E. C., & Hutchins, S. W. (2009). The Physiological Cost Index of walking with an isocentric reciprocating gait orthosis among patients with T (12)–L(1) spinal cord injury. *Prosthetics and Orthotics International*, *33*(1), 61–68.

Manna, S. K., Sarkar, S., Barr, J., Wise, K., Barrera, E. V., Jejelowo, O., et al. (2005). Single-walled carbon nanotube induces oxidative stress and activates nuclear transcription factor-kappaB in human keratinocytes. *Nano Letters*, *5*(9), 1676–1684.

Mertens, H., Biteen, J. S., Atwater, H. A., & Polman, A. (2006). Polarization-selective plasmon-enhanced silicon quantum-dot luminescence. *Nano Letters*, *6*(11), 2622–2625.

Modi, G., Pillay, V., Choonara, Y. E., Ndesendo, V. M., du Toit, L. C., & Naidoo, D. (2009). Nanotechnological applications for the treatment of neurodegenerative disorders. *Progress in Neurobiology*, *88*(4), 272–285. Epub 2009 May 30.

Moghimi, S. M. (2006). Recent developments in polymeric nanoparticle engineering and their applications in experimental and clinical oncology. *Anti-Cancer Agents in Medicinal Chemistry*, *6*(6), 5535–5561.

Moghimi, S. M., Hunter, A. C., & Murray, J. C. (2001). Long-circulating and target specific nanoparticles: Theory and practice. *Pharmacological Reviews*, *53*, 283–318.

Mills, N. L., Amin, N., Robinson, S. D., Anand, A., Davies, J., Patel, D., et al. (2006). Do inhaled carbon nanoparticles translocate directly into the circulation in humans? *American Journal of Respiratory and Critical Care Medicine*, *173*(4), 426–431. Epub 2005 Dec 9.

Mistry, A., Stolnik, S., & Illum, L. (2009). Nanoparticles for direct nose-to-brain delivery of drugs. *International Journal of Pharmaceutics*, *379*(1), 146–157. Epub 2009 Jun 23.

Na, K., Lee, K. H., Lee, D. H., Bea, U. H., (2006). Biodegradable thermo-sensitive nanoparticles from poly(l-lactic acid)/poly(ethylene glycol) alternating multi-block copolymer for potential anti-cancer drug carrier. *European Journal of Pharmaceutical Science*, *27*, 115–122.

Neretti, N., Wang, P. Y., Brodsky, A. S., Nyguyen, H. H., White, K. P., Rogina, B., et al. (2009). Long-lived Indy induces reduced mitochondrial reactive oxygen species production and oxidative damage. *Proceedings of the National Academy of Sciences of the United States of America*, *106*(7), 2277–2282. Epub 2009 Jan 21.

Nicolini, C., Lanzi, M., Accossato, P., Fanigliulo, A., Mattioli, F., & Martelli, A. (1995, Fall). A silicon-based biosensor for real-time toxicity testing in normal versus cancer liver cells. *Biosensors and Bioelectronics*, *10*(8), 723–733.

Niidome, T., Yamagata, M., Okamoto, Y., Akiyama, Y., Takahashi, H., Kawano, T., Katayama, Y., & Niidome, Y. (2006). PEG-modified gold nanorods with a stealth character for in vivo applications. *Journal of Controlled Release*, *114*, 343–347.

Oberdörster, G. (1996). Significance of particle parameters in the evaluation of exposure-dose-response relationships of inhaled particles. *Inhal Toxicol*, *8 Suppl*, 73–89. Review.

Oberdorster, G., Maynard, A., Donaldson, K., Castranova, V., Fitzpatrick, J., Ausman, K., et al., ILSI Research Foundation/Risk Science Institute Nanomaterial Toxicity Screening Working Group. (2005). Principles for characterizing the potential human health effects from exposure to nanomaterials: Elements of a screening strategy. *Particle and Fibre Toxicology*, *2*, 8.

Oberdorster, G., Elder, A., & Rinderknecht, A. (2009). Nanoparticles and the brain: Cause for concern? *Journal of Nanoscience and Nanotechnology*, *9*, 4996–5007.

Olivier, J. C., Fenart, L., Chauvet, R., et al. (1999). Indirect evidence that drug brain targeting using polysorbate

80-coated polybutylcyanoacrylate nanoparticles is related to toxicity. *Pharmaceutical Research, 16*, 1836–1842.

Olsson, Y., Sharma, H. S., Nyberg, F., & Westman, J. (1995). The opioid receptor antagonist naloxone influences the pathophysiology of spinal cord injury. *Progress in Brain Research, 104*, 381–399.

Olsson, Y., Sharma, H. S., & Pettersson, C. A. (1990). Effects of p-chlorophenylalanine on microvascular permeability changes in spinal cord trauma. An experimental study in the rat using 131I-sodium and lanthanum tracers. *Acta Neuropathologica, 79*(6), 595–603.

Olsson, Y., Sharma, H. S., Pettersson, Å, & Cervós-Navarro, J. (1992). Endogenous release of neurochemicals may increase vascular permeability, induce edema and influence on cell changes in trauma to the spinal cord. *Progress in Brain Research, 91*, 197–203.

Peracchia, M. T., Fattal, E., Desmaële, D. Besnard, M. Noël, J. P., Gomis, J. M., Appel, M., d'Angelo, J. Couvreur, P. (1999). Stealth PEGylated polycyanoacrylate nanoparticles for intravenous administration and splenic targeting. *Journal of Controlled Release, 60*, 121–128.

Pettersson, C. A., Sharma, H. S., & Olsson, Y. (1990). Vascular permeability of spinal nerve roots. A study in the rat with Evans blue and lanthanum as tracers. *ACTA Neuropathologica, 81*(2), 148–154.

Ren, T., Xu, N., Cao, C., Yuan, W., Yu, X., Chen, J., et al. (2009). Preparation and therapeutic efficacy of polysorbate-80-coated amphotericin B/PLA-b-PEG nanoparticles. *Journal of Biomaterials Science, Polymer Edition, 20*(10), 1369–1380.

Ricci-Junior, E., & Marchetti, J. M. (2006). Preparation, characterization, photocytotoxicity assay of PLGA nanoparticles containing zinc(II) phthalocyanine for photodynamic therapy use. *Journal of Microencapsulation, 23*, 523–538.

Saez, A., Guzmán, M., Molpeceres, J., Aberturas, M. R. (2000). Freeze-drying of polycaprolactone and poly(D,L-lactic-glycolic) nanoparticles induce minor particle size changes affecting the oral pharmacokinetics of loaded drugs. *European Journal of Pharmaceutics and Biopharmaceutics, 50*, 379– 387.

[SCENIHR] EU Scientific Committee on Emerging and Newly Identified Health Risks. (2008). Modified Opinion (after public consultation) on the appropriateness of existing methodologies to assess the potential risks associated with engineered and adventitious products of nanotechnologies. 2006. SCENIHR/002/05. Retrieved June 23, 2008, from http://ec.europa.eu/health/ph_risk/committees/04_scenihr/docs/scenihr_o_003b.pdf.

Schins, R. P., Duffin, R., Höhr, D. Knaapen, A. M., Shi, T., Weishaupt, C., Stone, V., Donaldson, K., Borm, P. J. (2002). Surface modification of quartz inhibits toxicity, particle uptake, and oxidative DNA damage in human lung epithelial cells. *Chemical Research in Toxicology, 15*, 1166–1173.

Seki, J., Sonoke, S., Saheki, A., Fukui, H., Sasaki, H., Mayumi, T. (2004). A nanometer lipid emulsion, lipid nano-sphere (LNS), as a parenteral drug carrier for passive drug targeting. *International Journal of Pharmaceutics, 273*, 75–83.

Senior, J., Delgado, C., Fisher, D., Tilcock, C., Gregoriadis, G. (1991). Influence of surface hydrophoilicity of liposomes on their interaction with plasma protein and clearance from the circulation: Studies with poly(ethylene glycol)-coated vesicles. *Biochimica et Biophysica Acta (BBA)-Biomembranes, 1062*(1), 77–82.

Senior, J., & Gregoriadis, G. (1982). Is half-life of circulating small unilamellar liposomes determined by changes in their permeability? *FEBS Letters, 145*, 109–114.

Sharma, H. S. (2004). Pathophysiology of the blood-spinal cord barrier in traumatic injury. In H. S. Sharma & J. Westman (Eds.), *The blood-spinal cord and brain barriers in health and disease* (pp. 437–518). San Diego: Elsevier Academic Press.

Sharma, H. S. (2005). Pathophysiology of blood-spinal cord barrier in traumatic injury and repair. *Current Pharmaceutical Design, 11*(11), 1353–1389. Review.

Sharma, H. S. (2006). Hyperthermia influences excitatory and inhibitory amino acid neurotransmitters in the central nervous system. An experimental study in the rat using behavioural, biochemical, pharmacological, and morphological approaches. *Journal of Neural Transmission, 113*(4), 497–519.

Sharma, H. S. (2007a). Neurotrophic factors in combination: A possible new therapeutic strategy to influence pathophysiology of spinal cord injury and repair mechanisms. *Current Pharmaceutical Design, 13*(18), 1841–1874. Review.

Sharma, H. S. (2007b). Nanoneuroscience: Emerging concepts on nanoneurotoxicity and nanoneuroprotection. *Nanomedicine, 2*(6), 753–758. Editorial.

Sharma, H. S. (2007c). A select combination of neurotrophins enhances neuroprotection and functional recovery following spinal cord injury. *Annals of the New York Academy of Sciences, 1122*, 95–111.

Sharma, H. S. (2008). New perspectives for the treatment options in spinal cord injury. *Expert Opinion on Pharmacotherapy, 9*(16), 2773–2800. Review.

Sharma, H. S. (2009a). Blood–central nervous system barriers: The gateway to neurodegeneration, neuroprotection and neuroregeneration. In A. Lajtha, N. Banik, & S. K. Ray (Eds.), *Handbook of neurochemistry and molecular neurobiology* (pp. 363–457). New York: Springer US.

Sharma, H. S. (2009b). Nanoneuroscience: Nanoneurotoxicity and nanoneuroprotection. Editorial, *Journal of Nanoscience and Nanotechnology, 9*, 4992–4995.

Sharma, H. S., Ali, S. F., Dong, W., Tian, Z. R., Patnaik, R., Patnaik, S., et al. (2007a). Drug delivery to the spinal cord tagged with nanowire enhances neuroprotective efficacy and functional recovery following trauma to the rat spinal cord. *Annals of the New York Academy of Sciences, 1122*, 197–218.

Sharma, H. S., Ali, S. F., Hussain, S. M., Schlager, J. J., & Sharma, A. (2009d). Influence of engineered nanoparticles from metals on the blood-brain barrier permeability, cerebral blood flow, brain edema and neurotoxicity. An experimental study in the rat and mice using biochemical and morphological approaches. *Journal of Nanoscience and Nanotechnology, 9*, 5055–5072.

Sharma, H. S., Ali, S. F., Tian, Z. R., Hussain, S. M., Schlager, J. J., Sjöquist, P.-O., et al. (2009c). Chronic treatment with nanoparticles exacerbate hyperthermia induced blood-brain barrier breakdown, cognitive dysfunction and brain pathology in the Rat. Neuroprotective effects of nanowired-antioxidant compound H-290/51. *Journal of Nanoscience and Nanotechnology*, *9*, 5073–5090.

Sharma, H. S., Ali, S., Tian, Z. R., Patnaik, R., Patnaik, S., Lek, P., et al. (2009b). Nano-drug delivery and neuroprotection in spinal cord injury. *Journal of Nanoscience and Nanotechnology*, *9*, 5014–5037. Review.

Sharma, H. S., Gordh, T., Wiklund, L., Mohanty, S., & Sjöquist, P. O. (2006, April). Spinal cord injury induced heat shock protein expression is reduced by an antioxidant compound H-290/51. An experimental study using light and electron microscopy in the rat. *Journal of Neural Transmission*, *113* (4), 521–536.

Sharma, H. S., & Kiyatkin, E. A. (2009). Rapid morphological brain abnormalities during acute methamphetamine intoxication in the rat: An experimental study using light and electron microscopy. *Journal of Chemical Neuroanatomy*, *37*(1), 18–32. Epub 2008 Aug 19.

Sharma, H. S., Lundstedt, T., Flärdh, M., Skottner, A., & Wiklund, L. (2006b). Neuroprotective effects of melanocortins in CNS injury. *Current Pharmaceutical Design*, *13*(19), 1929–1941. Review.

Sharma, H. S., & Olsson, Y. (1990). Edema formation and cellular alterations following spinal cord injury in rat and their modification with p-chlorophenylalanine. *Acta Neuropathologica (Berlin)*, *79*, 604–610.

Sharma, H. S., Olsson, Y., & Cervós-Navarro, J. (1993a). Early perifocal cell changes and edema in traumatic injury of the spinal cord are reduced by indomethacin, an inhibitor of prostaglandin synthesis. *Acta Neuropathologica (Berlin)*, *85*, 145–153.

Sharma, H. S., Olsson, Y., & Cervós-Navarro, J. (1993b). p-Chlorophenylalanine, a serotonin synthesis inhibitor, reduces the response of glial fibrillary acidic protein induced by trauma to the spinal cord. *Acta Neuropathologica (Berlin)*, *86*, 422–427.

Sharma, H. S., Olsson, Y., & Dey, P. K. (1990). Early accumulation of serotonin in rat spinal cord subjected to traumatic injury. Relation to edema and blood flow changes. *Neuroscience*, *36*, 725–730.

Sharma, H. S., Olsson, Y., Nyberg, F., & Dey, P. K. (1993c). Prostaglandins modulate alterations of microvascular permeability, blood flow, edema and serotonin levels following spinal cord injury: An experimental study in the rat. *Neuroscience*, *57*(2), 443–449.

Sharma, H. S., Olsson, Y., & Westman, J. (1995). A serotonin synthesis inhibitor, p-chlorophenylalanine reduces the heat shock protein response following trauma to the spinal cord: An immunohistochemical and ultrastructural study in the rat. *Neuroscience Research*, *21*(3), 241–249.

Sharma, H. S., Patnaik, R., Sharma, A., Sjöquist, P.-O., & Lafuente, L. V. (2009b). Silicon dioxide nanoparticles (SiO2, 40–50 nm) exacerbate pathophysiology of traumatic spinal cord injury and deteriorate functional outcome in the rat. An experimental study using pharmacological and morphological approaches. *Journal of Nanoscience and Nanotechnology*, *9*, 4970–4980.

Sharma, H. S., & Sharma, A. (2007). Nanoparticles aggravate heat stress induced cognitive deficits, blood-brain barrier disruption, edema formation and brain pathology. *Progress in Brain Research*, *162*, 245–273. Review.

Sharma, H. S., & Sharma, A. (2008). Antibodies as promising novel neuroprotective agents in the central nervous system injuries. *Central Nervous System Agents in Medicinal Chemistry (Formerly Current Medicinal)*, *8*(3), 143–169. DOI: 10.2174/187152408785699640.

Sharma, H. S., & Sjöquist, P. O. (2002). A new antioxidant compound H-290/51 modulates glutamate and GABA immunoreactivity in the rat spinal cord following trauma. *Amino Acids*, *23*(1–3), 261–272.

Sharma, H. S., Sjöquist, P. O., & Alm, P. (2003). A new antioxidant compound H-290151 attenuates spinal cord injury induced expression of constitutive and inducible isoforms of nitric oxide synthase and edema formation in the rat. *ACTA Neurochirurgica Supplementum*, *86*, 415–420.

Sharma, H. S., Sjöquist P.-O., & Westman, J. (2001). Pathophysiology of the blood-spinal cord barrier in spinal cord injury. Influence of a new antioxidant compound H-290/51. In D. Kobiler, S. Lustig, & S. Shapra (Eds.), *Blood-brain barrier. Drug delivery and brain pathology* (pp. 401–416). New York: Kluwer Academic/Plenum Publishers.

Sharma, H. S., & Westman, J. (2004a). *The blood-spinal cord and brain barriers in health and disease* (pp. 1–617). San Diego: Academic Press (Release date: Nov. 9, 2003).

Sharma, H. S., & Westman, J. (2004b). The heat shock proteins and hemeoxygenase response in central nervous system injuries. In H. S. Sharma & J. Westman (Eds.), *The blood-spinal cord and brain barriers in health and disease* (pp. 329–360). San Diego: Elsevier Academic Press.

Sharma, H. S., Westman, J., Cervós-Navarro, J., Dey, P. K., & Nyberg, F. (1998a). Blood-brain barrier in stress: A gateway to various brain diseases. In A. Levy, E. Grauer, D. Ben-Nathan, & E. R. de Kloet (Eds.), *New frontiers of stress research: Modulation of brain function* (pp. 259–276). Amsterdam: Harwood Academic Publishers Inc.

Sharma, H. S., Westman, J., & Nyberg, F. (1998b). Pathophysiology of brain edema and cell changes following hyperthermic brain injury. *Progress in Brain Research*, *115*, 351–412.

Sharma, H. S., Wiklund, L., Badgaiyan, R. D., Mohanty, S., & Alm, P. (2006). Intracerebral administration of neuronal nitric oxide synthase antiserum attenuates traumatic brain injury-induced blood-brain barrier permeability, brain edema formation, and sensory motor disturbances in the rat. *ACTA Neurochirurgica Supplementum*, *96*, 288–294.

Sharma, H. S., Winkler, T., Stålberg, E., Olsson, Y., & Dey, P. K. (1991). Evaluation of traumatic spinal cord edema using evoked potentials recorded from the spinal epidural space. An experimental study in the rat. *Journal of the Neurological Sciences*, *102*, 150–162.

Shimizu, M., Tainaka, H., Oba, T., Mizuo, K., Umezawa, M., & Takeda, K. (2009). Maternal exposure to nanoparticulate titanium dioxide during the prenatal period alters gene expression related to brain development in the mouse. *Particle and Fibre Toxicology*, *6*, 20.

Shvedova, A. A., Kisin, E. R., Mercer, R., Murray, A. R., Johnson, V. J., Potapovich, A. I., et al. (2005). Unusual inflammatory and fibrogenic pulmonary responses to single-walled carbon nanotubes in mice. *American Journal of Physiology. Lung Cellular and Molecular Physiology, 289*(5), L698–L708.

Silva, G. A., Czeisler, C., Niece, K. L., Beniash, E., Harrington, D. A., Kessler, J. A., et al. (2004). Selective differentiation of neural progenitor cells by high-epitope density nanofibers. *Science, 303*(5662), 1352–1355.

Stålberg, E., Sharma, H. S., & Olsson, Y. (1998). *Spinal cord monitoring. Basic principles, regeneration, pathophysiology and clinical aspects* (pp. 1–527). New York: Springer Wien.

Teeguarden, J. G., Hinderliter, P. M., Orr, G., Thrall, B. D., & Pounds, J. G. (2007). Particokinetics in vitro: Dosimetry considerations for in vitro nanoparticle toxicity assessments. *Toxicological Sciences, 95*(2), 300–12. Epub 2006 Nov 10. Review. Erratum in *Toxicological Sciences* (2007) *97*(2), 614.

Tomazic-Jezic, V. J., Merritt, K., & Umbreit, T. H. (2001). Significance of the type and size of biomaterial particles on phagocytosis and tissue distribution. *Journal of Biomedical Materials Research, 55*, 523–529.

Vierikko, T., Kivistö, S., Järvenpää, R., Uitti, J., Oksa, P., Virtema, P., et al. (2009). Psychological impact of computed tomography screening for lung cancer and occupational pulmonary disease among asbestos-exposed workers. *European Journal of Cancer Prevention, 18*(3), 203–206.

Wagner, V., Dullaart, A., Bock, A. K., & Zweck, A. (2006). The emerging nanomedicine landscape. *Nature Biotechnology, 24* (10), 1211–1217.

Wang, Z. G., Ma, X. Y., Su, Y. B., Feng, B., Zhang, F. R., & Li Z. Q. (1994). Effects of vitamins and methionine on blastogenic response of splenocytes and lipid peroxidation in lung and liver of mice exposed to SiO2 dust. *Journal of Environmental Pathology, Toxicology and Oncology, 13*(3), 205–207.

Warheit, D. B., Laurence, B. R., Reed, K. L., Roach, D. H., Reynolds, G. A., & Webb, T. R. (2004). Comparative pulmonary toxicity assessment of single-wall carbon nanotubes in rats. *Toxicological Sciences, 77*(1), 117–125.

Williams, D. (2006). Quantum dots in medical technology. *Medical Device Technology, 17*(4), 8–9.

Winkler, T., Sharma, H. S., Stålberg, E., Westman, J. (1998) Spinal cord bioelectrical activity, edema and cell injury following a focal trauma to the spinal cord. An experimental study using pharmacological and morphological approach. In E. Stålberg, H. S. Sharma, & Y. Olsson (Eds.), *Spinal cord monitoring. basic principles, regeneration, pathophysiology and clinical aspects* (pp. 281–348). New York: Springer Wien.

Xia, T., Kovochich, M., Brant, J., Hotze, M., Sempf, J., Oberley, T., et al. (2006). Comparison of the abilities of ambient and manufactured nanoparticles to induce cellular toxicity according to an oxidative stress paradigm. *Nano Letters, 6*(8), 1794–1807.

Xiong, Y., & Hall, E. D. (2009). Pharmacological evidence for a role of peroxynitrite in the pathophysiology of spinal cord injury. *Experimental Neurology, 216*(1), 105–114. Epub 2008 Dec 11.

Yang, H., Liu, C., Yang, D., Zhang, H., & Xi, Z. (2009). Comparative study of cytotoxicity, oxidative stress and genotoxicity induced by four typical nanomaterials: The role of particle size, shape and composition. *Journal of Applied Toxicology, 29*(1), 69–78.

Yao, C., Doose, D. R., Novak, G., & Bialer, M. (2006). Pharmacokinetics of the new antiepileptic and CNS drug RWJ-333369 following single and multiple dosing to humans. *Epilepsia, 47*(11), 1822–1829.

Yin, Q. Q., Wu, L., Gou, M. L., Qian, Z. Y., Zhang, W. S., & Liu, J. (2009). Long-lasting infiltration anaesthesia by lidocaine-loaded biodegradable nanoparticles in hydrogel in rats. *ACTA Anaesthesiologica Scandinavica, 53*(9), 1207–1213. Epub 2009 Jul 29.

Zhang, W. W., Cao, Q. Q., Xie, J. L., Ren, X. M., Lu, C. S., Zhou, Y., et al. (2003). Structural, morphological, and magnetic study of nanocrystalline cobalt-nickel- copper particles. *Journal of Colloid and Interface Science, 257*(2), 237–243.

H.S. Sharma (Ed.)
Progress in Brain Research, Vol. 180
ISSN: 0079-6123

CHAPTER 10

Solid lipid nanoparticles and microemulsions for drug delivery: the CNS

Maria Rosa Gasco[1,*], Lorenzo Priano[2,3], and Gian Paolo Zara[4]

[1]*Nanovector s.r.l, Torino, Italy*
[2]*Department of Neurosciences, University of Turin, Torino, Italy*
[3]*IRCCS — Istituto Auxologico Italiano, Ospedale S. Giuseppe — Piancavallo, Verbania, Italy*
[4]*Department of Anatomy Pharmacology and Forensic Medicine, University of Turin, Torino, Italy*

Abstract: The chapter examined solid lipid nanoparticles (SLN) and microemulsions, chosen as carriers of drugs, administered *in vivo* to be transported to the central nervous system. Drugs of different structures and for different therapies have been studied such as doxorubicin SLN stealth and nonstealth administered in rats by intravenous route, apomorphine SLN administered in rats by duodenal route, melatonin SLN administered by transdermal and oral routes in humans, and apomorphine microemulsion administered by transdermal route in Parkinson's patients. The pharmacokinetics of the drug, followed in most studies, put in evidence that the many important pharmacokinetic parameters were notably improved versus the drug alone or in a commercial formulation.

Keywords: solid lipid nanoparticles; microemulsions; drug delivery system; central nervous system

Introduction

The brain homeostasis is of primary importance for survival so that specific interfaces, also referred to as barriers, tightly regulate the exchange between the peripheral blood circulation and the cerebrospinal fluid (CSF) circulatory system. These barriers are represented by the choroid plexus epithelium, the arachnoid epithelium, and the blood–brain barrier (BBB). The concentration and clearance of endogenous and exogenous molecules, essential for the normal brain functions or dangerous because of their toxicity, are strictly regulated by the anatomic and physiologic features of each barrier (Abbott, 2002; Segal, 2000).

The presence of the BBB is certainly the most critical issue encountered in brain drug delivery. Among the possible strategies to deliver therapeutic molecules into the brain, namely, intracerebral, intraventricular, and intravascular delivery, the latest represents the most reliable one because of its potential efficacy, safety, and compliance (Silva, 2007).

Brain capillaries, differently from the peripheral capillaries, present no fenestrae, a low amount of pinocytosis vesicles and particular tight junctions also known zonula occludens. Tight junctions are structures that form a narrow and continuous seal surrounding each endothelial and epithelial cell at the apical border and are at strictly regulating the movements the molecules through the paracellular

* Corresponding author.
E-mail: maria.gasco@nanovechor.it

DOI: 10.1016/S0079-6123(08)80010-6

pathway. These structures, together with the brain endothelial cells, make an almost impermeable barrier for drugs administered through the peripheral circulation (Kniesel & Wolburg, 2000; Lapierre, 2000).

A further contribution to the peculiar BBB functions is given by the periendothelial structures represented by astrocytes, pericytes, and the basal membrane (Balahanov & Dore-Duffy, 1998; Lay & Kuo, 2005).

The presence of BBB transport systems further complicates the scenario. In fact, these transporters may assist or hinder the drug delivery to the brain. The carrier-mediated transport may be able to shuttle drugs or prodrugs into the brain in therapeutic concentrations, mimicking nutrients or endogenous compounds (Conford & Hyman, 1999; Pardrige, 1998).

Unfortunately, the presence of active efflux transporters to the BBB also limits the therapeutic efficacy of drugs virtually able to access the brain. The P-glycoprotein (P-gp) is an ATP-dependent drug transport protein present at the apical membranes of different epithelial cell types including those forming the BBB.

Recently, it has been demonstrated, either *in vitro* or *in vivo*, that BBB P-gp can prevent the accumulation of many molecules including a variety of drugs in the brain (Stouch & Gudmundsson, 2002), and P-gp inhibition has been proposed as a possible strategy to enhance the drug penetration (Skinkel, 1999).

Different strategies have been studied for the delivery of drugs to the brain. Indeed most part of the small drug molecules and of large molecules such as recombinant proteins or gene-based molecules are not able to penetrate the BBB and many efforts have been spent in the previous years toward delivery and targeting of drugs to the brain (de Boer & Gaillard, 2007). Many investigations have been carried out in the previous years to improve brain tumors therapy with nanoparticulates; there are less number of studies regarding colloidal carriers of drugs for neurological diseases or of diagnostics. Liposomes, polymeric nanoparticles, and solid lipid nanoparticles (SLN) have been studied, with different approaches, and the problems of overcoming the BBB.

In this chapter, we consider SLN and microemulsions as carriers for the delivery only of drugs active on the central nervous system (CNS). In particular, examining drugs used for therapy in neurological diseases, as many times their administration gives problems, such as high amount of drug administered by parenteral route, short half-life, high hydrophilicity, and poor transport through the BBB. The aim of all the researchers is to study if some improvements in pharmacokinetic parameters in laboratory animals and/or in humans could be achieved using colloidal formulations; the review considers studies on SLN and microemulsions carrying only drugs active on CNS.

Solid lipid nanoparticles

Different approaches are followed for the SLN preparation.

They can be prepared by high-pressure homogenization at elevated or low temperatures, via warm microemulsions, by solvent emulsification–evaporation–diffusion, by high-speed stirring, and/or sonication (Muller, Kader, & Gohla, 2000).

Here we refer only about SLN carrying drugs active on CNS (at brain level).

SLN carrying the lipophilic antipsychotic drug clozapine were prepared by hot homogenization followed by ultrasonication method. Clozapine has a very poor bioavailability (Manjunath & Venkateswarlu, 2005). The SLN were administered by intravenous (IV) and duodenal routes to Swiss albino mice. For the intravenous administration, stearylamine was entrapped with clozapine in SLN; the area under curve (AUC) in the brain increased up to 2.91-fold the one of clozapine suspension.

The same authors (Manjunath & Venkateswarlu, 2006) developed SLN as carriers of the highly lipophilic drug nitrendipine, using different triglycerides for the lipid matrix, soy lecithin, and Poloxamer 188. Positive and negative charged nitrendipine SLN were also produced and then examined to explore the influence of the charge on oral bioavailability. The different kinds of SLN were administered by IV and intraduodenal routes to rats; pharmacokinetic parameters of nitrendipine SLN were examined, tissue distribution studies were carried out in Swiss albino mice, against that of a nitrendipine suspension. Following IV administration nitrendipine-loaded SLN were found to be taken up to a greater extent in tested organs than nitrendipine suspension. The AUC and Mean Residence Time (MRT) of nitrendipine SLN were higher than those of nitrendipine suspension especially in brain and heart. Positively charged SLN were better taken up by the brain and moderately taken up by the heart. Reticuloendothelial system (RES) organs such as liver and spleen were compared with the ones after nitrendipine suspension administration. The higher levels of the drug were maintained for over 6 h in confront to only 3 h with nitrendipine suspension.

SLN were investigated for their ability to deliver quinine dihydrochloride for the management of cerebral malaria (Gupta, Jain, & Jain, 2007). Quinine was incorporated in SLN and successively coupling of SLN with transferrin (Tf) was achieved by a cross-linker. IV administration of Tf-conjugated SLN enhanced the brain uptake of quinine in confront to the SLN loaded of quinine alone.

In order to enhance the delivery of atazanavir, a HIV protease inhibitor, spherical SLN carrying the drug were tested at first using a well-characterized human brain microvessel endothelial cell line (hCMEC/D3). Cell viability experiments demonstrated that SLN exhibit no toxicity on hCMEC/D3 cells up to a concentration corresponding to 200 nM of the drug. Delivery of ^{3}H-atazanavir by SLN led to a significantly higher accumulation by the endothelial cell monolayer as compared to the drug aqueous solution (Chattopadhyay, Zastre, Wong, Wu, & Bendayan, 2008).

The transport *in situ* of lipid nanoparticles to the brain was evaluated by Koziara, Lockman, Allen, and Mumper (2003); the lipidic nanoparticles were prepared by warm microemulsion precursors followed by hot homogenization technique. Their components were emulsified wax (E wax) or Brij 72 as matrix, and water and Brij 78 as surfactant. The warm microemulsion was cooled upon stirring and the lipid SLN were obtained and homogenized. The SLN were labelled with ^{3}H cetyl alcohol. The transport of the nanoparticles was measured by an "*in situ*" rat brain perfusion method; significant uptake of SLN was obtained . suggesting CNS uptake. The same group studied also the effect that the addition of a thiamine ligand to nanoparticles (NPs), obtained by microemulsion as precursors, causes association with the BBB thiamine transporter (Lockman et al., 2003).

Muller and coworkers studied the preferential adsorption of blood protein onto intravenously injected particulate carriers from different origins (Luck, Paulke, Schroder, Blunk, & Muller, 1998); in particular, Apolipoprotein E (Apo E) on the surface of P80-coated SLN after their incubation in human plasma citrate. Delivery to the brain using nanoparticulate drug carriers in combination with the targeting principles of "differential protein adsorption" has been proposed (Dehouck et al., 1997). The Pathfinder technology (Muller & Schmidt, 2002) exploits proteins present in the blood which absorb onto the surface of intravenously injected carriers for targeting nanoparticles to the brain. Apo E is one of such targeting molecules for the delivery of nanoparticles to the endothelial cells of the BBB. Apo E can play an important role in the transport of lipoprotein into brain via the low-density lipoprotein receptor present on the BBB. Atoquavone (Muller & Keck, 2004; Scholler et al., 2001) is a drug poorly adsorbed after oral administration, showing poor therapeutic efficacy against toxoplasma encephalitis (TE). Nanocrystals of the drug were produced, their surface was modified with Tween 80 leading to *in vivo* preferential absorption of Apo E; the nanosuspension was IV administered to a murine model of TE, obtaining the disappearance of parasites and of cysts at dose 10-fold smaller than the one of atoquavone administered by oral route.

Solid lipid nanoparticles from warm microemulsions

SLN can be achieved from warm microemulsions.

Warm microemulsions are prepared at temperature ranging from 60°C to 80°C by using melted lipids (such as triglycerides/fatty acids) as oil, surfactants such as lecithin, and cosurfactants (such as short-chain carboxylates, biliar salts); the warm microemulsions are subsequently dispersed in cold water. The nanodroplets of warm microemulsion, using this procedure, become SLN; they are successively washed by tangential flow filtration. SLN are spherical in shape and with a narrow size distribution .The zeta potential is normally high (30/40 mV) being positive or negative depending on the starting formulation.

Hydrophilic and lipophilic molecules (drugs or diagnostics) can be incorporated in SLN using different methods.

SLN are able to carry drugs of different structure and lipophilicity, such as cyclosporine A (Ugazio, Cavalli, & Gasco, 2002), paclitaxel (Cavalli, Caputo, & Gasco, 2000), doxorubicin (Fundaro, Cavalli, Bargoni, Vighetto, & Gasco, 2000), tobramycin (Cavalli et al., 2003), short fatty acids (Dianzani et al., 2006), peptides (Morel, Cavalli, & Gasco, 1996), antisense oligonucleotides (Brioschi et al., 2008), and melatonin (MT) (Rezzani et al., 2009). Also diagnostic compounds such as iron oxides (Peira, 2003) have been incorporated into SLN.

SLN can be internalized within 2–3 min into all the tested cell lines (Miglietta, Cavalli, Bocca, Gabriel, & Gasco, 2000; Serpe et al., 2006); administered by duodenal route and are targeted to lymph (Bargoni et al., 1998). SLN stealth can also be prepared to avoid their recognition by the RES, thus prolonging their residence time (Podio, 2001). SLN drug, unloaded or loaded, stealth/or nonstealth, are transported through the BBB (Podio, 2001; Zara et al., 2002).

Drug-loaded solid lipid nanoparticles

In the late 1990s SLN were proposed for brain drug targeting by several groups (Yang, Zhu, Lu, & Liang, 1999; Zara et al., 1999), which studied the pharmacokinetics of two anticancer agents: camptothecin and doxorubicin. After oral and IV administration, they observed drug accumulation into the brain.

Both stealth and nonstealth stearic acid unloaded labelled SLN were found in rat CSF 20 min after IV administration even though low amount of radioactivity was found in the CSF samples collected from cysterna magna (Podio, Zara, Carazzone, Cavalli, & Gasco, 2000).

When the same kind of SLN were loaded with doxorubicin, significantly higher drug concentrations were found in the brain of the animals treated with stealth SLN as compared to nonstealth SLN and doxorubicin solution. The overall plasma pharmacokinetics of stealth and nonstealth SLN provided to be significantly different from that of the doxorubicin solution (Fundaro et al., 2000).

R-apomorphine (10,11-dihydroxyapomorphine) is a well-known potent short-acting dopamine agonist at D1 and D2 dopamine receptors and it was proposed as an antiparkinsonian drug more than a century ago. It significantly reduces the severity and duration of "off" periods and it is able to reverse bradykinesia when administered alone. Despite these favorable clinical effects, the drug's clinical use is somewhat limited by its pharmacokinetic profile: short half-life (~30 min), rapid clearance from the plasma, lack of storage and retention in brain regions, poor oral bioavailability (5%), and first-pass hepatic metabolism are significant limitations to chronic oral administration. Our group evaluated a new formulation of apomorphine in SLN; the study was designed to investigate the pharmacokinetics and biodistribution of apomorphine incorporated in SLN, injected orally or intravenously in rats.

In vitro the release over time of apomorphine from the SLN dispersion was almost linear After IV administration the peak plasma concentration was higher after apomorphine solution administration than after apomorphine SLN. However, the total area under curve (AUC_{tot}) was nonsignificantly different after SLN than apomorphine solution. The terminal half-life was significantly longer following apomorphine SLN.

Following intraduodenal administration we found that the C_{max} and AUC_{tot} were significantly higher with apomorphine SLN compared to apomorphine solution; on the contrary, the clearance was shorter after apomorphine solution than after the SLN formulation.

In the brain the apomorphine concentration was significantly higher 30 min after apomorphine SLN IV administration versus solution; it was detected only at 4 h after apomorphine SLN injection.

After duodenal administration the drug was detectable in brain only at 30 min after apomorphine SLN administration. No drug was found neither at 4 h nor at 24 h after injection of either apomorphine SLN or the solution.

Furthermore, the free drug concentration was measured in human plasma and we showed that the release started after the absorption of the apomorphine SLN. We also measured the free apomorphine concentration in human blood over time. The amounts in question are relatively low, but may be sufficient to expect clinical effects when administered to parkinsonian patients. After apomorphine solution administration, the amounts of apomorphine determined in the plasma were by far lower than those from SLN, confirming previous studies on the duodenal administration of drug loaded and unloaded SLN (Fig. 1).

In order to furnish a general model for SLN-based delivery systems of drugs devoid of favorable pharmacokinetics, we have recently incorporated MT (melatonin) in SLN (MT-SLN). MT has been chosen for our *in vivo* study because of its safeness in humans even at high dosages.

MT is a hormone produced by the pineal gland at night, involved in the regulation of circadian rhythms. For clinical purposes (mainly disorders of the sleep–wake cycle and insomnia in the elderly), exogenous MT administration should mimic the typical nocturnal endogenous MT levels, but its pharmacokinetics is not favorable due to its short half-life of elimination (DeMuro, Nafziger, Blask, Menhinick, & Bertino, 2000; Mallo et al., 1990). The pharmacokinetics of MT-SLN has been examined in humans after administration by oral and transdermal route (Priano et al., 2007). Three kinds of freeze-dried MT-SLN containing different amounts of MT were prepared and characterized: (a) MT-SLN: MT = 1.8% for *in vitro* experiments (average diameter: 85 nm, polydispersity index = 0.135); (b) MT-SLN: MT = 2% for transdermal application (average diameter = 91 nm, polydispersity index = 0.140); and (c) MT-SLN: MT = 4.13% for oral route (average diameter = 111 nm, polydispersity index = 0.189).

In vitro, MT-SLN produced a flux of MT of 1 μg/h/cm^2 through hairless mice skin, following a pseudo-zero-order kinetics Priano et al. (2007). At the same time, *in vivo* study produced very interesting results,

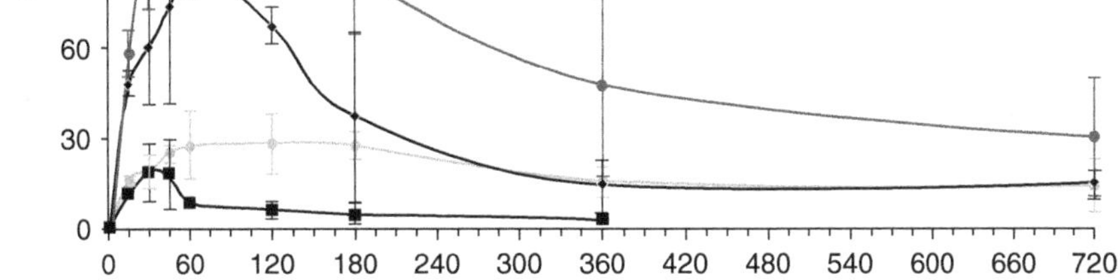

Fig. 1. Plasma levels of free apomorphine and total apomorphine after duodenal administration of apomorphine solution or apomorphine SLN in rats.

confirming in humans that SLN can act as a reservoir that allows a constant and prolonged release of the included drugs (Peira et al., 2003). MT (3 mg) incorporated in SLN was orally administered at 8.30 a.m. to seven healthy subjects; for control purposes, 1 week later the same subjects received orally a standard formulation of MT at the same dose (3 mg) and again at 8.30 a.m. Compared to the MT standard solution, T_{max} observed after MT-SLN administration was delayed of about 20 min, while mean AUC and mean half-life of elimination were significantly higher (respectively $169{,}944.7 \pm 64{,}954.4$ pg/mL × hour vs. $85{,}148.4 \pm 50{,}642.6$ pg/mL × hour, $p = 0.018$; and 93.1 ± 37.1 min vs. 48.2 ± 8.9 min, $p = 0.009$). Even more, standard formulation and MT-SLN after oral administration produced similar peak plasma levels of MT, even if delayed of about half an hour in the case of MT-SLN. More interestingly, detectable and clinically significant MT plasma levels after MT-SLN oral administration were maintained for a longer period of time, suggesting that SLN orally administered to humans can yield a sustained release of the incorporated drug, a feature that could be particularly useful for molecules, such as MT, characterized by unfavorable kinetics (Priano et al., 2007). Previous studies in laboratory animals indicated a probable targeting of SLN — either drug-loaded or unloaded — to lymph, after duodenal administration (Bargoni et al., 1998). Similarly, the significantly longer half-life of MT observed in the study of Priano et al. (2007) may suggest a targeting of MT-SLN to human lymph, even though the capsules used to administer SLN were not gastro-resistant. In fact, MT half-life of elimination has been calculated in about 40 min after an intravenous bolus and following oral administration low bioavailability and rapid clearance from plasma have been shown, primarily due to a marked first-pass hepatic metabolism. Moreover, pharmacokinetic analysis following transdermal administration of MT-SLN demonstrated that plasma levels of MT similar to those produced by oral administration may be achieved for more than 24 h (50). In 10 healthy subjects, SLN incorporating MT were administered transdermally by applying a patch at 8.30 a.m. and leaving it in place for 24 h. In this delivery system, MT absorption and elimination were slow (mean half-life of absorption = 5.3 ± 1.3 h; mean half-life of elimination = 24.6 ± 12.0 h) so that MT plasma levels above 50 pg/mL were maintained for at least 24 h (Figs. 2 and 3). Tolerability of MT-SLN administered transdermally or by oral

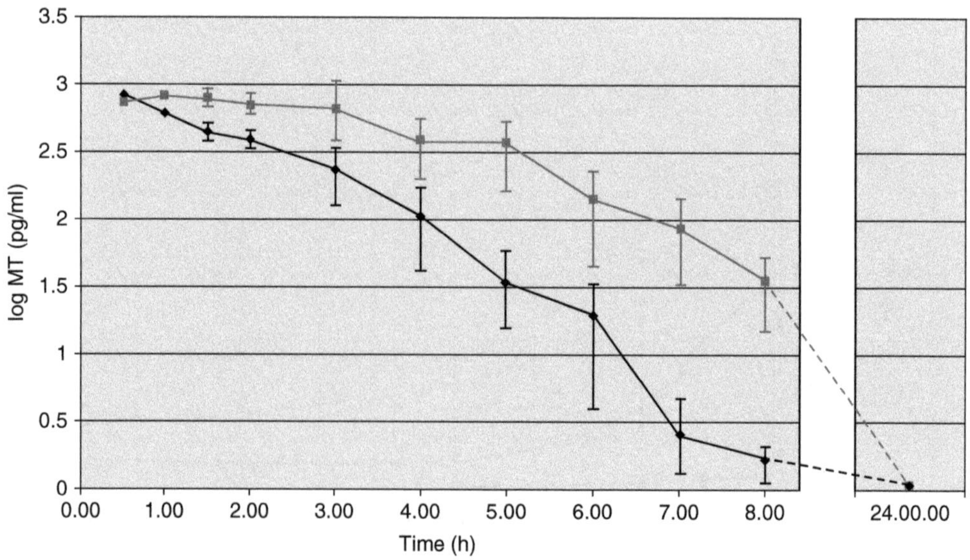

Fig. 2. MT plasma profile in humans after MT (♦) and MT-SLN (■) oral administration.

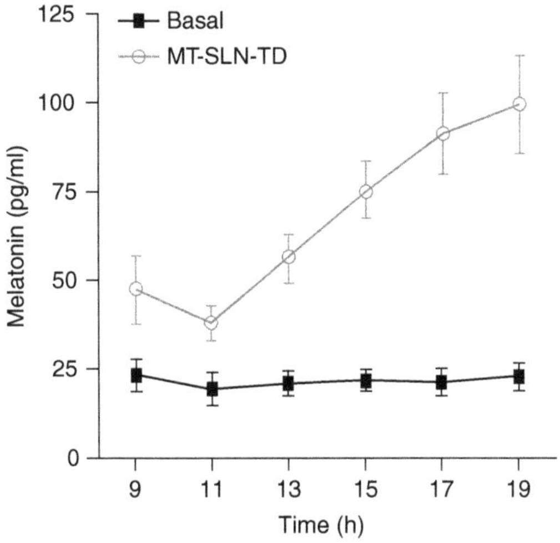

Fig. 3. MT plasma levels in humans at baseline and after MT-SLN transdermal administration (MT-SLN-TD).

route was good and no adverse effect occurred, apart from a predictable mild somnolence and transient erythema after gel application. This means that, at least at the doses used in that study, SLN administration via the oral or transdermal routes is safe.

In this context, we also tested transdermal MT-SLN for three consecutive nights in five patients suffering from delayed sleep phase syndrome (unpublished data), confirming the safeness of this formulation. Due to the small sample, however, the tendency of clinical benefits was present but statistical significance could not be reached, so that further investigations in larger samples are needed in order to evaluate the impact of this new formulation in clinical practice.

However, these very favorable results, obtained in humans administering MT-loaded SLN, clearly suggest that SLN can be considered effective *in vivo* delivery systems that could be suitably applied to different drugs, and in particular to those requiring prolonged high plasma levels but that have unfavorable pharmacokinetics. Finally, it must be stressed that, since doses and concentrations of drugs included in SLN can be varied, different plasma level profiles could be obtained, thus disclosing new chances for sustained delivery systems adaptable to a variety of clinical conditions (Priano et al., 2007).

Suitability of SLN to convey drugs into CNS is also confirmed by studies regarding baclofen included in SLN. Intrathecal baclofen administration represents the reference treatment for spasticity of spinal or cerebral origin. Nevertheless, surgical involvement together with risk of infection or catheter dysfunction may limit the number of potentially treatable patients (Dario & Tomei, 2004; Perot & Almeida-Silveira, 1994). In order to explore alternative and efficacious routes of administration, we studied a new pharmaceutical preparation characterized by SLN incorporating baclofen (baclofen-SLN). Baclofen concentration, after reconstitution with water of freeze-dried SLN, was 1.7 mg/mL. Groups of Wistar rats were injected intraperitoneally with physiological solution and unloaded SLN at 10 mL/kg (control groups), with baclofen-SLN (baclofen-SLN group), and baclofen solution (baclofen-sol group) at increasing dosages of 2.5, 5, 7.5, 8.5, and 10 mg/kg. At different times up to the fourth hour, efficacy testing was performed by means of H-reflex, while behavioral characterization was obtained using two scales validated for motor symptoms due to spinal lesions and sedation in rat models (Nemethy, Paroli, Williams-Russo, & Blanck, 2002; Tsunoda, Kuang, Tolley, Whitton, & Fujinami, 1998). Rats were sacrificed for detecting baclofen concentration in blood and tissue. Compared to baclofen-sol and control group, *H*/*M* amplitude curve after baclofen-SLN injection was characterized by a dose-dependent reduction at the first and second hours, so confirming efficacy, and a rebound increase at the fourth hour, indicating an unexpected belated spinal hyperexcitability (Fig. 4). Similarly, baclofen-SLN effect on behavioral scales was stronger compared to baclofen-sol group, with the maximum effects obtained at the first hour. Moreover, clinical effects were detectable after low dosages of baclofen-SLN (2.5 mg/kg) but only after higher dosages of baclofen-sol (7.5 mg/kg). After 4 h from the injection, only the rats treated with the higher dosages of baclofen-SLN still presented clinical signs consisting in sedation (8.5 mg/kg) or complete paralysis and piloerection (10 mg/kg). On the whole, these data suggest a dose-dependent modulation of spinal reflex excitability, which is not

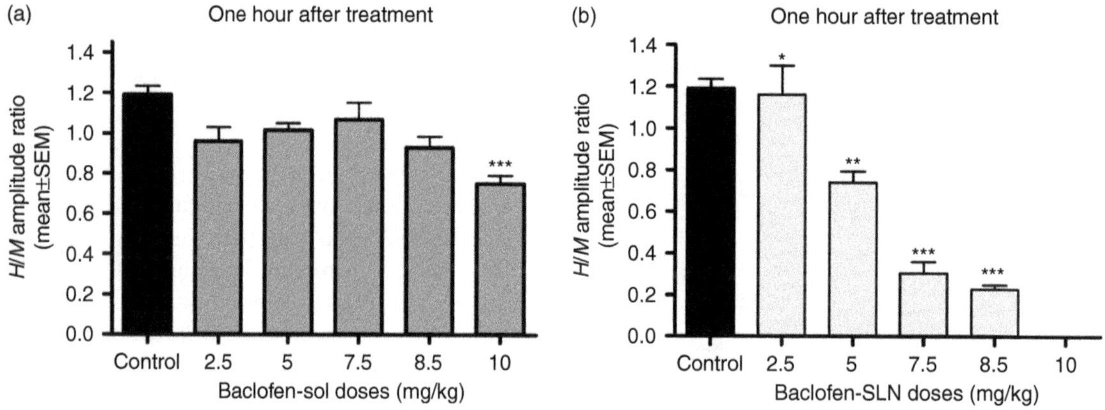

Fig. 4. H/M amplitude ratios after baclofen-solution (a) and after baclofen-SLN (b), at increasing doses, compared to control animal group. $*p < 0.05$; $**p < 0.01$; $***p < 0.001$.

so evident after administration of standard formulation of baclofen. Nevertheless, important cortical effects were also present. Clinical data were related with plasma and tissue concentrations. In fact, after 2 and 4 h only baclofen-SLN administration produced measurable baclofen plasma concentrations, with an almost linear decrease of baclofen appreciable for 4 h. On the contrary, undetectable amount of baclofen in plasma were noticed 2 h after administration of baclofen-sol. In brain, both the two formulations (baclofen in solution and in SLN) gave a maximum after 2 h but concentrations after SLN were almost twice the ones after solution. This last data might be due partly to the free drug already released and to baclofen-SLN overcoming the BBB. We realize that for clinical purposes this effect of baclofen-SLN is undesiderable, as it is responsible for sedation. However, baclofen-sol injections also produced sedation, even if weaker and corresponding to lower plasma concentrations, compared to baclofen-SLN. In conclusion, higher spinal and cortical effects of baclofen-SLN, compared to equivalent dosages of baclofen-sol, seem attributable to higher and more prolonged concentrations of drugs in plasma and brain.

As previously noted, unloaded SLN administered by duodenal route are targeted to lymph and the incorporated drug can be partly distributed in the brain; moreover, SLN can also be prepared stealth for increasing their residence time (Bargoni et al., 1998; Fundaro et al., 2000; Podio, 2001; Zara et al., 2002). Other new studies will be directed toward a duodenal administration of baclofen-SLN stealth, not only for prolonging their residence time but also to target them to lymph, enhancing their bioavailability. Further research should also be directed toward the optimization of dosages and concentrations of baclofen included in SLN, in order to preserve the prolonged antispastic effect, peculiar of this new formulation, but devoid of clinically significant cortical effects.

Solid lipid nanoparticles as potential diagnostics

Superparamagnetic iron oxides are classified as contrast agents for magnetic resonance imaging (MRI). They are able to affect the water relaxation times T_1 and T_2; their ability in altering such properties is quantified by the parameter relaxivity. Iron oxides are able to affect preferentially the T_2 relaxation times of tissues (and are called T_2-relaxing agents) while paramagnetic contrast agents such as Gd complexes affect mainly T_1 and are called T_1-relaxing agents.

Iron oxides are insoluble in water; therefore, to be clinically used they must be transformed in modified colloids while their magnetic properties should remain unchanged. The surface of the iron oxide nanoparticles can be modified, covering them by hydrophilic macromolecules; such as dextran in the case of Endorem.

A research was performed in order to know whether SLN can load iron oxides and whether they are able to reach the brain. Two different SLN, SLN-Fe^A and SLN-Fe^B containing iron oxides were prepared from warm microemulsions and studied at first *in vitro* (1). The comparison of Fe-SLN was performed with Endorem. Both the Fe-SLN preparations showed relaxometric properties similar to the ones of Endorem. The good T_2-relaxation-enhancing properties allow an *in vivo* study of their distribution by MRI. Fe-SLN^B, at the higher Fe concentration, were administered IV to rats; the comparison was performed with Endorem. Images obtained after Endorem IV administration show early modification, but soon return to baseline; these findings are consistent with short Endorem retention time in blood. Results from SLN-Fe^B show a different behavior. For each part of the brain, maximal SS (Signal Suppression) is reached in the last images (135 min after administration). SS increase from the first to the last acquisition. This study shows that after inclusion in SLN, Endorem becomes a new type of contrast agent: Endorem is taken by the liver and does not cross the BBB, while Endorem containing SLN-Fe^B shows CNS uptake. This means that SLN-Fe kinesis is related to SLN and not to their iron oxide content as already seen.

Microemulsions

Microemulsions are transparent, thermodynamically stable dispersions of water and oil, usually stabilized by a surfactant and a cosurfactant. They contain particles smaller than 0.1 μm. Microemulsions are often defined as thermodynamically stable liquid solutions; the stability of microemulsions is a consequence of the ultralow interfacial tension between the oil and water phases. A clear distinction exists between microemulsion and coarse emulsions. The latter are thermodynamically unstable, droplets of their dispersed phase are generally larger than 0.1 μm and consequently their appearance is normally milky rather than transparent.

The limits in the use of microemulsions in the pharmaceutical field are chiefly from the need of all the components to be acceptable, particularly surfactants and cosurfactants — the amounts of surfactants and cosurfactants required to form microemulsions are usually higher than those required for emulsions.

Recently, apomorphine was incorporated into microemulsions to study whether they are a feasible vehicle for transdermal transport of this drug. In the preparatory *in vitro* study (Peira, Scolari, & Gasco, 2001), two different microemulsions whose components were all biocompatible were studied: the concentration of apomorphine was 3.9% in each. Since apomorphine is highly hydrophilic, to increase its lipophilicity, apomorphine–octanoic acid ion pairs were formed. At pH 6.0, log P_{app} of apomorphine increased from 0.3 in the absence of octanoic acid to log P_{app} = 2.77 for a molar ratio 1:2.5 (apomorphine: octanoic acid). The flux of drug from the two thickened microemulsions through hairless mouse skin was, respectively, 100 and 88 μg/h/cm^2. The first formulation, having the higher flux, was chosen for *in vivo* administration to Parkinson's patients.

For the *in vivo* study, 21 patients with idiopathic Parkinson's disease who presented long-term L-DOPA syndrome, motor fluctuations and prolonged "off" periods were selected (Priano et al., 2004). Here, 10 g of apomorphine hydrochloride (3.9%), included in microemulsion for transdermal delivery (Apo-MTD), was applied to a 100 cm^2 skin area on the chest; the area was delimited by 1-mm-thick biocompatible foam tape and covered with a polyester-based membrane and an occlusive membrane to prevent evaporation. In these conditions, a single layer of microemulsion (1 mm thick) was directly in contact with the skin surface and acted as a reservoir of apomorphine. Apo-MTD was applied at 8.00 a.m. and left for 12 h. In all patients, except two, apomorphine was detected in blood samples after a variable lag time. Pharmacokinetic analysis revealed that epicutaneous–transdermal apomorphine absorption was rapid (mean half-life of absorption = 1.03 h) with a variability among patients (half-life of absorption, SD = 1.39 h). Mean C_{max} was above the therapeutic range (mean C_{max} = 42.81 ± 11.67 ng/mL), with a mean T_{max} of 5.1 ± 2.24 h. Therapeutic concentrations of apomorphine were reached after a mean latency of 45 min (range 18–125), and stable concentrations, above the therapeutic range, continued for as long as Apo-MTD was maintained in place. At the 12th hour, Apo-MTD was removed, and the apomorphine plasma concentration then decreased at a rate comparable to that described for subcutaneous administration (mean half-life of elimination equal to 10.8 ± 1.93 h). C_{max} and AUC showed good correlations with the reduction of "off" periods duration and with the improvement of clinical scores evaluating motor performances (r values ranging form 0.49 to 0.56, with p values ranging from 0.02 to 0.04). Apo-MTD overall tolerability was good: systemic side effects were similar to those caused by subcutaneous apomorphine injection (sleepiness, mild orthostatic hypotension, and transient nausea), and in the case of nausea, they were strictly related to the highest plasma level of apomorphine. Moreover, regarding local side effects, the large majority of patients (71.4%) presented a transient mild erythema at the site of Apo-MTD application, with a complete regression within 48 h, whereas only in two cases the erythema lasted more than 3 days and required local therapy. This study clearly demonstrated that in most Parkinson's patients Apo-MTD is absorbed by the epicutaneous–transdermal route. This result is in contrast with other reports, where the transdermal route did not produce detectable plasma levels of apomorphine, or in which no apomorphine was transported passively through the skin (Gancher, Nutt, & Woodward, 1991; van der Geest, van Laar, Gubbens-Stibbe, Boddé, & Danhof, 1997). Probably, this difference was mainly due to the peculiar pharmaceutical preparation used. Even if pharmacokinetic parameters are variable, Apo-MTD demonstrated the feasibility of providing therapeutic apomorphine plasma levels for much longer periods of time than previously tested apomorphine preparations (several hours), allowing a more constant dopaminergic stimulation. These results are encouraging and Apo-MTD might become of clinical value in some parkinsonian patients suffering from uncontrolled "wearing-off" and prolonged "off" phenomena. On the contrary, because of the lag time of about 1 h before therapeutic concentrations are reached, Apo-MTD may not be the "ideal" preparation for rapid relief of "off" periods.

Since Apo-MTD was found to provide constant drug release over several hours, other studies have been addressed to its use for the nocturnal sleep disorders of Parkinson's patients. Twelve parkinsonian patients underwent standard polysomnography on basal condition and during one night treatment with Apo-MTD (applied to 100 cm^2 from 10 p.m. until 8 a.m.; Priano et al., 2003). Sleep analysis during APO-MTD treatment in comparison to basal condition showed very favorable findings: 16% increment of total sleep time, 12% increment of sleep efficiency, 16% increment of stage 3 and 4 nonrapid eye movement (NREM), 15% reduction of periodic limb movements index, 22% reduction of arousal index, and 23% reduction of the "cycling alternating pattern" rate, an objective measure of disruption and fragmentation of NREM sleep. Pharmacokinetic analysis confirmed the absorption of apomorphine and the maintenance of therapeutic plasma levels for several hours (mean C_{max} = 31.8 ± 9.7 ng/mL; mean T_{max} = 3.1 ± 1.6 h; mean half-life of absorption = 1.2 ± 1.4 h; mean half-life of elimination = 8.8 ± 1.9 h). On the whole, this study confirmed that APO-MTD in Parkinson's disease might be able to reduce nocturnal anomalous movements, akinesia, and rigidity, and might be efficacious for reducing the instability of sleep maintenance typical of parkinsonian sleep.

References

Abbott, N. J. (2002). Astrocyte-endothelial interactions and blood-brain-barrier permeability. *Journal of Anatomy*, *200*, 629–638.

Balahanov, R., & Dore-Duffy, P. (1998). Role of the CNS microvascular pericyte in the blood-brain barrier. *Journal of Neuroscience Research*, *53*, 637–644.

Bargoni, A., Cavalli, R., Caputo, O., Fundaro, A., Gasco, M. R., & Zara, G. P. (1998). Solid lipid nanoparticles in lymph and plasma after duodenal administration to rats. *Pharmaceutical Research*, *15*, 745–750.

Brioschi, A., Calderoni, S., Pradotto, L. G., Guido, M., Strada, A., Zenga, F., et al. (2008). Solid lipid nanoparticles carrying oligonucleotides inhibit vascular endothelial grow factor expression in rat glioma models. *Journal of Nanoneuroscience, 1*, 1–10.

Cavalli, R., Caputo, O., & Gasco, M. R. (2000). Preparation and characterization of solid lipid nanospheres containing paclitaxel. *European Journal of Pharmaceutical Science, 10*, 305–308.

Cavalli, R., Bargoni, A., Podio, V., Muntoni, E., Zara, G. P., & Gasco, M. R. (2003). Duodenal administration of solid lipid nanoparticles loaded with different percentages of tobramycin. *Journal of Pharmaceutical Sciences, 92*, 1085–1094.

Chattopadhyay, N., Zastre, J., Wong, H. L., Wu, X. Y., & Bendayan, R. (2008). Solid lipid nanoparticles enhance the delivery of the HIV protease inhibitor, atazanavir, by a human brain endothelial cell line. *Pharmaceutical Research, 25*, 2262–2271.

Conford, F. M., & Hyman, S. (1999). Blood-brain barrier permeability to small and large molecules. *Advanced Drug Delivery Reviews, 36*, 145–163.

Dario, A., & Tomei, G. (2004). A benefit-risk assessment of baclofen in severe spinal spasticity. *Drug Safety, 27*, 799–818.

de Boer, A. G., & Gaillard, P. J. (2007). Drug targeting to the brain. *Annual Review of Pharmacology and Toxicology, 47*, 323–355.

Dehouck, R., Ferrari, L., Dehouck, M. P., Pierce, A., Torpier, G., & Cecchelli, R. (1997). A new function for the LDL receptor. Transcytosis of LDL across the blood brain barrier. *Journal of Cell Biology, 38*, 877–889.

DeMuro, R. L., Nafziger, A. N., Blask, D. E., Menhinick, A. M., & Bertino, J. S. (2000). The absolute bioavailability of oral melatonin. *Journal of Clinical Pharmacology, 40*, 781–784.

Dianzani, C., Cavalli, R., Zara, G. P., Gallicchio, M., Lombardi, G., Gasco, M. R., et al. (2006). Cholesteryl butyrate solid lipid nanoparticles inhibit adhesion of human neutrophils to endothelial cells. *British Journal of Pharmacology, 148*, 648–656.

Fundaro, A., Cavalli, R., Bargoni, A., Vighetto, G. P., & Gasco, M. R. (2000). Non-stealth and stealth solid lipid nanoparticles (SLN) carrying doxorubicin: Pharmacokientics and tissue distribution after i.v. administration to rats. *Pharmacological Research, 42*, 337–343.

Gancher, S. T., Nutt, J. G., & Woodward, W. R. (1991). Absorption of apomorphine by various routes in parkinsonism. *Movement Disorders, 6*, 212–216.

Gupta, Y., Jain, A., & Jain, S. K. (2007). Transferrin conjugate solid lipid nanoparticles for enhanced delivery of quinine dihydrochloride to the brain. *Journal of Pharmacy and Pharmacology, 59*, 935–940.

Kniesel, K., & Wolburg, H. (2000). Tight junctions of the blood brain barrier. *Cellular and Molecular Neurobiology, 20*, 57–76.

Koziara, J. M., Lockman, P. R., Allen, D. D., & Mumper, R. J. (2003). In situ blood-brain barrier transport of nanoparticles. *Pharmaceutical Research, 20*, 1772–1777.

Lapierre, L. A. (2000). The molecular structure of the tight junctions. *Advanced Drug Delivery Reviews, 41*, 255–264.

Lay, C. H., & Kuo, K. H. (2005). The critical component to establish in vitro BBB model: Pericyte. *Brain Research Reviews, 50*, 258–265.

Lockman, P. R., Oyewumi, M. O., Koziara, J. M., Roder, K. E., Mumper, R. J., & Allen D. D. (2003). Brain uptake of thiamine-coated nanoparticles. *Journal of Controlled Release, 93*, 271–282.

Luck, M., Paulke, B. R., Schroder, W., Blunk, T., & Muller, R. H. (1998). Analysis of plasma protein adsorption on polymeric nanoparticles with different surface characteristics. *Journal of Biomedical Materials Research, 39*, 478–485.

Mallo, C., Zaidan, R., Galy, G., Vermeulen, E., Brun, J., Chazot, G., et al. (1990). Pharmacokinetics of melatonin in man after intravenous infusion and bolus injection. *European Journal of Clinical Pharmacology, 38*, 297–301.

Manjunath, K., & Venkateswarlu, V. (2005). Pharmacokinetics, tissue distribution and bioavailability of clozapine solid lipid nanoparticles after intravenous and intraduodenal administration. *Journal of Controlled Release, 107*, 215–228.

Manjunath, K., & Venkateswarlu, V. (2006). Pharmacokinetics, tissue distribution and bioavailability of nitrendipine solid nanoparticles after intravenous and intraduodenal administration. *Journal of Drug Targeting, 14*, 632–645.

Miglietta, A., Cavalli, R., Bocca, C., Gabriel, L., & Gasco, M. R. (2000). Cellular uptake and cytotoxicity of solid lipid nanospheres (SLN) incorporating doxorubicin or paclitaxel. *International Journal of Phamaceutics, 210*, 61–67.

Morel, S., Cavalli, R., & Gasco, M. R. (1996). Thymopentin in solid lipid nanoparticles. *International Journal of Phamaceutics, 132*, 259–262.

Muller, R. H., Kader, K., & Gohla, S. (2000). Solid lipid nanoparticles (SLN) for controlled drug delivery – a review of the state of the art. *European Journal of Pharmaceutics and Biopharmaceutics, 50*, 161–177.

Muller, R. H., & Schmidt, S. (2002). Path. Finder technology for the delivery of drugs to the brain. *New Drugs, 2*, 38–42.

Muller, R. H., & Keck, C. M. (2004). Challenges and solutions for the delivery of biotech. Drugs – a review of drug nanocrystal technology and lipid nanoparticle. *Journal of Biotechnology, 113*, 151–170.

Nemethy, M., Paroli, L., Williams-Russo, P. G., & Blanck, T. J. (2002).Assessing sedation with regional anesthesia: Inter-rater agreement on a modified Wilson sedation scale. *Anesthesia and Analgesia, 94*, 723–728.

Pardrige, W. M. (1998). CNS drug design based on principles of blood-brain barrier transport. *Journal of Neurochemistry, 70*, 1781–1792.

Peira, E., Scolari, P., & Gasco, M. R. (2001). Transdermal permeation of apomorphine through hairless mouse skin from microemulsions. *International Journal of Phamaceutics, 226*, 47–51.

Peira, E., Marzola, P., Podio, V., Aime, S., Sbarbati, A., & Gasco, M. R. (2003). In vitro and in vivo study of solid lipid nanoparticles loaded with superparamagnetic iron oxide. *Journal of Drug Targeting, 11*, 19–24.

Perot, C., & Almeida-Silveira, M. I. (1994). The human H and T reflex methodologies applied to the rat. *Journal of Neuroscience Methods, 51*, 71–76.

Podio, V., Zara, G. P., Carazzone, R., Cavalli, R., & Gasco, M. R. (2001). Biodistribution of stealth and non-stealth solid lipid nanoparticles after intravenous administration in rats. *Journal of Pharmacy and Pharmacology, 52*, 1057–1063.

Priano, L., Albani, G., Brioschi, A., Guastamacchia, G., Calderoni, S., Lopiano, L., et al. (2003). Nocturnal anomalous movement reduction and sleep microstructure analysis in parkinsonian patients during 1-night transdermal apomorphine treatment. *Neurological Science, 24*, 207- 208.

Priano, L., Albani, G., Brioschi, A., Calderoni, S., Lopiano, L., Rizzone, M., et al. (2004). Transdermal apomorphine permeation from microemulsions: A new treatment in Parkinson's disease. *Movement Disorders, 19*, 937–942.

Priano, L., Esposti, D., Esposti, R., Castagna, G., De Medici, C., Fraschini, F., et al. (2007). Solid lipid nanoparticles incorporating melatonin as new model for sustained oral and transdermal delivery systems. *Journal of Nanoscience and Nanotechnology, 7*, 3596–3601.

Rezzani, R., Fabrizio Rodella, L., Fraschini, F., Gasco, M. R., Demartini, G., Musicanti, C., et al. (2009). Melatonin delivery in solid lipid nanoparticles: Prevention of cyclosporin A induced cardiac damage. *Journal of Pineal Research, 46*, 255–261.

Segal, M. B. (2000). The choroid plexuses and the barriers between the blood and the cerebrospinal fluid. *Cellular and Molecular Neurobiology, 20*, 183–196.

Scholler, N., Krause, K., Kayser, O., Muller, R. H., Borner, K., Hahn, H., et al. (2001). Atovaquone nanosuspensions show excellent therapeutic effect in a new murine model of reactivated toxoplasmosis. *Antimicrobial Agents and Chemotherapy, 45*, 1771–1779.

Skinkel, A. H. (1999). P-glycoprotein, a gatekeeper in the blood-brain barrier. *Advanced Drug Delivery Reviews, 36*, 179–194.

Serpe, L., Cavalli, R., Gasco, M. R., Muntoni, E., Cavalli, R., Panzanelli, P., et al. (2006). Intracellular accumulation and cytotoxicity of doxorubicin with different pharmaceutical formulations in human cancer cells. *Journal of Nanoscience and Nanotechnology, 6*, 3062–3069.

Silva, A. S. (2007). Nanotechnology approaches for drug and small molecule delivery across the blood brain barrier. *Surgical Neurology, 67*, 113–116.

Stouch, T. R., & Gudmundsson, O. (2002). Progress in understanding the structure-activity relationship pf P-glycoprotein. *Advanced Drug Delivery Reviews, 54*, 315–328.

Tsunoda, I., Kuang, L. Q., Tolley, N. D., Whitton, J. L., & Fujinami, R. S. (1998). Enhancement of experimental allergic encephalomyelitis (EAE) by DNA immunization with myelin proteolipid protein (PLP) plasmid DNA. *Journal of Neuropathology & Experimental Neurology, 57*, 758–767.

Ugazio, E., Cavalli, R., & Gasco, M. R. (2002). Incorporation of cyclosporin A in solid lipid nanoparticles in solid lipid nanoparticles. *International Journal of Phamaceutics, 241*, 341–344.

van der Geest, R., van Laar, T., Gubbens-Stibbe, J. M., Boddé, H. E., & Danhof, M. (1997). Iontophoretic delivery of R- apomorphine — II: An in vivo study in patients with Parkinson's disease. *Pharmaceutical Research, 14*, 1804–1810.

Yang, S. C., Zhu, J. B., Lu, Y., & Liang, C. Z. (1999). Body distribution of camptothecin solid lipid nanoparticles after oral administration. *Pharmaceutical Research, 16*, 751–757.

Zara, G. P., Cavalli, R., Fundaro, A., Bargoni, A., Caputo, O., & Gasco, M. R. (1999). Pharmacokinetics of doxorubicin incorporated in solid lipid nanospheres (SLN). *Pharmaceutical Research, 40*, 281–286.

Zara, G. P., Cavalli, R., Bargoni, A., Fundaro, A., Vighitto, D., & Gasco, M. R. (2002). Intravenous administration to rabbits of non-stealth and stealth doxorubicin loaded solid lipid nanoparticles at increasing concentration of stealth agent: Pharmacokinetics and distribution of doxorubicin in brain and in other tissues. *Journal of Drug and Targeting, 10*, 327–335.

H.S. Sharma (Ed.)
Progress in Brain Research, Vol. 180
ISSN: 0079-6123

CHAPTER 11

Solid lipid nanoparticles for brain tumors therapy: state of the art and novel challenges

Andrea M. Brioschi[1,*], Sara Calderoni[1], Gian Paolo Zara[2], Lorenzo Priano[3,4], Maria Rosa Gasco[5] and Alessandro Mauro[1,3]

[1]*Department of Neurology and Laboratory of Clinical Neurobiology, Ospedale S. Giuseppe, Istituto Auxologico Italiano, IRCCS, Verbania, Italy*
[2]*Department of Anatomy Pharmacology Forensic Medicine, University of Torino, Torino, Italy*
[3]*Departement of Neurosciences, University of Torino, Torino, Italy*
[4]*Department of Neurology, Ospedale S. Giuseppe, Istituto Auxologico Italiano, IRCCS, Verbania, Italy*
[5]*Nanovector S.r.l., Torino, Italy*

Abstract: Malignant gliomas, despite aggressive multimodal therapies and adequate supportive care, still maintain poor prognosis. Solid lipid nanoparticles (SLN) are colloidal carriers that could be regarded as a highly flexible platform for brain tumor imaging and therapeutical purposes. In this chapter we will first describe brain tumors characteristics and conventional therapeutical approaches. In the subsequent sections, we will analyze SLN properties, effectiveness, and future perspectives in both imaging and targeted treatment of malignant gliomas.

Keywords: solid lipid nanoparticles; brain tumors; drug targeting; brain tumor imaging; cholesterylbutyrate; angiogenesis

Brain tumors

Brain tumors constitute a complex of heterogeneous clinico-pathological diseases, often characterized by poor prognosis and associated with low quality of life (Buckner et al., 2007; DeAngelis, 2001; Louis, Pomeroy, & Cairncross, 2002).

Central nervous system (CNS) tumors are classified by the World Health Organization according to their presumed cell of origin as well as to their localization, histopathological appearance, and lineage markers (Louis et al., 2007).

Primary brain tumors show in the United States an average annual incidence rate of 14.4 per 100,000 persons (Fisher, Schwartzbaum, Wrensch, & Wiemels, 2007) and about half of them are histologically malignant, showing an annual gender incidence rate of 7.0/100,000 in men and 5.2/100,000 in women (Fisher et al., 2007).

Primary malignant brain tumors account for the first cause of death for solid tumors in children and the third cause of death for all cancer types in

* Corresponding author.
Phone No. +39 323 514337; Fax No. +39 323 514364
E-mail: a.brioschi@auxologico.it

DOI: 10.1016/S0079-6123(08)80011-8

adolescents and young adults (Buckner et al., 2007). Primary malignant CNS neoplasms show a relative survival probability at 2 years of 37.7% and at 5 years of 30.2% (Fisher et al., 2007). These epidemiological data clearly suggest that primary malignant brain tumors – despite a lower incidence rate compared to all cancers (about 2%) – display high morbidity and mortality rate and consequently they could be numbered among the most devastating human neoplastic diseases (Buckner et al., 2007).

Gliomas are the most common type among primary CNS tumors and account for an average annual incidence rate in the United States of 6.42/100,000 persons (Fisher et al., 2007). Gliomas include different histopathological entities: astrocytomas, oligodendrogliomas, mixed gliomas (a combination of oligodendroglial and astrocytic features), and ependimomas (Fisher et al., 2007; Norden & Wen, 2006). According to the degree of differentiation and anaplasia, gliomas could receive a histopathological grade that, in turn, strictly correlates with prognosis. High-grade gliomas (otherwise defined malignant gliomas), accounting for more than half of all gliomas in adults and for 78% of all primary malignant CNS tumors, include glioblastomas, anaplastic astrocytomas, anaplastic oligodendrogliomas, and anaplastic oligoastrocytomas. Survival time after diagnosis of malignant glioma depends on both the histological subtype and the age at onset (Fisher et al., 2007; Sathornsumetee, Rich, & Reardon, 2007). The 2-year relative survival probability value (according to the histology and age group at diagnosis) in the United States is variable: 1.4–29.8% for glioblastomas, 4.1–71.4% for anaplastic astrocytomas, 4.9–76.5% for anaplastic oligodendrogliomas, and 37.6–84.7% for mixed gliomas (Fisher et al., 2007).

Genetic factors and molecular markers were recently identified as prognostic indicators for malignant gliomas, in addition to previously known clinical, histological, and neuroradiological factors such as age and functional status at diagnosis, extent of surgical resection, degree of necrosis, and pre- and postsurgery tumor size. (Sathornsumetee et al., 2007).

Different glioma subtypes and grades exhibit a set of peculiar genetic alterations, mainly occurring in genes encoding proteins involved in signal transduction pathways and cell-cycle regulation of tumor initiation and progression. These genetic changes frequently involve growth factors that can be overexpressed (i.e., epidermal growth factor – EGF, platelet-derived growth factor – PDGF, fibroblast growth factor, ciliary neurotrophic factor), or show activating mutations like those commonly occurring (40% of glioblastomas) in the EGF receptor – EGFR gene. Other common molecular changes include tumor suppressor loss (i.e., Phosphatase and Tensin (PTEN) mutations, occurring in nearly 25% of glioblastomas) or deletions, ink4a/arf locus deletions, Rb and p53 mutations (Cavaliere, Wen, & Schiff, 2007; Fisher et al., 2007; Fomchenko & Holland, 2006; Martin-Villalba, Okuducu, & von Deimling, 2008; Sanson, 2008).

For instance, altered EGFR expression inversely correlates to survival increasing proliferation rates, resistance to chemotherapy, invasion, and apoptosis (Rich & Bigner, 2004). Moreover, PDGF ligands are highly expressed in malignant gliomas and the activation of PDGF receptors stimulates proliferation, resistance to apoptosis, cellular motility, and angiogenesis (Rich & Bigner, 2004).

In addition to the aforementioned variability in the pattern of genetic alterations, during tumor progression glioma cells could display additional mutations and epigenetic changes that yield these tumors to become genetically and phenotypically different from the cancer-initiating focus and perhaps sharing variable levels of chemo- and/or radiosensitivity (Cavaliere et al., 2007; Fomchenko & Holland, 2006; Wong, Bendayan, Rauth, Li, & Wu, 2007). Among these latter additional changes we could mention the hypermethylation of methylguanine-DNA-methyltransferase (MGMT) promotor (that results in reduced MGMT expression and consequently in a better response to alkylating drugs) and contrariwise the possible development of a multidrug resistance phenotype by the activation of membrane-associated transporters (such as P-glycoprotein) that actively expel from the cytoplasm a broad range of cytotoxic agents (Criniere et al., 2007; Martin-Villalba et al., 2008; Wong et al., 2007). In conclusion, gene expression profile could help to differentiate glioma subtypes in order to identify tumor indistinguishable on morphological ground (such as primary and secondary glioblastomas) and thereby to predict clinical course (Martin-Villalba et al., 2008).

Taken together, previously reported data suggest that malignant gliomas could be regarded as a group of different diseases, each of them showing distinctive clinical–pathological behavior (Fisher et al., 2007; Louis et al., 2002). Consequently, the recognition of specific prognostic factors may be crucial to identify different subgroup of patients who could be more sensitive to differing schedules of conventional treatment options (i.e., combining chemotherapeutics and/or radiotherapy with treatment sensitizers) (Cavaliere et al., 2007). Furthermore, the identification of these prognostic factors will open a new therapeutical way (the so-called molecular chemotherapy) both by using new treatment agents (including, i.e., monoclonal antibodies, cytokines, synthetic molecules, gene constructs) and by targeting different extra- and/or subcellular pathways (such as cell-cycle control, cell migration, tissue invasion, and angiogenesis) (Cavaliere et al., 2007; Sanson, 2008).

To date, despite aggressive multimodal therapeutic approaches (such as surgery, radiation therapy, and chemotherapy) and adequate supportive care, malignant gliomas still maintain poor prognosis. Among newly diagnosed glioblastoma patients that receive the best treatment schedule possible median survival rate is 14.6 months (Buckner et al., 2007; Carpentier, 2005; Fisher et al., 2007; Norden & Wen, 2006; Norden, Drappatz, & Wen, 2008a; Rich & Bigner, 2004; Stupp et al., 2005).

This substantial failure of conventional treatments could be ascribed to three main reasons, principally related to the peculiar characteristics of high-grade gliomas.

1. First of all, the inability to achieve effective intratumoral concentrations of common chemotherapeutic agents, mainly due to the presence of the blood–brain (BBB) and the brain–tumor barrier, as well as to the intrinsic properties of commonly used cytotoxic drugs (i.e., poor specificity, high systemic toxicity, and propensity to induce chemoresistance).
2. Furthermore, the characteristic early infiltrative behavior of these neoplasms that limits surgical aggressive resections and thereby negatively influence multimodal approaches.
3. Finally, the noticeable cellular and genetic intratumoral, spatial–temporal heterogeneity that modifies the individual response to chemo- and radiotherapy (Cavaliere et al., 2007; Fomchenko & Holland, 2006; Sanson, 2008; Sanson, Laigle-Donadey, & Benouaich-Amiel, 2006).

The normal BBB is a highly effective physical and physiological barrier that regulates the CNS homeostasis and thereby controls the delivery of drugs to the brain (Blakeley, 2008; Kaur, Bhandari, Bhandari, & Kakkar, 2008). Mechanical limitations are mainly carried out by endothelial cell tight junctions that in turn are supported by the absence of fenestration and the reduction of pinocytotic vesicles at endothelial level and by the presence of a composite anatomical barrier constituted by astrocytic end-feet, pericytes, and extracellular matrix. Physiological properties that characterize the normal BBB are formed by the presence of both high electrical resistance across the endothelial cell barrier (even reaching 2,000 ωcm^2) and effective efflux transporters (mainly members of the adenosine triphosphate-binding cassette – ABC), located on cell surface of endothelial and cancer cells (Blakeley, 2008; Kaur et al., 2008; Pardridge, 2007; Wong et al., 2007).

Several factors influence the specific ability of a given molecule to pass through the BBB, including size, water solubility, charge, plasma protein binding, and serum concentration (Blakeley, 2008; Kaur et al., 2008). Other factors, such as cerebral blood flow rate, influx and efflux values at the BBB and blood–CSF (cerebrospinal fluid) barrier, rate of metabolism, and interactions–binding of the drug in the brain, may influence drug cerebral distribution (Kaur et al., 2008). However, only less than 5% of all drugs proved active into CNS, and almost 100% of large molecule drugs – including recombinant proteins and enzymes, monoclonal antibodies, antisense agents, short interfering RNA, and gene products – under physiological conditions do not cross the BBB (Pardridge, 2007). Invasiveness and neoangiogenetic processes of malignant gliomas are accompanied by focal disruptions of the BBB and increased permeability of capillary endothelium. Nevertheless, this BBB disruption is not able to produce any effect on tumor therapy response probably because both intrinsic characteristics of tumor cells (i.e., high proliferation rate and chemo- and radiotherapy escape phenomena) and the not homogeneous localization of these vascular breakages into the tumor mass (Beduneau, Saulnier, & Benoit, 2007).

Nowadays, the most common chemotherapeutic agents in clinical use for malignant glioma treatment include DNA-alkylating cytotoxic drugs (such as carmustine), triple combination (often used at high dosages) of procabazine, cisplatin, and vincristine, and the more recently available temozolomide (TMZ), etoposide, and lomustine. Novel strategies to achieve effective intratumoral bioavailability of chemotherapeutic agents, regardless of their physical–chemical properties, partially escaping in a passive manner the BBB control, were proposed. Drug dose intensification, use of more lipophilic analogs, and intra-arterial delivery preceded by iatrogenic disruption of the BBB using osmotics, magnetic resonance imaging (MRI)-guided ultrasound, or radiotherapy showed debatable or uncertain results (Blakeley, 2008; Rich & Bigner, 2004). Postsurgical implantation into residual tumoral cavity of drug-embedded biodegradable polymers or catheters (for both convection-enhanced delivery or reservoir continuous release) are in clinical use but are still limited to a well clinically selected group of patients referring to much more selected neurosurgical teams (Beduneau et al., 2007; Blakeley, 2008; Kaur et al., 2008). Furthermore, active methods to cross the BBB are in study and will be evaluated in the following sections of this chapter.

Drug delivery systems – solid lipid nanoparticles

In order to obtain a better profile of drug stability, biodistribution, pharmacokinetics, and anticancer activity after parenteral administration, so allowing more targeted antitumoral activity, lower systemic toxicity and reduced adverse side effects, several passive and active carriers were developed. Among them, lipoplexes, dendrimers, cyclodextrins, liposomes, microspheres, niosomes, and nanoparticles were investigated in experimental models and some of them were even put on the market (Barratt, 2003; Cho, Wang, Nie, Chen, & Shin, 2008; Gaidamakova, Backer, & Backer, 2001; Kim et al., 2005; Koziara, Lockman, Allen, & Mumper, 2006; Kreuter, 2001; Lu et al., 2005; Mehnert & Mader, 2001; Muller & Keck, 2004; Olbrich, Bakowsky, Lehr, Muller, & Kneuer, 2001; Pardridge, 2007; Parveen & Sahoo, 2008; Rich & Bigner, 2004; Serikawa et al., 2006; Tiwari & Amiji, 2006; Wong et al., 2007).

An effective delivery system should display some of the following characteristics:

- ability to load a high amount of drugs,
- physical and chemical storage stability,
- low systemic toxicity (that is to say favorable *in vivo* fate of the carrier)
- easy and large-scale production process,
- low overall cost,
- chance of specifically target tumor tissue (Kaur et al., 2008; Mehnert & Mader, 2001).

Solid lipid nanoparticles (SLN) are colloidal (namely, submicron sized) carriers constituted by a solid lipid matrix at room and body temperature, composed of physiological lipids (lipid acids, mono-, di-, or triglycerides, glycerine mixtures, and waxes), and stabilized by biocompatible surfactants (nonionic or ionic) (Marcato & Duran, 2008; Wissing, Kayser, & Muller, 2004; Wong et al., 2007). SLN were shown to satisfy nearly all the aforementioned characteristics combining some advantages (mainly drug bioavailability, controlled release, and drug targeting) and avoiding disadvantages of other vehicles in a more simple and versatile way (Blasi, Giovagnoli, Schoubben, Ricci, & Rossi, 2007; Kaur et al., 2008).

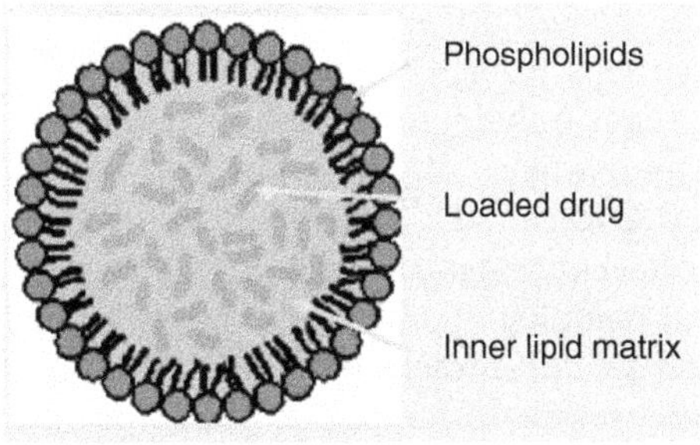

Fig. 1. Schematic structure of loaded SLN obtained by warm microemulsion method.

SLN could be prepared by different approaches such as high-pressure homogenization at high or low temperatures, warm microemulsions (Fig. 1), solvent emulsification–evaporation–diffusion, and high-speed stirring and/or sonication (Blasi et al., 2007; Muller, Mader, & Gohla, 2000). The first two processes show the most versatile technique (mainly in terms of avoidance of nonbiocompatible components, scale-up feasibility, and sterilization) and consequently are the most frequently used (Blasi et al., 2007).

SLN could carry different agents including both hydrophilic and lipophilic therapeutics and diagnostic tools. For parenteral administration benzodiazepines, antipsychotics, pilocarpine, steroids, timolol, antineoplastic agents, peptides, and more recently gene therapeutical agents, such as plasmid DNA and antisense oligonucleotides (AS-ODN), and MRI contrast agents were successfully incorporated into SLN (Dass, 2002; Gasco, 2007; Manjunath & Venkateswarlu, 2005; Muller et al., 2000; Peira et al., 2003; Wissing et al., 2004). Furthermore, SLN could be administered by different routes (such as parenteral, transdermal, oral, and ocular) (Gasco, 2007). The drug solubility in the lipid melt together with the structure and the polymorphic state of the lipid matrix are the main factors that influence the drug loading capacity (Wissing et al., 2004).

SLN are able to increase chemical stability and to protect from systemic degradation the vehiculated molecule (hence consequently increasing its plasma half life) by virtue of the presence of a solid hydrophobic core (the so-called solid high melting fat matrix) in which lipophilic compounds are dissolved or dispersed (Kaur et al., 2008; Wissing et al., 2004). The carried drug – according to its lipid ratio and solubility – could be mainly located into the core, into the shell or dispersed into the matrix of the SLN (Wissing et al., 2004). By modifying the composition of the lipid matrix, the type and the concentration of the surfactant, and the productions parameters it is possible to modulate the drug release profile: drug-enriched shell (burst release), solid solution (intermediate release), and drug-enriched core (sustained release, even for up to several weeks) (Cavalli et al., 2003; Wissing et al., 2004).

The SLN content of only well-tolerated, biocompatible, and biodegradable lipids and the avoidance of any organic solvent during the preparation process justify the common statement that SLN could be regarded as safe (Blasi et al., 2007; Fundaro et al., 2000; Zara et al., 1999, 2002a, 2002b).

Moreover, SLN could be easily sterilized and produced on a large industrial scale, so reducing the overall cost (Blasi et al., 2007; Gasco, 2007; Kaur et al., 2008).

Several *in vivo* experimental studies demonstrated that pharmacokinetics and body distribution profile of different drugs after parenteral administration are significantly changed if vehiculated by SLN. Among the tested agents we could count different compounds for which the passage through the BBB is usually troublesome: chemotherapeutics (i.e., doxorubicin, paclitaxel, idarubicin, camptothecin, etoposide, retinoic acid, TMZ), chemosensitizers (such as verapamil and cyclosporine-A), neuroleptics (such as clozapine), and contrast agents for MRI. These studies (that will be better analyzed in the following section) clearly demonstrated that SLN are able to significantly increase plasma peak, modify plasma concentration curve (raising the area under curve – AUC – from 3- to 20-fold and lowering the rate of clearance so increasing plasma half-lives), and reduce the volume of distribution (Fundaro et al., 2000; Huang, Zhang, Bi, & Dou, 2008; Manjunath & Venkateswarlu, 2005; Shenoy, Vijay, & Murthy, 2005; Wissing et al., 2004; Wong et al., 2007; Yang et al., 1999; Zara et al., 1999, 2002a, 2002b). Furthermore, a different body distribution pattern of the drug was shown: usually the highest concentrations and mean residence times (MRT) are found in the brain, the lowest ones are seen in lung, heart, and kidney, and variable results are obtained from liver and spleen (Fundaro et al., 2000; Huang et al., 2008; Manjunath & Venkateswarlu, 2005; Shenoy et al., 2005; Wissing et al., 2004; Yang et al., 1999; Zara et al., 1999, 2002a, 2002b). These properties, probably justified by the effect of SLN on both the rate of crossing biologic barriers and the pattern of drug release, coupled with the reduction of the drug total dose needed, could significantly contribute to decrease side effects of carried agents (Shenoy et al., 2005).

Compared to other vehicles, SLN show a higher ability to escape the reticuloendothelial system (RES), so bypassing liver and spleen filtration and consequently increasing the bioavailability of the carried agent (Cho et al., 2008; Kaur et al., 2008). SLN characteristics (mainly size and surface) could be easily modified in order to modulate the body distribution hence increasing bioavailability into CNS of the complex drug carrier. Size not exceeding a maximum diameter of 200 nm, sphericity, and adequate deformability are crucial peculiarities to ensure the escape from the sinusoidal spleens (Cho et al., 2008; Kaur et al., 2008). The coating of SLN surface with a hydrophilic or flexible polymer (such as polyethylene glycol, PEG) and/or the use of a surfactant (such as polysorbate and Epikuron) prevent opsonization [namely, the recognition by macrophage membrane of peculiar blood plasma proteins (opsonin) adsorbed onto the colloidal carrier] and the consequent phagocytosis carried out by macrophages in the liver (Cho et al., 2008; Kaur et al., 2008). This mechanism was summarized in the concept of the "differential protein adsorption" under that physical–chemical surface characteristics of nanoparticles induce qualitatively and quantitatively different adsorption patterns that in turn determine the *in vivo* fate of the carrier system (Muller & Keck, 2004). In addition to opsonins (mainly immunoglobulins and complement factors), that facilitate RES recognition, dysopsonins (such as albumin, apolipoprotein A-I, A-IV, C-III, and H) are contrariwise able to reduce the affinity of colloidal carriers to RES and perhaps to increase passive targeting to specific organs only by modifying the composition of the nanoparticle (W. Mehnert & Mader, 2001; Muller & Keck, 2004).

The mechanisms by which SLN cross the BBB are not completely understood but it is indisputable that a central role is played by the interactions between plasma proteins adsorbed onto the SLN surface and endothelial cells, hence facilitating or hindering nanoparticles adhesion and subsequently activating or not endocytotic process (Goppert & Muller, 2005; Kreuter, 2001; W. Mehnert & Mader, 2001). Among proteins adsorbed onto the SLN surface, ApoE, Apo C-II, albumin, and immunoglobulin G seem to be crucial in the site-specific targeting to the brain (Blasi et al., 2007; Goppert & Muller, 2005). Other mechanisms, namely, increased retentions of nanoparticles in the brain blood capillaries and transcytosis, could be advocated and could work together with the aforementioned endocytotic process (Blasi et al., 2007).

Furthermore, different surfactants (such as Polysorbate 80 and Poloxamer 188) were shown to facilitate the BBB crossing of different drugs (i.e., doxorubicin) vehiculated by both polybutylcyanoacrylate (PBCA) nanoparticles and SLN (Blasi et al., 2007; Cho et al., 2008; Dehouck et al., 1997; Goppert & Muller, 2005; Kaur et al., 2008; Petri et al., 2007; Steiniger et al., 2004).

For instance, our group showed that *in vivo* SLN containing stearic acid and PEG 2000 as stealthing agents, unloaded or loaded with different chemotherapeutics (i.e., doxorubicin), are able to facilitate the passage through the BBB and to increase the bioavailability of the drug into the brain tissue compared to nonstealth SLN or free drug solutions; moreover, stealth SLN show lesser degree of recognition by the RES so prolonging drug plasma half-life (Fundaro et al., 2000; Podio, Zara, Carazzonet, Cavalli, & Gasco, 2000b; Zara et al., 2002b).

Yang and Colleagues showed that camptothecin-loaded SLN stabilized by Poloxamer 188 compared to the free solution of this antineoplastic agent after i.v. administration induce a higher maximum concentration (corresponding to 180%

increase) and a better profile of the AUC/dose curve and MRT in the brain, heart, and RES (Yang et al., 1999).

Koziara and Colleagues evaluated the CNS uptake of two kinds of SLN composed by the emulsifying wax (E wax) or Brij 72 as matrix, and, respectively, Brij 78 and Tween 80 as surfactant. The SLN were labeled with [^{3}H]cetyl-alcohol and the transport of the SLN was measured by an "*in situ*" rat brain perfusion method. A significant increase in the CNS uptake of both types of SLN was observed compared to [^{14}C]sucrose (Koziara, Lockman, Allen, & Mumper, 2003, 2004). The same group confirmed the aforementioned results in an analogous experiment using paclitaxel-loaded polysorbate nanoparticles (Koziara et al., 2004).

Petri and Colleagues recently showed in a *in vivo* rat intracranial glioblastoma model that both the surfactants Polysorbate 80 and Poloxamer 188 promote the adsorption onto PCBA–nanoparticles of various blood plasma proteins, including different classes of apolipoprotein (respectively Apo E and Apo A-I). These apolipoproteins in turn activate a specific receptor-mediated mechanism at the capillary brain endothelial cells: Polysorbate 80–Apo E complex activate a LDL-receptor mediated endocytosis and Poloxamer 188–Apo A-I stimulate a scavenger receptor class B type I (SR-BI)-mediated nanoparticle adhesion (Dehouck et al., 1997; Petri et al., 2007; Steiniger et al., 2004).

Moreover, the use of differently charged surfactants significantly influences the passage through the BBB. Lockman and Colleagues evaluated the effects of differently charged nanoparticles on both the BBB integrity and the brain permeability. The authors showed that only neutral and low concentrations of anionic nanoparticles warrant the BBB integrity and that the brain uptake is better for low concentration of anionic nanoparticles. These results suggest that neutral and low concentrations of anionic nanoparticles can be regarded as effective colloidal carriers to the brain (Lockman, Koziara, Mumper, & Allen, 2004).

In conclusion, the aforementioned data show that SLN could be effectively and easily tailored (mainly acting on the composition of their surface) in order to passively increase CNS targeting (passive targeting). Furthermore, SLN are able to allow a more specific targeting directed to genetic and phenotypic features displayed by brain tumors, the so-called active targeting, that will be discussed in the following sections of this chapter (Parveen & Sahoo, 2008).

Solid lipid nanoparticles and brain tumors

As introductory remarks we have to remind that *in vitro* and *in vivo* experimental glioma models are not able to fully reproduce the extremely complex characteristics of human gliomas, both phenotypically and genotypically. More in details, *in vitro* models based on established primary animal or human glial tumor cell cultures (Barth, 1998; Claes et al., 2008; Fomchenko & Holland, 2006; Martinez-Murillo & Martinez, 2007; Mathieu, Lecomte, Tsanaclis, Larouche, & Fortin, 2007) are useful to study biochemical and biological tumor cell properties (such as chemo- and radiosensitivity) but cannot recapitulate the interactions between the tumor and the host environment (i.e., neoangiogenesis and immunological reactions) as well as the genetic variability of human gliomas. Furthermore, *in vivo* models (mainly subcutaneous or brain orthotopic xenografts of selected primary glioma cell cultures), although influenced by tumor–host interactions, do not show some distinctive phenotypical features of naïve malignant gliomas – such as diffuse infiltrative behavior or angiogenesis. Moreover, these models are lacking genetic heterogeneity and native stromal support as well as provide a synchronous instead of a stepwise disease development paradigm (Claes et al., 2008; Fomchenko & Holland, 2006). In conclusion, experimental glioma models are not able to accurately reproduce the naturally occurring disease and consequently the findings obtained from these models have to be critically and with careful consideration translated into clinical phase I and II trials and hence to clinical practice (Claes et al., 2008; Fomchenko & Holland, 2006).

In the following sections we will mainly focus on SLN prepared by our group from warm microemulsions. Microemulsions are chemical–physical systems that are composed of oil, water, cosurfactant, and surfactant, and that show an interfacial tension near zero, thus accounting for their long-term stability. Microemulsion nanodroplets display a mean diameter below 80 nm. Warm microemulsions are prepared at temperature ranging from 60°C to 80°C by using melted lipids (such as fatty acids/triglycerides) and are subsequently dispersed in cold water. Nanodroplets obtained using this procedure become SLN, which are successively washed by tangential flow filtration. SLN display spherical shape and a narrow size distribution. The zeta potential is always high (30/40 mV), being negative or positive according to the starting formulation.

Drugs of different structure and lipophilicity, such as paclitaxel and doxorubicin, were loaded into SLN using different methods. Drug-loaded SLN show a mean diameter ranging from 80 to 200 nm, depending on the chemical characteristics and the amount of the incorporated molecules.

In vitro experimental models

Intracellular trafficking of nanoparticles

In previous sections of this chapter we already evaluated the properties that a systemically administered colloidal carrier must display in order to effectively reach brain tumor mass. At this level, the carrier should also display other abilities in order to pass through the selective plasma membrane and to reach effective concentrations in the cytoplasm of neoplastic cells as well as to display a controlled drug release.

In nonphagocytic cells, the preferential mechanism of nanocarriers cellular uptake is mediated by endocytosis (Rejman, Oberle, Zuhorn, & Hoekstra, 2004; Soldati & Schliwa, 2006). Endocytic pathway of nanoparticles starts at the plasma membrane level and it can be either clathrin-dependent or clathrin-independent, and the latter could be in turn divided into caveolar or clathrin–caveolae-independent (Mayor & Pagano, 2007). The first step of the clathrin-mediated endocytosis is a receptor-mediated process based on ligand–receptor recognition and interaction at membrane level. Subsequently, the pathway progress by generating clathrin-coated pits that invaginate into the cytoplasm and then detach from it, so forming the endocytic vesicles. Therefore, these vesicles undergo to both early and late endosomal transport and lastly to lysosomal digestion. On the other hand, caveolin-coated vescicles invaginate from plasma membrane domains which are especially enriched in cholesterol and sphingolipids (Simons & Ikonen, 1997). This caveolae-mediated uptake could be very advantageous in carrier-mediated drug delivery because nanoparticles can avoid the lysosomal degradation, so increasing their cytoplasmic half-life and consequently maintain for longer periods a sustained intracellular drug release (Shin & Abraham, 2001). However, Lai and Colleagues suggested that certain types of polymeric nanoparticles can exploit a nonclathrin, noncaveolae, and cholesterol-independent pathway in order to undergo nondegradative trafficking into HeLa cells (Lai et al., 2007). Among other physiological mechanisms perhaps involved in nanocarrier intake spontaneous fluid-phase macropinocytosis could be included. This process, commonly occurring in all eukaryotic cells during their life time, is characterized by cyclic internalization of plasma membrane bits in which substances dissolved in the extracellular fluids are entrapped. Through this physiological recycling pathway it is possible that positively charged or neutral nanoparticles glued to the negative outer surface of the cell membrane could reach the cytoplasm (Partlow, Lanza, & Wickline, 2008).

One or more of the aformentioned pathways could be involved in neoplastic cellular uptake of nanoparticulate carriers, depending on their size, z-potential, chemical characteristics, and coated molecules (Cho et al., 2008; Edetsberger, Gaubitzer, Valic, Waigmann, & Kohler, 2005; Vasir & Labhasetwar, 2007; Vijayaraghavalu, Raghavan, & Labhasetwar, 2007).

Several studies demonstrated that SLN (including also both polymer–lipid hybrid nanoparticles and nanostructured lipid carriers) are able to easily enter into the cytoplasm of different tumoral cells including U373 (human astrocytoma), U87 MG (human glioblastoma–astrocytoma), Lipari (human glioblastoma), and C6 (rat glioma) (Brioschi et al., 2009; Lim, Lee, & Kim, 2004; Miglietta, Cavalli, Bocca, Gabriel, & Gasco, 2000; Serpe et al., 2006; Stevens, Sekido, & Lee, 2004; Wong, Bendayan, Rauth, Wu, 2004; Wong et al., 2006a, 2006b). Nevertheless, the exact mechanism by which SLN cross the cell membrane is till now poorly understood (Muller & Olbrich, 1999). We observed (unpublished data) that fluorescent SLN are rapidly uptaken from human U373 tumoral cells (in 5 min) and accumulate into cytoplasm. Moreover, SLN do not enter either into the nucleus or into cytoplasmic organelles such as mitochondria or Golgi apparatus. Furthermore, after 4 h we observed a lysosomal entrapment of the greater part of the uptaken nanoparticles. The persistence of few SLN into the cytoplasm could lead us to suppose the existence of either alternative concomitant endocytotic entrance pathways or a mechanism for lysosomal escape. Moreover, SLN seem to display a biphasic drug release profile: from 10 to 30 min we observed a characteristic "burst release" phase while from 30 min to 24 h SLN produced a till robust and sustained drug delivery. These data seem to suggest that at early times, corresponding to the cytoplasmic localization, SLN rapidly release loaded compounds until an equilibrium with the environment is reached. Subsequently, the prolonged and till sustained drug delivery could be probably ascribed to lysosomal digestion of SLN lipid matrix by resident acidic lipases (Du, Sheriff, Bezerra, Leonova, & Grabowski, 1998).

Further studies will clarify both the role played by different endocytotic processes possibly involved in SLN tumor cell uptake and the contribute given by the lysosomal escape mechanism in the delayed drug release phase.

Antineoplastic drugs loaded into SLN

Several chemotherapeutics were incorporated in SLN, such as doxorubicin, idarubicin, paclitaxel, camptothecin, etoposide, SN-38 (irinotecan analog), retinoic acid, 5-fluorouracil (5-FU), and TMZ.

Anthracyclin antibiotics such as doxorubicin, idarubicin, and daunorubicin have general anticancer properties that include interaction with DNA in a variety of different ways such as intercalation, DNA strand breakage, and inhibition operated by topoisomerase II. Most of these compounds at effective dosages produce significant toxicity. Doxorubicin (Di Marco, 1978; Stan, Casares, Radu, Walter, & Brumeanu, 1999) is currently in clinical use for the treatment of several solid tumors, while daunorubicin and idarubicin are exclusively used for the treatment of leukemia. However, due to their chemical properties, anthracyclines do not easily pass the BBB and hence they do not achieve effective intracerebral concentrations for the treatment of brain tumors. Moreover, the dose-related characteristic systemic side effects such as cardiomyopathy, congestive heart failure (Minotti, Menna, Salvatorelli, Cairo, & Gianni, 2004; Singal, Li, Kumar, Danelisen, & Iliskovic, 2000), bone marrow depression, and alopecia (Minow, Benjamin, Lee, & Gottlieb, 1977) have to be taken into account.

Paclitaxel is a diterpenoid isolated from *Taxus brevifolia* that shows anticancer effects against both hematopoietic and solid tumors (von Holst et al., 1990). Because of its high hydrophilicity, paclitaxel does not easily cross the BBB. Moreover, its clinical use is highly limited by systemic side effects, such as peripheral neuropathy and cardiac arrhythmia (Rowinsky & Donehower, 1995) as well as alopecia and bone marrow depression (Minow et al., 1977).

Camptothecin, an alkaloid plant isolated from *Camptotheca acuminata* (Wall, Wani, Natschke, & Nicholas, 1986), is the prototype of antitumor agents that display a peculiar mechanism of action. These compounds target the nuclear enzyme topoisomerase I that physiologically transiently breaks and rejoins DNA strands in order to facilitate their replication, recombination, and transcription. Because of poor water solubility, instability at biological pH, and severe toxicity of the carboxylated form, camptothecin is not used in clinical applications (Potmesil, 1994).

Etoposide, a semisynthetic derivative of podophyllotoxin, a substance extracted from the mandrake root *Podophyllum peltatum*, shows potent antineoplastic properties (Xu, Lv, & Tian, 2009). More in details, etoposide binds to and inhibits topoisomerase II main function of ligating cleaved DNA molecules and consequently induces accumulation of single- or double-strand DNA breaks, inhibition of DNA replication and transcription, and apoptotic cell death. Moreover, etoposide do not readily penetrate the CNS. In the first 24 h during i.v. administration of etoposide at various dosages, the concentration of this drug in the CSF ranges from undetectable to less than 5% of that concurrently found in the plasma.

SN-38, an irinotecan analog derived from camptothecin (O'Dwyer & Catalano, 2006), is a topoisomerase I inhibitor primarily used in the treatment of colorectal cancer.

Tretinoin, also known as all-*trans*-retinoic acid, is a natural derivative of vitamin A. Retinoids are important regulators of cell reproduction, proliferation, and differentiation and are commonly used to treat dermatological disorders (Cheepala, Syed, Trutschl, Cvek, & Clifford, 2007). Furthermore, tretinoin could also be regarded as a pro-differentiating antineoplastic agent and it is used in the treatment of acute promyelocytic leukemia (Cornic et al., 1992).

5-FU – an analog of uracil – is converted to a fraudulent nucleotide that impair thymidylate synthesis. The result is inhibition of DNA synthesis but not RNA or protein production (Cohen, Flaks, Barner, Loeb, & Lichtenstein, 1958).

TMZ is an orally available methylating agent at specific DNA sites and thus affects DNA synthesis and consequently triggers apoptosis. TMZ became a new standard-of-care treatment of patients affected by glioblastoma, both as adjuvant and as concurrent chemotherapy during radiotherapy (Friedman, Kerby, & Calvert, 2000; Sathornsumetee et al., 2007). This alkylating agent is able to pass through the BBB and to achieve in the CSF approximately 40% of the corresponding plasma concentration. Unfortunately, TMZ displays adverse effects such as hematological toxicity and oral ulceration and an unusual cardiomyopathy, directly due to the accumulation of the drug in the heart (Sathornsumetee et al., 2007).

Cytotoxicity of SLN loaded with chemotherapeutic agents

In vitro, the efficacy of doxorubicin- or paclitaxel-loaded SLN compared to drug-free solutions were evaluated in different neoplastic cells, including glioma and astrocytoma cell lines.

In a recent study our group compared the intracellular accumulation and toxicity in human tumoral cell lines (including U373 astrocytoma) of different doxorubicin formulations: loaded into SLN (Doxo-SLN), carried by pegylated liposomes (Caelyx), and administered as free solutions (Serpe et al., 2006). Doxo-SLN were significantly more efficient in inhibiting cell growth in comparison to both pegylated liposomes and free solutions, suggesting that the intrinsic characteristics of the delivery system by itself may improve the uptake and accumulation of doxorubicin into the cells.

Doxorucicin- and paclitaxel-SLN showed an improved cytotoxicity when compared to the free solutions at same concentrations; moreover, in human glioma cell lines (U87 and U373) drug-loaded SLN were able to induce consistent cell death at lower concentrations (from 10- to 100-fold) and at shorter exposure times if compared to drug-free solutions (Mauro & Brioschi, unpublished data, Miglietta et al., 2000).

Recently, polymer–lipid hybrid nanoparticle and lipid nanoparticles loaded with paclitaxel or doxorubicin and SLN loaded with vinorelbine bitartrate were prepared and their cytotoxicity was tested on tumoral cell lines of nonglial origin, showing promising results.

Wong and Colleagues developed a new polymer–hybrid nanoparticle system able to load and release water-soluble doxorubicin; they obtained a complex between cationic doxorubicin and soybean oil-based anionic polymer, dispersed together with a lipid in water to form Doxo-loaded SLN (Wong et al., 2006b). Treatment of Multidrug Resistant (MDR) cells (human breast cancer) with this formulation induced an increase of cell death when compared to the drug-free solution.

Stevens and Colleagues synthesized and incorporated paclitaxel-7-carbonyl-cholesterol, a paclitaxel prodrug, into lipid nanoparticles that also contained folate–PEG–cholesterol as ligand for targeting folate receptor (FR) expressing tumoral cells (Stevens et al., 2004). The FR-targeted lipid nanoparticles showed greater uptake and cytotoxicity than the nontargeted ones in FR(+) cell lines (M109 and KB) than in FR(–) cell lines (CHO).

Wan and Colleagues evaluated the uptake and cytotoxicity of PEG 2000–stearic acid SLN loaded with vinorelbine bitartrate in RAW26 (mouse macrophages), MCF-7 (human breast cancer), and A549 (human alveolar basal epithelial) cell lines (Wan et al., 2008). They demonstrated that the phagocytic uptake of SLN by RAW26 are progressively inhibited by the addition of increasing concentrations PEG 2000; inversely, high quantities of PEG 2000 promote the intracellular uptake of SLN by tumoral cell lines such as MCF-7 and A549, in accordance with previously reported data (Bocca et al., 1998).

Moreover, the essay of anticancer activity *in vitro* demonstrated that, due to the increased cellular internalization of drug, the cytotoxicity of vinorelbine bitartrate is enhanced by encapsulation in pegylated SLN.

Jain and Colleagues prepared plain SLN and SLN loaded with 5-FU subsequently targeted with ferritin (Fr-SLN) using the ethanol injection method (S. K. Jain et al., 2008). The cellular uptake and IC_{50} values of the Fr-SLN formulation were determined *in vitro* in MDA-MB-468 breast cancer cells. *In vitro* cell binding of Fr-SLN exhibits 7.7-fold higher binding of Fr-SLN to cancer cells in comparison to SLN; moreover, cytotoxicity assays on Fr-SLN gave IC_{50} of 1.25 μM and 3.56 μM for plain SLN.

In vivo experimental models

SLN pharmacokinetics in healthy animals

Pharmacokinetics of SLN were first studied by us in healthy rats treated intravenously with Doxo-SLN or with the free drug solution (Zara et al., 1999). This study demonstrated the superior efficacy of SLN in achieving and maintaining doxorubicin plasma concentration in comparison to the free solution. Moreover, SLN were able to modify the drug biodistribution, in particular, decreasing heart and liver drug concentration while improving cerebral accumulation.

Afterward, Fundarò and Colleagues compared doxo-SLN, stealth doxo-SLN and doxorubicin-free solution confirming the ability of SLN (stealth more than nonstealth) to increase plasma half-life and brain accumulation of doxurubicin (Fundaro et al., 2000). On the contrary, the free doxorubicin solution was very rapidly cleared from the blood stream (in 2.5 h) and is not able to enter the brain parenchyma. In all rat tissues examined, except the brain, the amount of doxorubicin was always lower after injection of the two types of SLN in comparison to the commercial solution; in particular, SLN significantly decreased heart concentrations, thus decreasing the characteristics side effects of the drug.

Furthermore, to confirm tissue distribution and transport across the BBB of modified SLN, both drug-free and drug-loaded stealth (pegylated) and nonstealth SLN were administered intravenously to rats (Podio, Zara, Carazzonet, Cavalli, & Gasco, 2000a). In the first part of the experiment, rats were injected with labeled (with 17-[^{131}]iodo-heptadecanoic acid) nonstealth or stealth SLN and radioactivity tissue accumulation measured after 60 min. This study showed that in liver and lungs the radioactivity was much lower after stealth-SLN formulation administration compared to the nonstealth counterpart, confirming that there is a difference in body distribution among the two SLN types, perhaps due to the stealthing agents (stearic acid and PEG 2000). In the second part of this work, rats were injected with unlabeled stealth and nonstealth SLN and after 20′ both types of SLN were detected in the brain thus confirming the BBB passage, as proved by CSF samples transmission electron microscopy analysis.

Finally, we demonstrated that pegylated doxorubicin SLN reach the brain in larger amounts than the nonstealth SLN and that the brain drug concentrations increase proportionally to the percentage of stealth agent used in the formulation (Zara et al., 2002b). Favorable pharmacokinetics and tissue distribution were demonstrated also for idarubicin-loaded SLN after i.v. or duodenal administration routes (Zara et al., 2002a). SLN-based formulation was more effective in maintaining plasma drug concentrations than the idarubicin-free solution (improved AUC). Tissue distribution was significantly modified by the encapsulation: after SLN administration, idarubicin and idarubicinol concentrations are lower in heart, lung, spleen, and kidneys, while brain accumulation was enhanced. Duodenal route further on improved idarubicin pharmacokinetics compared to i.v. injection, thus suggesting that SLN can be considered for oral delivery of antineoplastic drugs in both systemic and brain tumors.

Yang and Colleagues evaluated the body distribution of intravenously injected camptothecin SLN (CA-SLN) in C57BL/6J mice (Yang et al., 1999). SLN were obtained from high-pressure homogenization technique using camptothecin, stearic acid, soybean lecithin and Poloxamer 188. The results of this study showed that the AUC/dose and the mean residence times of CA-SLN were much higher than those of camptothecin solutions, especially in brain, heart, and reticuloendothelial cells containing organs.

Chen and Colleagues by using the emulsification–evaporation technique, prepared stearic acid–lecithin SLN containing paclitaxel, coated with either Brij78 or Poloxamer F_{68} surfactants (D. B. Chen, Yang, Lu, & Zhang, 2001). Evaluation of drug pharmacokinetics in Kunming (KM) mice showed that encapsulation of paclitaxel in both kind of SLN produce noticeable differences compared to the free drug (Cremophor EL) pharmacokinetics.

Huang and Colleagues prepared temozolomide SLN (TMZ-SLN) by emulsification and low-temperature solidification method (Huang et al., 2008). The AUC/dose and MRT of the TMZ-SLN i.v. injected in healthy rabbits demonstrated much higher and longer than those obtained with TMZ solution, especially in brain and in reticuloendothelial cells-containing organs; moreover the AUC ratio between the TMZ-SLN and TMZ solution in the brain was the highest among the tested organs.

Taken together, these studies demonstrate that SLN can modify the distribution of loaded drugs. In particular, doxorubicin, when vehiculated by SLN, is able to achieve lower concentrations in lung, liver, and heart compared to the free solution, thus being able to reduce its systemic toxicity. At the same time, doxorubicin brain accumulation is greatly enhanced if carried by SLN, allowing cerebral targeting for drug delivery in brain tumors. Overall, the addition of stealth agents to SLN seems to improve the aforementioned properties, mainly by decreasing the recognition of SLN by the RES in liver and spleen, thus increasing drug–SLN plasma half-life.

SLN pharmacokinetics in rats bearing glioma cell subcutaneous or intracerebral orthotopic xenografts

In order to assess SLN-mediated drug delivery in an *in vivo* glioma model, we established in Wistar rats orthotopic intracerebral stereotactic C6 cell implants. At day 14, rats were intravenously injected with either doxo-SLN or doxorubicin-free solution. Doxo-SLN achieved intratumoral drug concentrations ranging from 12- (after 30 min) to 50-fold (after 24 h) higher compared to free solutions at same times. Furthermore, in the contralateral healthy hemisphere, only doxorubicin vehiculated by SLN was able to reach subtherapeutic concentrations, ranging from 3.2 (after 30 min) to 12 μg/g (after 24 h), compared to free drug solution (Mauro & Guido, unpublished data).

Williams and Colleagues showed that the SLN formulation of SN-38 is able to increase drug plasma half-life in nude mice bearing subcutaneously xenografted human HT29 cells, a model of chemoresistant colon adenocarcinoma (Williams et al., 2003).

Jain and Colleagues in the second *in vivo* part of the aforementioned study, treated i.v. nude Balb/c mice bearing MDA-MB-468 breast cancer cells subcutaneous xenografts with either 5-FU solution, plain 5-FU-SLN, or Fr-5-FU-SLN (S. K. Jain et al., 2008). The authors showed that administration of Fr-5-FU-SLN formulation results in effective reduction of tumor growth as compared with free 5-FU and plain 5-FU-SLN (delay in tumor growth and increase in life span). Furthermore, Fr-5-FU-SLN allow an increased drug level in the tumor and decreased systemic drug accumulation as well as a reduced IC_{50} compared to both plain 5-FU-SLN and 5-FU solution.

Further suggestions on how nanoparticles work *in vivo* came from three recent studies.

Steininger and Colleagues used an experimental animal model based on intracerebral implanted 101/8 glioblastoma cells in rats (Steiniger et al., 2004). Implanted rats were injected at days 2, 5, and 8 with the following formulations: blank PBCA nanoparticles coated with polysorbate 80 (NP+PS), doxorubicin in saline (DOX), doxorubicin in 1% polysorbate solution (DOX+PS), doxorubicin bound to NP (DOX-NP), or doxorubicin bound to NP coated with polysorbate (DOX-NP+PS). Rats treated with DOX-NP+PS showed a significantly higher survival times compared to all other groups and a 20% rate of long-term remission without any evidence of neurotoxicity.

Xu and Colleagues produced PEG-coated PBCA nanoparticles loaded with paclitaxel and targeted with transferrin (ATN, actively targetable nanoparticles) or not targeted (NTN, nonactively targeted nanoparticles) (Xu et al., 2005). Pharmacokinetics and biodistribution studies of ATN, NTN, and paclitaxel solution were performed in KM strain mice bearing S-180 tumor nodules of about 10 mm, while the evaluation of antitumor activity *in vivo* were done in S-180-bearing KM mice. The authors showed that ATN exhibited a markedly delay in blood clearance in mice and higher paclitaxel levels at 24 h after ATN injection compared to that obtained after free drug solution administration. The distribution profiles of ATN showed that after i.v injection, the tumor accumulation of paclitaxel increases with time, and its concentration at 6 h was about 4.8–2.1 fold higher than those from, respectively, free paclitaxel and NTN administration. A significant tumor regression was observed and complete tumor remission was evident in five out of nine KM mice treated i.v. with ATN.

Ambruosi and Colleagues investigated the biodistribution of blank [^{14}C]-PBCA uncoated and coated with Polysorbate 80 as well as doxorubicin-loaded Polysorbate 80-coated [^{14}C]-PBCA in glioblastoma 101/8-bearing rats after i.v. injection (Ambruosi et al., 2006). The authors showed that the overcoating of [^{14}C]-PBCA-Polysorbate 80 decreased their concentrations in RES organs, while the addition of doxorubicin to the pegylated formulation counteracts the coating effects perhaps by increasing the positive charge of the particles and consequently by altering their adsorption properties both to plasma proteins and to other cells in the body. However, the accumulation of [^{14}C]-PBCA-Polysorbate 80 nanoparticles in the tumor site and in contralateral hemisphere of glioma-bearing rats demonstrated the efficacy of the enhanced permeability and retention effect on brain delivery of nanoparticles. Despite the reduced rate of BBB passage displayed by doxorubicin-loaded Plysorbate 80-coated nanoparticles (perhaps due to the interaction between the drug itself and the surfactant), the concentration of doxorubicin at the tumor site was still higher than in contralateral hemisphere and in brains from healthy rats.

Taken together, the aforementioned results clearly showed that nanoparticles, and in particular SLN, are able to significantly increase intracellular and intratumoral bioavailability of various chemotherapeutics potentially highly effective for brain tumors. Furthermore, in comparison to free solutions, SLN allow a noteworthy reduction in the amount of incorporated drug required to produce cytotoxic effects (as showed by the significant reduction of IC_{50}). In this manner, drug-related and dose-dependent systemic side effects could be avoided.

New therapeutical strategies

Prodrugs, solid lipid nanoparticles, and brain tumors

The use of prodrugs was proposed to overcome pharmacokinetics limitations of otherwise potentially effective drugs. Till now, the decreased cytotoxicity rate, the increased serum opsonization (limiting the passage through the BBB), and the reduced mobility within the brain of the new synthesized lipophilic analogs compared to the referring drugs were advocated as the cause of the disappointing results of prodrugs treatment of brain tumors (Blakeley, 2008; J. X. Wang, Sun, & Zhang, 2002).

Wang and Colleagues synthesized from the cytotoxic agent 5-fluoro-2′-deoxyuridine (FUdR) a lipophilic prodrug, the 3′,5′-dioctanoyl-5-fluoro-2′-deoxyuridine (DO-FUdR), that in turn was incorporated into SLN prepared by thin-layer ultrasonication technique (DO-FUdR-SLN) (J. X. Wang et al., 2002). A comparative study in mice showed that DO-FUdR-SLN allow the best AUC–time curve, MRT and $t_{1/2}$ value in brain tissue compared to both DO-FUdR and free FUdR. More in details, the brain AUC–time curve of DO-FUdR-SLN was 2.06-fold higher than that of DO-FUdR and both these curves were, respectively, 10.97- and 5.32-fold higher compared to free FUdR. Furthermore, the brain $t_{1/2}$ value of DO-FUdR-SLN was 1.49-fold higher that DO-FUdR. The overall drug targeting efficiency (TE^C) of DO-FUdR-SLN to the brain was about threefold higher compared to free FUdR solutions (respectively 29.84 and 11.77). Moreover, the TE^C of DO-FUdR-SLN was decreased in the hearth and kidney compared FUdR free solutions. These data clearly suggest that SLN are able to further enhance the brain targeting of even lipophilic prodrugs and perhaps to partially reduce systemic undesired effects.

Our group chose cholesterylbutyrate (Chol-but) (Fig. 2) – the ester of cholesterol and butyric acid – as another matrix to prepare, from warm oil-in-water microemulsions, Chol-but SLN as a prodrug of butyric acid (Bach Knudsen, Serena, Canibe, & Juntunen, 2003, Brioschi, Zara, Calderoni, Gasco, & Mauro, 2008). This molecule belongs to the family of short-chain fatty acids, physiological compounds produced in the colon of all mammalian organisms (Bach Knudsen et al., 2003; Miller, 2004; Santini, Gozzini, Scappini, Grossi, & Rossi Ferrini, 2001), and could be regarded as a prototype of an effective *in vitro* anti-inflammatory and anticancer drug whose clinical use is heavily limited by its poor

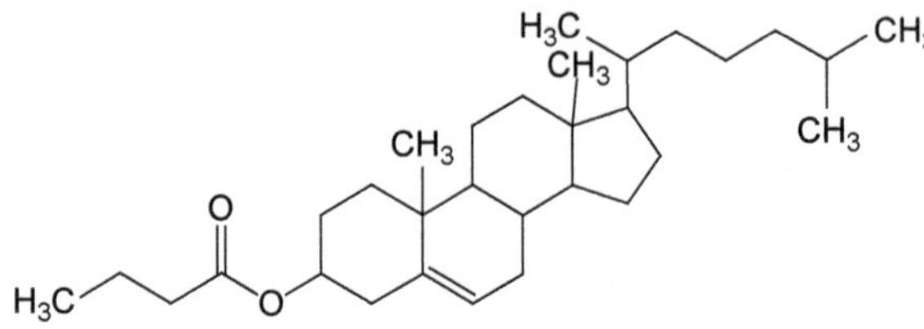

Fig. 2. Structure of cholesterylbutyrate.

pharmacokinetics (Egorin, Yuan, Sentz, Plaisance, & Eiseman, 1999; Miller, 2004; Pouillart, 1998).

Butyrate acts as an anticancer agent by inhibiting proliferation, by stimulating differentiation, and by inducing apoptosis in a wide panel of neoplastic cell lines (including colorectal, breast, gastric, lung, pancreas, and brain districts) (J. S. Chen, Faller, & Spanjaard, 2003; Miller, 2004; Santini et al., 2001). Butyrate could be also numbered as an endogenous member of the family of histone deacetylases (HDAC) inhibitors (HDACI). Disequilibrium in the balance between histone acetyltransferases and HDAC and altered expression of HDAC are involved in the development and the progression of cancer (Balakin, Ivanenkov, Kiselyov, & Tkachenko, 2007; Bolden, Peart, & Johnstone, 2006; J. S. Chen et al., 2003; J. M. Mehnert & Kelly, 2007). HDACI acts as antineoplastic agents by increasing acetylation of both nuclear histones and nonhistone proteins, so inducing transcriptional and nontranscriptional effects and consequently gene expression modulation and activation–inhibition of different pathways (Balakin et al., 2007; Entin-Meer et al., 2005; J. M. Mehnert & Kelly, 2007; Minucci & Pelicci, 2006). HDACI anticancer activities could also include the regulation of the host immune responses and tumor angiogenesis (Bhalla, 2005; Bolden et al., 2006) as well as – in particular for butyrate – mRNA stabilization and direct action on gene transcription (Jiang & Sharfstein, 2008; Lee, Kim, Kummar, Giaccone, & Trepel, 2008; Miller, 2004). Butyrate displays *in vitro* a broad and diversified antineoplastic activity, partially similar to other HDACI, suggesting a possible use of this drug as an effective alternative and/or synergic chemotherapeutic agent. However, *in vivo* studies on butyrate were disappointing (J. S. Chen et al., 2003; Conley et al., 1998; Miller, Kurschel, Osieka, & Schmidt, 1987; Miller, 2004; Patnaik et al., 2002; Pouillart, 1998; Santini et al., 2001), mainly because of poor pharmacokinetics (such as rapid plasma clearance and high liver first pass metabolism) and adverse events (Miller et al., 1987; Pouillart, 1998; Patnaik et al., 2002; Santini et al., 2001; Chen, Faller, & Spanjaard, 2003; Miller, 2004; Conley et al., 1998). For these reasons, more stable and safer prodrugs of butyrate, such as acyloxymethyl esters (tributyrin, AN-1, AN-9), were developed (Entin-Meer et al., 2007; Nudelman et al., 2001; Reid et al., 2004; Rephaeli et al., 2006).

Antineoplastic effects of Chol-but SLN were analyzed *in vitro* on several cancer cell lines (Pellizzaro et al., 1999; Salomone et al., 2000; Serpe et al., 2004; Ugazio et al., 2001) and compared to sodium-butyrate (Na-but).

In nonsmall-cell lung carcinoma cell line (NIH-H460) cultures Pellizzaro and Colleagues showed that Chol-but SLN are able to induce 90% cell growth inhibition at concentration six times lower than Na-but. Complete growth inhibition was obtained at a concentration (0.25 mM) at which Na-but causes only about 55% growth reduction (Pellizzaro et al., 1999; Salomone et al., 2000; Ugazio et al., 2001).

In melanoma cell lines (human MELTO1 and mouse B16) Salomone and Colleagues found that Chol-but SLN compared to Na-but exert antiproliferative and proapoptotic effects at lower doses and shorter treatment times. Furthermore, these effects of Chol-but SLN are time- and dose-dependent within the first 24 h, whereas at prolonged times they become strictly dose-dependent. Moreover, a significant decrease of proliferating cells and an increase of cells blocked in the G0/G1 to S transition phase were seen after 24 h of Chol-but SLN treatment (Salomone et al., 2000).

In three human leukemic cell lines (Jurkat from lymphoid, U937, and HL-60 from myeloid origin) Serpe and Colleagues confirmed that Chol-but SLN (0.25, 0.5, and 1 mM) compared to Na-but (same concentrations) are able to induce a greater cell growth inhibition. Furthermore, the authors showed that *c-myc* expression is rapidly and transiently downregulated in all the three cell lines after Chol-but SLN treatment (0.25 mM) whilst it is slightly decreased only in U937 cells after Na-but treatment at higher concentrations (1 mM). Cell-cycle arrest caused by Chol-but SLN is different among the two groups of cells: block in G1 phase for myeloid (U937 and HL-60) and mainly in G2 phase for lymphoid cells (Jurkat). This result could suggest a different mechanism of action of Chol-but SLN in the various cell types (Serpe et al., 2004).

Antineoplastic effects of Chol-but SLN were analyzed *in vitro* on several cancer cell lines and compared to Na-but, showing that Chol-but SLN exert cell growth inhibition and proapoptotic at lower doses and shorter treatment times in all the cell lines tested (Figs. 3 and 4). However, in these studies the effect of Chol-but SLN on the cell cycle of the various cell lines appeared different, suggesting that the mechanisms of the antineoplastic Chol-but SLN effects may be differently modulated in different cellular contexts (Serpe et al., 2004).

Moreover, in a pilot study we i.v. treated Wistar rats bearing intracerebral stereotactic C6 cell implants with Chol-but SLN 30 mg/kg or with saline every day from day 15th to 21st after implant and then sacrificed. Morphological and immunohistochemical analyses showed a significant shrinkage in tumors of treated animals. The implanted area was replaced by large cysts surrounded by residual tumor cells mostly displaying apoptotic or monstrous (multinucleated) features. Confocal microscopy studies showed that Fluorescent Chol-but SLN, labeled with 6-coumarin were rapidly internalized into tumor cells and persisted for few days into their cytoplasms.

In summary, our *in vitro* studies on different neoplastic cell lines and preliminary *in vivo* study in a rat glioma model convincingly indicate that Chol-but SLN are able to induce consistent antiproliferative and proapoptotic effects earlier and at significantly lower concentrations compared to Na-but. The mechanisms of action of Chol-but SLN and of butyrate are similar but do not completely overlap. For instance, after Chol-but SLN treatment, a significant increase in G2/M block compared to Na-but is observed in some, but not

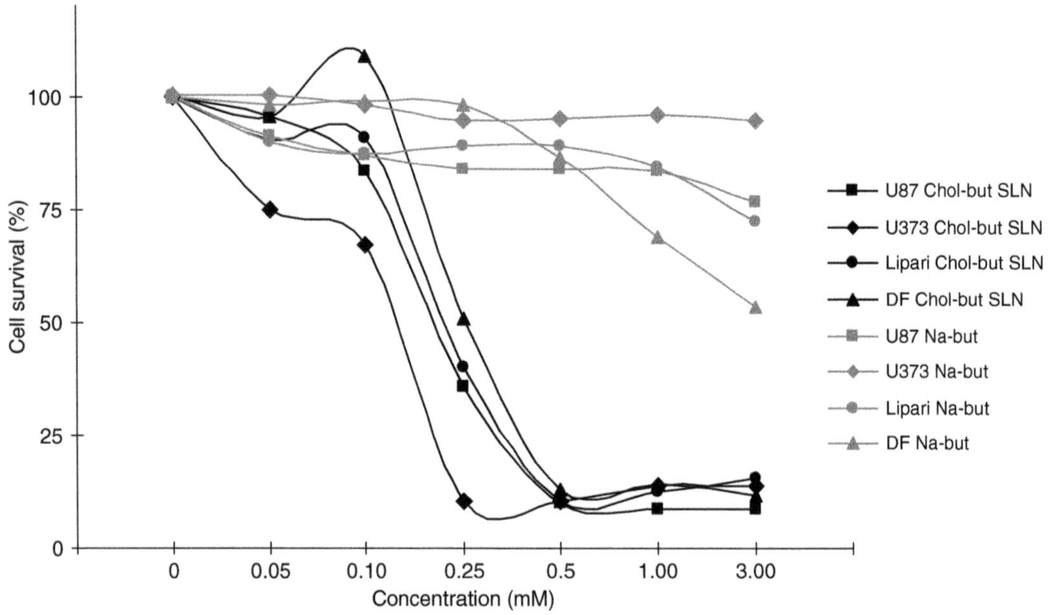

Fig. 3. Cytotoxic activity analysis Methylthiazolyldiphenyl-tetrazolium bromide (MTT test) performed on four human glioma cell cultures (U87, U373, Lipari, DF) treated with Na-but and Chol-but SLN at different concentrations after 72 h (reproduced from Brioschi et al., Molecules, 2008).

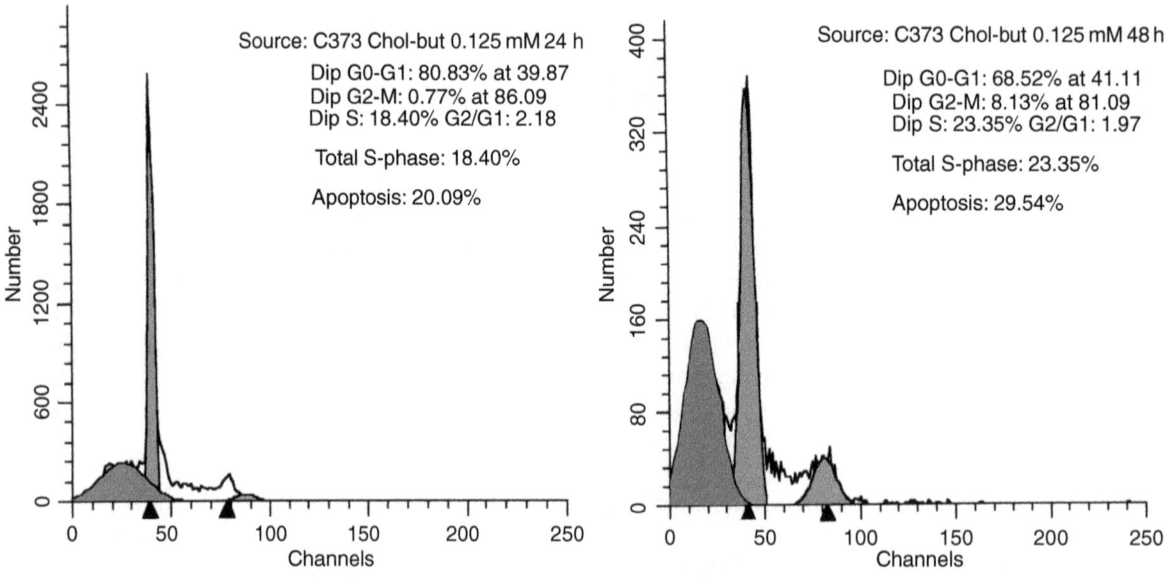

Fig. 4. Flow cytometry DNA analysis performed on U373 human glioma cell cultures treated with Chol-but SLN 0.125 mM after 24 (left) and 48 (right) h (reproduced from Brioschi et al., Molecules, 2008).

in all, of cancer cell lines and the percentage of apoptotic cells are found higher, even at later times (unpublished data).

Therefore, Chol-but SLN could be regarded as suitable and highly effective prodrug of butyric acid, still maintaining chemical–physical, pharmacokinetics, and pharmacodynamic properties of other SLN formulations.

Chol-but SLN seem to act as other nonselective HDACI (Peart et al., 2005) by modulating in a nonspecific manner different pathways, mainly involved in cell survival and proliferation. In malignant gliomas, Chol-but SLN, that proved to modify spontaneous immune response, could act also as a nonspecific stimulating agent of an already present but weak host immune reaction (Carpentier & Meng, 2006).

According to previously reported *in vivo* data, Chol-but SLN are able to effectively reach the CNS and thereby the implanted tumors, and at the same time to achieve significantly lower concentrations in other organs, hence decreasing systemic toxicity.

Antiangiogenetic agents, solid lipid nanoparticles, and brain tumors

Histological hallmarks of malignant gliomas include extensive neovascularization. Among CNS neoplasms primary malignant gliomas show the highest new vessel formation rate that is strictly connected to aggressive clinical behavior (Birner et al., 2003; R. K. Jain et al., 2007; Johansson, Brannstrom, Bergenheim, & Henriksson, 2002; Lamszus et al., 2003; Mischel et al., 2003; Rong, Durden, Van Meir, & Brat, 2006; Toi, Matsumoto, & Bando, 2001; Zhou, Tan, Hess, & Yung, 2003). Among the different mechanisms recruiting new blood vessels in brain tumors, neoangiogenesis is regarded as the major player because of its direct correlation with tumor progression and hence with prognosis. Endothelial proliferations within newly sprouted vessels (a hallmark of human glioblastomas) is probably a direct effect of central tumor hypoxia and necrosis that in turn induce pseudopalisading cells secreting proangiogenic factors (Rong et al., 2006).

Vascular endothelial growth factor A (VEGF-A) and its receptor VEGFR2 are considered crucial players among several known angiogenic cytokines (Ferrara & Davis-Smyth, 1997; R. K. Jain et al., 2007; Ke, Shi, Im, Chen, & Yung, 2000; Ke, Shi, & Yung, 2002; Lamszus et al., 2003; Rong et al., 2006).

The VEGF family consists of 34- to 45-kDa dimeric glycosylated protein isoforms. VEGF165 – the predominant isoform – is produced in most normal tissues, including the brain, and in both low- and high-grade gliomas in which its expression rate directly correlates to grading, vascularity, clinical behavior, and inversely to prognosis (Ferrara & Davis-Smyth, 1997; Ferrara, Gerber, & LeCouter, 2003; Jansen, de Witt Hamer, Witmer, Troost, & van Noorden, 2004; Rong et al., 2006; Rosenstein & Krum, 2004a; Toi et al., 2001).

In brain tumors, VEGF-A expression is mainly and independently regulated by both hypoxia (trough the hypoxia-inducible factor-1α, HIF-1α) and acidosis. In addition, different oncogenes and tumor suppressor genes, hormones, cytokines, and signaling molecules are able to modify VEGF expression pattern (R. K. Jain et al., 2007). Furthermore, not only malignant cells but also various host cells (such as stromal cells) and extracellular matrix could express VEGF in response to toxic insults.

Several experimental attempts to turn off the HIF/VEGF signaling pathway using different class of drugs and genes constructs were successful *in vitro* and *in vivo* to reduce tumor progression and angiogenesis. Various compounds such as angiostatin, anti-VEGF and anti-VEGF receptor (VEGFR) antibodies, inhibitors of VEGFR-2 tyrosine kinase activity, ribozymes, AS-ODN, and small interfering RNA constructs were tested to interfere with the VEGF signaling pathway (Breyer et al., 2000; Farhadi, Capelle, Erber, Ullrich, & Vajkoczy, 2005; Jansen et al., 2004; Kim et al., 2005; Lamszus et al., 2003; Niola et al., 2006; Peoch et al., 2002; Rich & Bigner, 2004). For instance, several clinical studies indicated that treatment of recurrent glioblastomas patients with a combination of bevacizumab (anti-VEGF humanized monoclonal antibody), various chemotherapeutics (i.e., TMZ, irinotecan) and radiotherapy significantly increases progression-free survival and reduces the need for steroids (Norden et al., 2008a; Norden, Drappatz, & Wen 2008b). Till now, different VEGFR inhibitors are under phase I and II clinical study (Norden et al., 2008a).

However, these approaches showed significant limits in the ability of overcome systemic degradation, reach effective bioavailability in the tumor target, and avoid systemic toxicity. Common adverse effects, such as hypertension, proteinuria, and increased risk of thromboembolism and hemorrhage, and the evidence that about 50% of patients develop antiangiogenic therapy resistance and time-variable response further on limit the clinical use of these class of therapeutics (Norden et al., 2008b). These drawbacks could be firstly related to both the physical–chemical properties of the drugs and the routes of administration (Barratt, 2003; Jansen et al., 2004; Koziara et al., 2006; Muller & Keck, 2004; Pardridge, 2007; Rich & Bigner, 2004; Tiwari & Amiji, 2006).

Furthermore, recent preclinical studies surprisingly showed that the blocking of VEGF-mediated neoangiogenesis could promote both tumor infiltration (perhaps by overexpression of proinvasive molecules or by co-option of existing cerebral blood vessels) and recruitment of circulating endothelial cells into the neoplasm (Norden et al., 2008a, 2008b). These data clearly suggest that at this moment anti-VEGF signaling pathway inhibition could optimally work only if combined to other cytotoxic chemotherapeutics, to non-VEGF-mediated antiangiogenetic factors, or to radiotherapy (Norden et al., 2008a, 2008b).

We specifically designed — always from warm oil-in-water microemulsions — SLN vehiculating VEGF antisense oligonucleotides (VEGF-AS-ODN SLN) in order to downregulate VEGF expression in a rat glioma model. We studied the effectiveness of VEGF-AS-ODN SLN both *in vitro* in C6 glioma cell cultures under hypoxic conditions (Fukumura et al., 2001; Serganova et al., 2004; Tan et al., 2005), and *in vivo* in the intracerebral rat C6 glioma model (Brioschi et al., 2009).

In vitro, rat C6 glioma cells under both normal and hypoxic conditions were treated with VEGF phosphorothioate AS-ODN, either free or vehiculated by SLN for 24 and 48 h. VEGF phosphorothioate sense-ODN (S-ODN) as free solution and carried by SLN (VEGF-S-ODN SLN) as well as Fluorescent SLN (Flu-SLN) carrying 6-coumarin instead of ODN were also tested as control.

At confocal microscopy observation within 5 min after treatment with Flu-SLN a sharp and homogeneous cytoplasmic green fluorescence reached the maximum and persisted unchanged for almost 90 min. Western blot analysis of cell homogenates from untreated normoxic cultures depicted a pattern of VEGF expression similar to that found in rat normal heart and brain homogenates when compared to that of human controls. Under hypoxia cotreatment with VEGF-AS-ODN, VEGF-S-ODN or VEGF-S-ODN SLN did not produce any appreciable VEGF expression modulation at both 24 and 48 h. VEGF 120, VEGF164, and VEGF188 expression at 24 and 48 h increased under hypoxia, as expected, while progressively decreased after VEGF-AS-ODN SLN treatment. A statistically significant reduction ($p < 0.01$) was evident for all the VEGF isoforms after 48 h when compared not only to the hypoxic but also to the basal conditions (Fig. 5).

In experiments *in vivo* with the Wistar rat C6 glioma model, the implanted rats were randomized into four main groups, each one treated for three consecutive days with free VEGF-S-ODN, free VEGF-AS-ODN, VEGF-S-ODN SLN or VEGF-AS-ODN SLN, at different concentrations. Three days after treatment all the animals were sacrificed. In control animals, tumor cells, mainly in the perinecrotic areas and tumor borders, showed a clear cytoplasmic VEGF immunostaining. Interestingly, a similar VEGF immunoreactivity was found in hippocampal neurons as well as in large pyramidal cortical and cerebellar Purkinje neurons in both the hemispheres.

In animals treated with VEGF-S-ODN- and VEGF-AS-ODN-free solutions as well as with VEGF-S-ODN SLN any appreciable modification in the VEGF expression in both tumor and normal brain tissue was found. Only treatment with VEGF-AS-ODN SLN induced a great reduction of VEGF expression in both central and peripheral regions of the tumors. VEGF expression was also decreased in normal brain tissue, but to a lesser extent than in tumors (Fig. 6).

In summary, *in vitro* findings clearly indicate that our SLN allow highly effective, quick, and sustained ODN delivery into tumor cells (at least 500-fold more efficient than the free solutions).

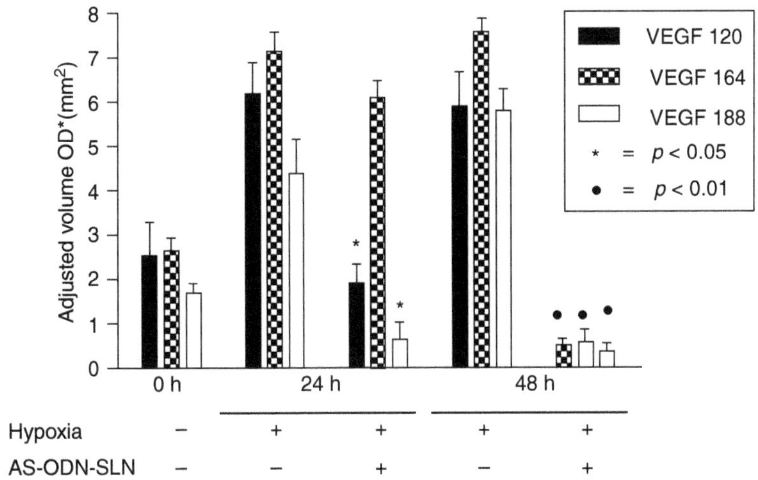

Fig. 5. VEGF expression statistical analysis and semiquantification of western blot data from cell homogenates (reproduced with permission from Brioschi et al., Journal of Nanoneuroscience, 2009).

Similar results have been previously described by Tondelli and Colleagues using *c-myb* AS-ODN incorporated in polymeric nanospheres (Saleh, Stacker, & Wilks, 1996; Tondelli, Ricca, Laus, Lelli, & Citro, 1998). Taken together these data demonstrate that SLN could be regarded as a good carrier not only for chemotherapeutic drugs but also for gene therapeutical agents. Furthermore, *in vivo* study showed that VEGF AS-ODN SLN efficiently downregulate VEGF expression in neoplastic cells, effectively reaching every part of the implanted tumors (Brioschi et al., 2009).

Future perspectives and novel challenges

SLN compared to other colloidal carriers display more versatile structural properties and hence could be potentially modified in order to vehiculate simultaneously more than one therapeutical compound. This goal could be reached acting on both the preparation process and lipid composition as well as surfactants and cosurfactants use. In this manner it will be possible to design SLN able to carry two or more therapeutical agents having different molecular structure and physical–chemical characteristics. In addition to lipophilic molecules it could be supposed that more hydrophilic and/or ionic compounds could be simultaneously loaded into SLN. Furthermore, SLN may be planned to allow a different release profile of the carried drugs, for instance by acting on the preferential location of the dispersed molecule into the core or into the shell portion of the nanoparticle. This could lead to time specific release profile of each carried drug. These statements suggest a future possible scenario in which a single carrier could be adapted to different requirements. For instance, SLN could be tailored to the clinical course of the disease, and constructed in order to act as "sensitizer" or preparatory to other therapies (i.e., surgery and radiotherapy), to take into account the genetic temporal and spatial heterogeneity of the tumor, and to reciprocally enhance the effects of the vehiculated drugs otherwise individually poorly effective.

Targeting to the brain

In the previous part of this chapter we already showed that SLN could be passively targeted to the brain by modifying both lipid composition and production processes. This passive targeting could be also sustained by the so-called enhanced permeability and retention effect, commonly found in systemic solid neoplasms (Parveen & Sahoo, 2008; Wong et al., 2007). The not homogeneous BBB disruption coupled with the secretion of vascular

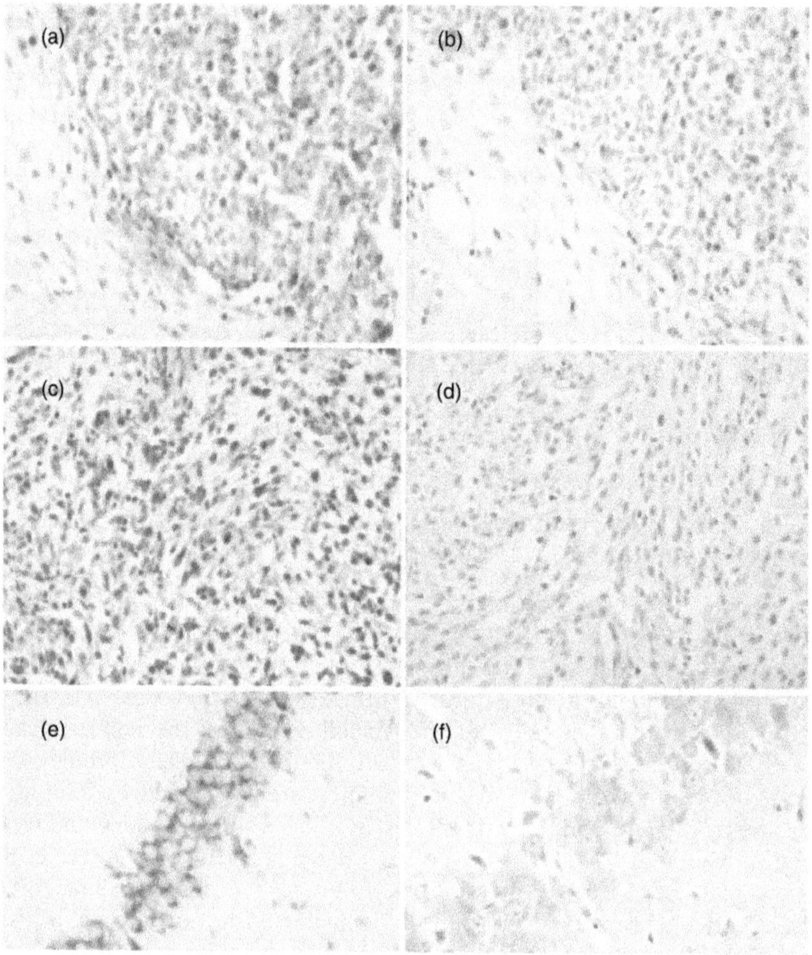

Fig. 6. VEGF immunohistochemistry on xenografted tumor sections from: control rats (a), corresponding negative (without primary antibody) control rats (b), animals treated with 2 mg/kg AS-ODN-free solution (c), and 2 mg/kg AS-ODN-SLN (d). VEGF immunohistochemistry on brain sections (hippocampus) from AS-ODN (e)- and AS-ODN-SLN (f)-treated animals (reproduced with permission from Brioschi et al., Journal of Nanoneuroscience, 2009).

mediators facilitating extravasation and the raised pressure exerted by tumor mass and surrounding edema contribute to slow the drainage of macromolecules and to facilitate the accumulation within the tumor of particulate carriers within the tumor (Parveen & Sahoo, 2008; Wong et al., 2007).

In addition to passive processes, active targeting is one of the most promising and potentially effective results of the use of nanoparticulate carriers. Active targeting could facilitate the SLN transport into brain tumors but also increase the specificity of the drug delivery into a peculiar neoplastic cell population. This could consequently further reduce the total amount of the drug effectively needed (and hence the possible systemic toxicity) and allow a better selected and temporal defined antineoplastic effect. Active targeting implies that SLN surface is suitably designed to specifically recognize peculiar tissues or cancer cells. Furthermore, active targeting to brain tumor cells may facilitate the BBB passage (Beduneau et al., 2007) addressing different influx–efflux transport systems displayed by brain endothelial cells that include carrier-mediated transport (i.e., D-glucose), receptor-mediated endocytosis [such as insulin, insulin-like growth factor, folic acid, and transferrin (Tf)], and adsorptive-mediated endocytosis (Beduneau et al., 2007; Pardridge, 2007). Therefore, active targeting of nanoparticulate carriers to brain tumor cells could be achieved by three main mechanisms: ligand–receptor interaction, antibody–antigen recognition, and use of aptamers [namely, DNA or RNA fragments that bind with high affinity and specificity molecular intracellular and/or membrane-bound targets (Parveen & Sahoo, 2008)]. All these processes implies that different classes of targeted compounds are displayed by SLN surface by covalent or notcovalent linkages (Beduneau et al., 2007).

Ligand–receptor interaction

Tf receptor is 2- to 10-fold overexpressed in most of the tumor cells compared to normal cells and hence could be regarded as the prototype of potential targets useful in order to enhance carrier–tumor cell interaction. Furthermore, Tf receptors are also expressed on the luminal membrane of brain endothelial cells and through receptor-mediated endocytosis allow the internalization of iron-saturated Tf (Fishman, Rubin, Handrahan, Connor, & Fine, 1987; Gupta, Jain, & Jain, 2007; Qian, Li, Sun, & Ho, 2002). This result, that seems to suggest the effectiveness of active BBB crossing, was disputed because of the demonstration of Tf retroendocytosis after dissociation from the iron, the latter compound being the only one completely transcytosed by endothelial cells (Beduneau et al., 2007). However, Gupta and Colleagues developed SLN conjugated with transferrin (Tf-SLN) and loaded with the antimalarial quinine dihydrochloride in order to increase the delivery of these agents to the brain. The authors found that Tf-SLN show the lowest plasma concentration and the highest brain uptake compared to nonconjugated SLN and free drug solution (Gupta et al., 2007). FR could be regarded as another possible useful system to actively target brain tumor cells, because of its high expression rate at the level of both endothelial and brain tumor cells (Beduneau et al., 2007). Stevens and Colleagues showed that paclitaxel prodrug-loaded SLN conjugated to folic acid are effective in mice bearing subcutaneously engrafted murine lung carcinoma cell tumors (Stevens et al., 2004).

Antibody–antigen recognition

Receptor-specific peptidomimetic monoclonal antibodies (MAb) could act as "molecular Trojan horse" and allow BBB crossing by any given attached compound. Till now, various types of so-called Trojan horse liposomes proved effective in brain targeting (Pardridge, 2007).

Tsutsui and Colleagues designed bionanocapsules to specifically *in vitro* and *in vivo* target human glioma cells. These nanoparticles, composed of antibody directed against human EGFR-VIII, were specifically uptaken by human Gli36 cells in culture and *in vivo* in a mouse orthotopic Gli36 intracerebral glioma model by direct intratumoral injection (Tsutsui et al., 2007).

Yang and Colleagues showed that a boronated monoclonal antibody directed against the EGFR-VIII linked to polyamido amine dendrimers is effective in a rat glioblastoma model (Yang et al., 2006b, 2008). This report confirms previous preclinical studies showing that monoclonal antibodies directed against the extracellular portion of EGFR-VIII and PDGFR could be effective in the treatment of gliomas and hence could be used to specifically target SLN carrying different drugs to glioma cells (Rich & Bigner, 2004). Similarly antibodies directed toward different target involved in the glioma VEGF signaling pathway (i.e., VEGFR) could be used to target SLN to glioma endothelial cells possibly interfering with the angiogenic process (Rich & Bigner, 2004).

Nevertheless, both ligand–receptor and antibody–antigen recognition could interact and activate systemic and local host biological reactions, potentially interfering with physiological antitumoral activities, such as immunological response. For this reason MAb directed to sites of the targeted molecule not involved in the endogenous ligand recognition were developed (Beduneau et al., 2007). MAb-conjugated liposomes, also known as immunoliposomes, proved effective as brain drug delivery systems. For instance, Zhang and Colleagues developed immunoliposomes, carrying a plasmid DNA encoding the EGF receptor antisense mRNA, conjugated with two MAb directed to mouse Tf receptor (in order to pass through the BBB) and to human insulin receptor (in order to intratumor cell delivery). This study showed that these immunoliposomes are effective after i.v. administration in mice bearing U87 brain tumors (Zhang, Jeong Lee, Boado, & Pardridge, 2002). A similar result was obtained in the same brain glioma model using immunoliposomes carrying a short hairpin RNA targeting EGFR mRNA (Boado, 2005; Zhang et al., 2004). Till now, various colloidal carriers (including pegylated nanoparticles and NLC) conjugated with a murine MAb antirat Tf (OX26) are under study and show promising results as brain drug delivery systems (Beduneau et al., 2007; Pardridge, 2007).

Aptamers

Aptamers (namely, DNA or RNA oligonucleotides targeted to specific antigens) display some advantages over antibodies such as lower immunogenicity, higher target specificity and affinity, and better tissue penetration. Poly(lactide-co-glycolide) (PLGA) nanoparticles carrying docetaxel functionalized with an RNA aptamer recognizing a plasma membrane-specific antigen were successfully tested in both *in vitro* and *in vivo* models of prostate cancer (Farokhzad et al., 2006a; Farokhzad, Karp, & Langer, 2006b; Martin-Villalba et al., 2008).

In conclusion, taken together these studies clearly suggest that SLN – due to their versatility – could be easily and variously engineered in order to successfully achieve active targeting to malignant brain tumors.

Gene therapy of brain tumors

Malignant gliomas display high genetic heterogeneity, as previously discussed ((62 Martin-Villalba,A. 2008)). In the recent years, many researches focused on the role of microRNAs (miRNAs) expression profile in diagnosis, staging, progression, prognosis, and response to therapy in brain tumors (Nicoloso & Calin, 2008). The miRNAs together with short interfering RNA belong to the family of small regulatory RNAs that act as riboregulators and are the crucial mediators of RNA interference strategy (Liu, Fortin, & Mourelatos, 2008). The miRNAs – namely, 19–25 nucleotides in length noncoding RNA – act at posttranscriptional level and are able to regulate gene expression by reducing the amount of transcribed mRNA and/or translated proteins (Liu et al., 2008). Consequently, alterations in miRNAs play a critical role in tumor initiation, progression, and metastasis (Liu et al., 2008) and, in fact, glioblastomas display an miRNAs expression pattern different from normal brain tissue (Ciafre et al., 2005; le Sage et al., 2007; Liu et al., 2008). MiR-21 is highly expressed (from 5- to 100-fold) in human glioblastomas and seems to act by interfering with the transcription of critical proapoptotic genes, probably reducing PTEN protein expression (Chan, Krichevsky, & Kosik, 2005). Furthermore, the cluster miR-221-222 is involved in cell-cycle regulation by targeting the CDK inhibitor $p27^{kip1}$ (le Sage et al., 2007; Visone et al., 2007). These findings suggest that miRNAs could represent potential targets in brain tumor therapy and different agents such as modified antisense single-stranded oligonucleotide complementary to specific miRNA (LNAs anti-miRNAs) and chemically modified and cholesterol-conjugated single-stranded RNA complementary to a given miRNA (antagomirs) are under study in order to achieve targeted miRNA inhibition (Nicoloso & Calin, 2008). However, antagomirs showed the same limitations displayed by other therapeutical compound in order to reach the CNS bypassing the BBB: these agents were not able to play any action in the brain when systemically administered in mice while they proved effective if directly injected into the mouse cortex (Krutzfeldt et al., 2007).

We believe that SLN could be suitable carriers for RNA interference approach as well as, in broader terms, for gene therapy of brain tumors, as proved for instance by our work on VEGF AS-ODN previously reported. Moreover, more specific and downstream or upstream located players belonging to various pathways involved in glioma initiation and progression, including neoangiogenesis, could be effectively targeted by using SLN. Nevertheless, till now active targeting did not show satisfactory results in clinical trials and antiangiogenic therapy proved significant effects only if combined with chemotherapy (Sanson, 2008; Vredenburgh et al., 2007). For instance, EGFR targeted inhibition in recurrent malignant gliomas proved dependent by the concomitant switching-off of the downstream PI3K/AkT pathway (Mellinghoff et al., 2005). However, clinical trials addressing both EGFR and PI3K/AkT (so called "vertical targeting") showed disappointing results, probably because of the concurrent activation of multiple tyrosine kinase pathways (Sanson, 2008). Therefore, the concomitant targeting ("horizontal targeting") of more than one tyrosine kinase receptors activated in malignant gliomas could be effective (Sanson, 2008; Stommel et al., 2007). Once more, SLN could offer a resourceful colloidal carrier platform for both vertical and horizontal targeted therapies of malignant gliomas.

Brain tumors imaging and thermotherapy

Till now, several types of nanoparticles targeted mainly to vascular epitopes (such as magnetic particles, quantum dots, immunotargeted nanoshells, liposomes, and dendrimers) have been developed for systemic cancer detection in order to increase both intratumor retention times and contrast enhancement. The use of these nanoparticles could allow not only earlier detection and prolonged observation times of the neoplastic lesions but also – according to the targeting molecules employed – more detailed phenotypical and/or genetic characterization of detected tumors (Parveen & Sahoo, 2008).

Superparamagnetic nanoparticles were developed as MRI contrast agents in order to increase the selectivity and detection abilities of brain tumor imaging (Chertok, David, Huang, & Yang, 2007; Muldoon, Sandor, Pinkston, & Neuwelt, 2005; Reddy et al., 2006). Different magnetic nanoparticles composed of a magnetic core (usually iron oxide) produce a MRI visible hypointense signal drop out on T2-weighted images (negative contrast) due to their ability to strongly enhance proton spin–spin relaxation (Y. X. Wang, Hussain, & Krestin, 2001). Polyacrylamide nanoparticles encapsulating iron oxide proved excellent tumor contrast enhancement after i.v. administration to rats bearing intracranial 9L gliomas (Moffat et al., 2003; Reddy et al., 2006).

Our group studied the pharmacokinetics and biodistribution of SLN loaded with superparamagnetic iron oxides (SLN-Fe^{A} and SLN-Fe^{B}) compared to Endorem® as contrast agents for MRI in rats. After parenteral administration both types of SLN-Fe showed a slower blood clearance and a more prolonged CNS retention time (up to 135′) compared to the commercial MRI contrast agent Endorem® (Peira et al., 2003).

Thermal ablation (namely, thermotherapy) often combined to chemotherapy could be potentially effective in the treatment of malignant gliomas but is highly limited by its nonfocalized field of action (Jain, 2007). Recent studies proved that both in animal glioma models and in selected patients (Maier-Hauff et al., 2007) the injection of magnetic nanoparticles into the tumors followed by exposure to an alternating magnetic field allows to a prolongation of survival, induces regression of tumor growth and is well tolerated. Moreover, several studies showed that magnetic nanoparticles, due to their magnetic responsiveness, could be retained at tumor sites for longer times after local application of external magnetic field (Chertok et al., 2007). Chertok and Colleagues recently demonstrated that – in rats bearing intracerebral orthotopic 9L-gliosarcoma tumors – the concomitant application of brain targeting magnetic field during i.v. administration of iron oxide nanoparticles induces a fivefold increase in tumor exposure to these nanoparticles compared to nontargeted neoplasms and a 3.6-fold rise in selective nanoparticles accumulation in tumors than in normal brain tissue (Chertok et al., 2007).

Taken together these data suggest that SLN – either specifically coated with glioma tumor-targeting agents or brain targeted using magnetic field exposure – could work as innovative tools in the field of brain tumors imaging by improving sensibility and specificity of commonly used techniques. Moreover, the feasibility of loading these tumor-targeted imaging-devoted SLN with different classes of antineoplastic agents could argue their possible use as neuroimaging and concomitantly therapeutic carriers, opening the way of a selective, localized, and concomitant or subsequent thermo- and chemotherapy.

Thereby the prolonged retention time allows long-lasting observation time of the tumor behavior and hence could be useful in noninvasive *in vivo* MRI monitoring of therapeutical effects produced by SLN-carried drugs.

Future SLN therapeutical applications

Furthermore, SLN could be used to enhance the effects of new attractive targeted therapeutical strategies of brain tumors, such as boron neutron capture therapy (BNCT) and photodynamic therapy (Jain, 2007).

- BNCT could allow highly localized radiotherapy, possibly limited to a range of a single neoplastic cell, by producing a nuclear reaction between thermal neutrons and ^{10}B, leading to generation of α particles and ^{7}Li nuclei (Jain, 2007). As previously reported, boronated antibodies directed against to specific glioma targets could be effective if conjugated to different carriers (Yang et al., 2006a). Thereby, it will be possible to design SLN carrying on their surface glioma targeted boronated antibodies and this could open the way to highly localized cotreatments. For instance, by including into these SLN different antineoplastic agents (i.e., radiosensitizers, chemotherapeutics, gene constructs) we could obtain concomitant or subsequent potentiation of localized radiotherapy.
- The use of a photosensitizer (such as Photofrin®) for photodynamic therapy is based on the ability of this class of compounds to generate, after exposure to a specific wavelength light, toxic oxygen species into target cells. This therapy is greatly limited by systemic (and in particular cutaneous) side effects (Jain, 2007; Parveen & Sahoo, 2008). Reddy and Colleagues incorporated into a polymeric nanoparticle targeted to tumoral neovasculature both Photofrin® and iron oxide, showing that, compared PDT delivered after treatment with either Photofrin® or nontargeted nanoparticles alone, these nanoparticles were more effective in prolonging survival in a rat intracerebral glioma model after Photodynamic Therapy (PDT) (Reddy et al., 2006). Better MRI resolution obtained by the use of these targeted nanoparticles vehiculating iron oxide could contribute to improve intracranial localization of the tumor and hence to facilitate PDT administration. Further on, superparamagnetic SLN could be successfully used for PDT if adequately coated with glioma tumor-targeting agent and loaded also with a photosensitizer molecule.

Conclusions

In the previous sections of this chapter we showed that SLN – due to their versatile properties – could be regarded as efficacious colloidal carriers for different classes of agents useful in both imaging and treatment of malignant gliomas. SLN allow the effective employment of otherwise toxic (and hence poorly efficacious at safer dosages) chemotherapeutics and could efficaciously vehiculate molecules having different chemical and ionic structure, including gene constructs, and acting on distinct pathways involved in tumor initiation–progression. Furthermore, these nanoparticles could be designed to escape the RES and thereby to passively target the brain, so increasing the AUC curve rate and prolonging the exposition time. Moreover, active targeting as well as the possibility of simultaneously vehiculate more than one compound and the feasibility of different release profiles are more than an attractive realistic perspective. These latter SLN properties could further on contribute to increase the drug selectivity contemporaneously reducing systemic toxicity. In summary, SLN could work as a highly flexible platform for brain tumor imaging and therapeutical purposes, allowing a more tailored approach to both genetically–phenotypically distinct malignant gliomas and patients' stratification.

Nevertheless, our data clearly showed that even the most passively brain targeted SLN at the lowest effective concentration used reach every part of the body where they could release the vehiculated drug. Furthermore, once crossed the BBB, nonactively targeted SLN carry the therapeutical agent in every CNS region where, on the contrary, it could be undesirable and/or dangerous. As previously described, VEGF-AS-ODN SLN can reduce VEGF expression not only in glioma cells but also in hippocampal neurons, potentially interfering with protective and repair processes involving VEGF (Krum & Rosenstein, 1998; Rosenstein & Krum, 2004b; Yano et al., 2005; Yasuhara et al., 2005).

This intra-CNS low selectivity claims on the one hand further *in vivo* studies directed to identify the minimal effective drug dose needed and on the other hand the development of more selective active-targeted SLN, in order to minimize undesired effects in normal brain but also in healthy systemic tissues.

Despite these latter suggested possible limitations anyway it would be advisable to plan well designed and controlled phase I and phase II clinical trials in humans with SLN carrying antiglioma drugs.

Acknowledgments

This chapter is in memory of our beloved friends and colleagues Cinzia Miscio and Simonetta D'Alisa.

References

Ambruosi, A., Khalansky, A. S., Yamamoto, H., Gelperina, S. E., Begley, D. J., & Kreuter, J. (2006). Biodistribution of polysorbate 80-coated doxorubicin-loaded [14C]-poly(butyl cyanoacrylate) nanoparticles after intravenous administration to glioblastoma-bearing rats. *Journal of Drug Targeting*, *14*(2), 97–105.

Bach Knudsen, K. E., Serena, A., Canibe, N., & Juntunen, K. S. (2003). New insight into butyrate metabolism. *The Proceedings of the Nutrition Society*, *62*(1), 81–86.

Balakin, K. V., Ivanenkov, Y. A., Kiselyov, A. S., & Tkachenko, S. E. (2007). Histone deacetylase inhibitors in cancer therapy: Latest developments, trends and medicinal chemistry perspective. *Anti-Cancer Agents in Medicinal Chemistry*, *7*(5), 576–592.

Barratt, G. (2003). Colloidal drug carriers: Achievements and perspectives. *Cellular and Molecular Life Sciences*, *60*(1), 21–37.

Barth, R. F. (1998). Rat brain tumor models in experimental neuro-oncology: The 9L, C6, T9, F98, RG2 (D74), RT-2 and CNS-1 gliomas. *Journal of Neuro-Oncology*, *36*(1), 91–102.

Beduneau, A., Saulnier, P., & Benoit, J. P. (2007). Active targeting of brain tumors using nanocarriers. *Biomaterials*, *28*(33), 4947–4967.

Bhalla, K. N. (2005). Epigenetic and chromatin modifiers as targeted therapy of hematologic malignancies. *Journal of Clinical Oncology*, *23*(17), 3971–3993.

Birner, P., Piribauer, M., Fischer, I., Gatterbauer, B., Marosi, C., Ambros, P. F., et al. (2003). Vascular patterns in glioblastoma influence clinical outcome and associate with variable expression of angiogenic proteins: Evidence for distinct angiogenic subtypes. *Brain Pathology*, *13*(2), 133–143.

Blakeley, J. (2008). Drug delivery to brain tumors. *Current Neurology and Neuroscience Reports*, *8*(3), 235–241.

Blasi, P., Giovagnoli, S., Schoubben, A., Ricci, M., & Rossi, C. (2007). Solid lipid nanoparticles for targeted brain drug delivery. *Advanced Drug Delivery Reviews*, *59*(6), 454–477.

Boado, R. J. (2005). RNA interference and nonviral targeted gene therapy of experimental brain cancer. *NeuroRx*, *2*(1), 139–150.

Bocca, C., Caputo, O., Cavalli, R., Gabriel, L., Miglietta, A., & Gasco, M. R. (1998). Phagocytic uptake of fluorescent stealth and non-stealth solid lipid nanoparticles. *International Journal of Pharmaceutics*, *175*, 185–193.

Bolden, J. E., Peart, M. J., & Johnstone, R. W. (2006). Anticancer activities of histone deacetylase inhibitors. *Nature Reviews. Drug Discovery*, *5*(9), 769–784.

Breyer, R., Hussein, S., Radu, D. L., Putz, K. M., Gunia, S., Hecker, H., et al. (2000). Disruption of intracerebral progression of C6 rat glioblastoma by in vivo treatment with anti-CD44 monoclonal antibody. *Journal of Neurosurgery*, *92*(1), 140–149.

Brioschi, A., Zara, G. P., Calderoni, S., Gasco, M. R., & Mauro, A. (2008). Cholesterylbutyrate solid lipid nanoparticles as a butyric acid prodrug. *Molecules*, *13*(2), 230–254.

Brioschi, A. M., Calderoni, S., Pradotto, L. G., Guido, M., Strada, A., Zenga, F., et al. (2009). Solid lipid nanoparticles carrying oligonucleotides inhibit vascular endothelial growth factor expression in rat glioma models. *Journal of Nanoneuroscience*, *1*(1), 65–74.

Buckner, J. C., Brown, P. D., O'Neill, B. P., Meyer, F. B., Wetmore, C. J., & Uhm, J. H. (2007). Central nervous system tumors. *Mayo Clinic Proceedings*, *82*(10), 1271–1286.

Carpentier, A. F. (2005). Neuro-oncology: The growing role of chemotherapy in glioma. *Lancet Neurology*, *4*(1), 4–5.

Carpentier, A. F., & Meng, Y. (2006). Recent advances in immunotherapy for human glioma. *Current Opinion in Oncology*, *18*(6), 631–636.

Cavaliere, R., Wen, P. Y., & Schiff, D. (2007). Novel therapies for malignant gliomas. *Neurologic Clinics*, *25*(4), 1141–1171.

Cavalli, R., Bargoni, A., Podio, V., Muntoni, E., Zara, G. P., & Gasco, M. R. (2003). Duodenal administration of solid lipid nanoparticles loaded with different percentages of tobramycin. *Journal Pharmaceutical Sciences, 92*(5), 1085–1094.

Chan, J. A., Krichevsky, A. M., & Kosik, K. S. (2005). MicroRNA-21 is an antiapoptotic factor in human glioblastoma cells. *Cancer Research, 65*(14), 6029–6033.

Cheepala, S. B., Syed, Z., Trutschl, M., Cvek, U., & Clifford, J. L. (2007). Retinoids and skin: Microarrays shed new light on chemopreventive action of all-trans retinoic acid. *Molecular Carcinogenesis, 46*(8), 634–639.

Chen, D. B., Yang, T. Z., Lu, W. L., & Zhang, Q. (2001). In vitro and in vivo study of two types of long-circulating solid lipid nanoparticles containing paclitaxel. *Chemical & Pharmaceutical Bulletin (Tokyo), 49*(11), 1444–1447.

Chen, J. S., Faller, D. V., & Spanjaard, R. A. (2003). Short-chain fatty acid inhibitors of histone deacetylases: Promising anticancer therapeutics? *Current Cancer Drug Targets, 3*(3), 219–236.

Chertok, B., David, A. E., Huang, Y., & Yang, V. C. (2007). Glioma selectivity of magnetically targeted nanoparticles: A role of abnormal tumor hydrodynamics. *Journal of Controlled Release, 122*(3), 315–323.

Cho, K., Wang, X., Nie, S., Chen, Z. G., & Shin, D. M. (2008). Therapeutic nanoparticles for drug delivery in cancer. *Clinical Cancer Research, 14*(5), 1310–1316.

Ciafre, S. A., Galardi, S., Mangiola, A., Ferracin, M., Liu, C. G., Sabatino, G., et al. (2005). Extensive modulation of a set of microRNAs in primary glioblastoma. *Biochemical and Biophysical Research Communications, 334*(4), 1351–1358.

Claes, A., Schuuring, J., Boots-Sprenger, S., Hendriks-Cornelissen, S., Dekkers, M., van der Kogel, A. J., et al. (2008). Phenotypic and genotypic characterization of orthotopic human glioma models and its relevance for the study of anti-glioma therapy. *Brain Pathology, 18*(3), 423–433.

Cohen, S. S., Flaks, J. G., Barner, H. D., Loeb, M. R., & Lichtenstein, J. (1958). The mode of action of 5-fluorouracil and its derivatives. *Proceedings of the National Academy of Sciences, USA, 44*(10), 1004–1012.

Conley, B. A., Egorin, M. J., Tait, N., Rosen, D. M., Sausville, E. A., Dover, G., et al. (1998). Phase I study of the orally administered butyrate prodrug, tributyrin, in patients with solid tumors. *Clinical Cancer Research, 4*(3), 629–634.

Cornic, M., Guidez, F., Delva, L., Agadir, A., Degos, L., & Chomienne, C. (1992). Mechanism of action of retinoids in a new therapeutic approach to acute promyelocytic leukemia. *Bulletin du Cancer, 79*(7), 697–704.

Criniere, E., Kaloshi, G., Laigle-Donadey, F., Lejeune, J., Auger, N., Benouaich-Amiel, A., et al. (2007). MGMT prognostic impact on glioblastoma is dependent on therapeutic modalities. *Journal of Neuro-Oncology, 83*(2), 173–179.

Dass, C. R. (2002). Vehicles for oligonucleotide delivery to tumours. *The Journal of Pharmacy and Pharmacology, 54*(1), 3–27.

DeAngelis, L. M. (2001). Brain tumors. *The New England Journal of Medicine, 344*(2), 114–123.

Dehouck, B., Fenart, L., Dehouck, M. P., Pierce, A., Torpier, G., & Cecchelli, R. (1997). A new function for the LDL receptor: Transcytosis of LDL across the blood-brain barrier. *The Journal of Cell Biology, 138*(4), 877–889.

Di Marco, A. (1978). Mechanism of action and mechanism of resistance to antineoplastic agents that bind to DNA. *Antibiotics and Chemotherapy, 23*, 216–227.

Du, H., Sheriff, S., Bezerra, J., Leonova, T., & Grabowski, G. A. (1998). Molecular and enzymatic analyses of lysosomal acid lipase in cholesteryl ester storage disease. *Molecular Genetics and Metabolism, 64*(2), 126–134.

Edetsberger, M., Gaubitzer, E., Valic, E., Waigmann, E., & Kohler, G. (2005). Detection of nanometer-sized particles in living cells using modern fluorescence fluctuation methods. *Biochemical and Biophysical Research Communications, 332*(1), 109–116.

Egorin, M. J., Yuan, Z. M., Sentz, D. L., Plaisance, K., & Eiseman, J. L. (1999). Plasma pharmacokinetics of butyrate after intravenous administration of sodium butyrate or oral administration of tributyrin or sodium butyrate to mice and rats. *Cancer Chemotherapy and Pharmacology, 43*(6), 445–453.

Entin-Meer, M., Rephaeli, A., Yang, X., Nudelman, A., Nudelman, A., & Haas-Kogan, D. A. (2007). AN-113, a novel prodrug of 4-phenylbutyrate with increased anti-neoplastic activity in glioma cell lines. *Cancer Letters, 253*(2), 205–214.

Entin-Meer, M., Rephaeli, A., Yang, X., Nudelman, A., VandenBerg, S. R., & Haas-Kogan, D. A. (2005). Butyric acid prodrugs are histone deacetylase inhibitors that show antineoplastic activity and radiosensitizing capacity in the treatment of malignant gliomas. *Molecular Cancer Therapeutics, 4*(12), 1952–1961.

Farhadi, M. R., Capelle, H. H., Erber, R., Ullrich, A., & Vajkoczy, P. (2005). Combined inhibition of vascular endothelial growth factor and platelet-derived growth factor signaling: Effects on the angiogenesis, microcirculation, and growth of orthotopic malignant gliomas. *Journal of Neurosurgery, 102*(2), 363–370.

Farokhzad, O. C., Cheng, J., Teply, B. A., Sherifi, I., Jon, S., Kantoff, P. W., et al. (2006a). Targeted nanoparticle-aptamer bioconjugates for cancer chemotherapy in vivo. *Proceedings of the National Academy of Sciences, USA, 103*(16), 6315–6320.

Farokhzad, O. C., Karp, J. M., & Langer, R. (2006b). Nanoparticle-aptamer bioconjugates for cancer targeting. *Expert Opinion on Drug Delivery, 3*(3), 311–324.

Ferrara, N., & Davis-Smyth, T. (1997). The biology of vascular endothelial growth factor. *Endocrine Reviews, 18*(1), 4–25.

Ferrara, N., Gerber, H. P., & LeCouter, J. (2003). The biology of VEGF and its receptors. *Nature Medicine, 9*(6), 669–676.

Fisher, J. L., Schwartzbaum, J. A., Wrensch, M., & Wiemels, J. L. (2007). Epidemiology of brain tumors. *Neurologic Clinics, 25*(4), 867–890, vii.

Fishman, J. B., Rubin, J. B., Handrahan, J. V., Connor, J. R., & Fine, R. E. (1987). Receptor-mediated transcytosis of transferrin across the blood-brain barrier. *Journal of Neuroscience Research, 18*(2), 299–304.

Fomchenko, E. I., & Holland, E. C. (2006). Mouse models of brain tumors and their applications in preclinical trials. *Clinical Cancer Research, 12*(18), 5288–5297.

Friedman, H. S., Kerby, T., & Calvert, H. (2000). Temozolomide and treatment of malignant glioma. *Clinical Cancer Research, 6*(7), 2585–2597.

Fukumura, D., Xu, L., Chen, Y., Gohongi, T., Seed, B., & Jain, R. K. (2001). Hypoxia and acidosis independently up-regulate vascular endothelial growth factor transcription in brain tumors in vivo. *Cancer Research, 61*(16), 6020–6024.

Fundaro, A., Cavalli, R., Bargoni, A., Vighetto, D., Zara, G. P., & Gasco, M. R. (2000). Non-stealth and stealth solid lipid nanoparticles (SLN) carrying doxorubicin: Pharmacokinetics and tissue distribution after i.v. administration to rats. *Pharmacological Research, 42*(4), 337–343.

Gaidamakova, E. K., Backer, M. V., & Backer, J. M. (2001). Molecular vehicle for target-mediated delivery of therapeutics and diagnostics. *Journal of Controlled Release, 74*(1–3), 341–347.

Gasco, M. R. (2007). Lipid nanoparticles: Perspectives and challenges. *Advanced Drug Delivery Reviews, 59*(6), 377–378.

Goppert, T. M., & Muller, R. H. (2005). Polysorbate-stabilized solid lipid nanoparticles as colloidal carriers for intravenous targeting of drugs to the brain: Comparison of plasma protein adsorption patterns. *Journal of Drug Targeting, 13*(3), 179–187.

Gupta, Y., Jain, A., & Jain, S. K. (2007). Transferrin-conjugated solid lipid nanoparticles for enhanced delivery of quinine dihydrochloride to the brain. *The Journal of Pharmacy and Pharmacology, 59*(7), 935–940.

Huang, G., Zhang, N., Bi, X., & Dou, M. (2008). Solid lipid nanoparticles of temozolomide: Potential reduction of cardial and nephric toxicity. *International Journal of Pharmaceutics, 355*(1–2), 314–320.

Jain, K. K. (2007). Use of nanoparticles for drug delivery in glioblastoma multiforme. *Expert Review of Neurotherapeutics, 7*(4), 363–372.

Jain, R. K., di Tomaso, E., Duda, D. G., Loeffler, J. S., Sorensen, A. G., & Batchelor, T. T. (2007). Angiogenesis in brain tumours. *Nature Reviews. Neuroscience, 8*(8), 610–622.

Jain, S. K., Chaurasiya, A., Gupta, Y., Jain, A., Dagur, P., Joshi, B., et al. (2008). Development and characterization of 5-fU bearing ferritin appended solid lipid nanoparticles for tumour targeting. *Journal of Microencapsulation, 25*(5), 289–297.

Jansen, M., de Witt Hamer, P. C., Witmer, A. N., Troost, D., & van Noorden, C. J. (2004). Current perspectives on antiangiogenesis strategies in the treatment of malignant gliomas. *Brain Research. Brain Research Reviews, 45*(3), 143–163.

Jiang, Z., & Sharfstein, S. T. (2008). Sodium butyrate stimulates monoclonal antibody over-expression in CHO cells by improving gene accessibility. *Biotechnology and Bioengineering, 100*(1), 189–194.

Johansson, M., Brannstrom, T., Bergenheim, A. T., & Henriksson, R. (2002). Spatial expression of VEGF-A in human glioma. *Journal of Neuro-Oncology, 59*(1), 1–6.

Kaur, I. P., Bhandari, R., Bhandari, S., & Kakkar, V. (2008). Potential of solid lipid nanoparticles in brain targeting. *Journal of Controlled Release, 127*(2), 97–109.

Ke, L. D., Shi, Y. X., Im, S. A., Chen, X., & Yung, W. K. (2000). The relevance of cell proliferation, vascular endothelial growth factor, and basic fibroblast growth factor production to angiogenesis and tumorigenicity in human glioma cell lines. *Clinical Cancer Research, 6*(6), 2562–2572.

Ke, L. D., Shi, Y. X., & Yung, W. K. (2002). VEGF(121), VEGF(165) overexpression enhances tumorigenicity in U251 MG but not in NG-1 glioma cells. *Cancer Research, 62*(6), 1854–1861.

Kim, W. J., Yockman, J. W., Lee, M., Jeong, J. H., Kim, Y. H., & Kim, S. W. (2005). Soluble flt-1 gene delivery using PEI-g-PEG-RGD conjugate for anti-angiogenesis. *Journal of Controlled Release, 106*(1–2), 224–234.

Koziara, J. M., Lockman, P. R., Allen, D. D., & Mumper, R. J. (2003). In situ blood-brain barrier transport of nanoparticles. *Pharmaceutical Research, 20*(11), 1772–1778.

Koziara, J. M., Lockman, P. R., Allen, D. D., & Mumper, R. J. (2004). Paclitaxel nanoparticles for the potential treatment of brain tumors. *Journal of Controlled Release, 99*(2), 259–269.

Koziara, J. M., Lockman, P. R., Allen, D. D., & Mumper, R. J. (2006). The blood-brain barrier and brain drug delivery. *Journal of Nanoscience and Nanotechnology, 6*(9–10), 2712–2735.

Kreuter, J. (2001). Nanoparticulate systems for brain delivery of drugs. *Advanced Drug Delivery Reviews, 47*(1), 65–81.

Krum, J. M., & Rosenstein, J. M. (1998). VEGF mRNA and its receptor flt-1 are expressed in reactive astrocytes following neural grafting and tumor cell implantation in the adult CNS. *Experimental Neurology, 154*(1), 57–65.

Krutzfeldt, J., Kuwajima, S., Braich, R., Rajeev, K. G., Pena, J., Tuschl, T., et al. (2007). Specificity, duplex degradation and subcellular localization of antagomirs. *Nucleic Acids Research, 35*(9), 2885–2892.

Lai, S. K., Hida, K., Man, S. T., Chen, C., Machamer, C., Schroer, T. A., et al. (2007). Privileged delivery of polymer nanoparticles to the perinuclear region of live cells via a non-clathrin, non-degradative pathway. *Biomaterials, 28*(18), 2876–2884.

Lamszus, K., Ulbricht, U., Matschke, J., Brockmann, M. A., Fillbrandt, R., & Westphal, M. (2003). Levels of soluble vascular endothelial growth factor (VEGF) receptor 1 in astrocytic tumors and its relation to malignancy, vascularity, and VEGF-A. *Clinical Cancer Research, 9*(4), 1399–1405.

le Sage, C., Nagel, R., Egan, D. A., Schrier, M., Mesman, E., Mangiola, A., et al. (2007). Regulation of the p27(Kip1) tumor suppressor by miR-221 and miR-222 promotes cancer cell proliferation. *The EMBO Journal, 26*(15), 3699–3708.

Lee, M. J., Kim, Y. S., Kummar, S., Giaccone, G., & Trepel, J. B. (2008). Histone deacetylase inhibitors in cancer therapy. *Current Opinion in Oncology, 20*(6), 639–649.

Lim, S. J., Lee, M. K., & Kim, C. K. (2004). Altered chemical and biological activities of all-trans retinoic acid incorporated in solid lipid nanoparticle powders. *Journal of Controlled Release, 100*(1), 53–61.

Liu, X., Fortin, K., & Mourelatos, Z. (2008). MicroRNAs: Biogenesis and molecular functions. *Brain Pathology, 18*(1), 113–121.

Lockman, P. R., Koziara, J. M., Mumper, R. J., & Allen, D. D. (2004). Nanoparticle surface charges alter blood-brain barrier integrity and permeability. *Journal of Drug Targeting, 12*(9–10), 635–641.

Louis, D. N., Ohgaki, H., Wiestler, O. D., Cavenee, W. K., Burger, P. C., Jouvet, A., et al. (2007). The 2007 WHO classification of tumours of the central nervous system. *Acta Neuropathologica, 114*(2), 97–109.

Louis, D. N., Pomeroy, S. L., & Cairncross, J. G. (2002). Focus on central nervous system neoplasia. *Cancer Cell, 1*(2), 125–128.

Lu, W., Zhang, Y., Tan, Y. Z., Hu, K. L., Jiang, X. G., & Fu, S. K. (2005). Cationic albumin-conjugated pegylated nanoparticles as novel drug carrier for brain delivery. *Journal of Controlled Release, 107*(3), 428–448.

Maier-Hauff, K., Rothe, R., Scholz, R., Gneveckow, U., Wust, P., Thiesen, B., et al. (2007). Intracranial thermotherapy using magnetic nanoparticles combined with external beam radiotherapy: Results of a feasibility study on patients with glioblastoma multiforme. *Journal of Neuro-Oncology, 81*(1), 53–60.

Manjunath, K., & Venkateswarlu, V. (2005). Pharmacokinetics, tissue distribution and bioavailability of clozapine solid lipid nanoparticles after intravenous and intraduodenal administration. *Journal of Controlled Release, 107*(2), 215–228.

Marcato, P. D., & Duran, N. (2008). New aspects of nanopharmaceutical delivery systems. *Journal of Nanoscience and Nanotechnology, 8*(5), 2216–2229.

Martinez-Murillo, R., & Martinez, A. (2007). Standardization of an orthotopic mouse brain tumor model following transplantation of CT-2A astrocytoma cells. *Histology and Histopathology, 22*(12), 1309–1326.

Martin-Villalba, A., Okuducu, A. F., & von Deimling, A. (2008). The evolution of our understanding on glioma. *Brain Pathology, 18*(3), 455–463.

Mathieu, D., Lecomte, R., Tsanaclis, A. M., Larouche, A., & Fortin, D. (2007). Standardization and detailed characterization of the syngeneic Fischer/F98 glioma model. *The Canadian Journal of Neurological Sciences, 34*(3), 296–306.

Mayor, S., & Pagano, R. E. (2007). Pathways of clathrin-independent endocytosis. *Nature Reviews. Molecular Cell Biology, 8*(8), 603–612.

Mehnert, J. M., & Kelly, W. K. (2007). Histone deacetylase inhibitors: Biology and mechanism of action. *Cancer Journal, 13*(1), 23–29.

Mehnert, W., & Mader, K. (2001). Solid lipid nanoparticles: Production, characterization and applications. *Advanced Drug Delivery Reviews, 47*(2–3), 165–196.

Mellinghoff, I. K., Wang, M. Y., Vivanco, I., Haas-Kogan, D. A., Zhu, S., Dia, E. Q., et al. (2005). Molecular determinants of the response of glioblastomas to EGFR kinase inhibitors. *The New England Journal of Medicine, 353*(19), 2012–2024.

Miglietta, A., Cavalli, R., Bocca, C., Gabriel, L., & Gasco, M. R. (2000). Cellular uptake and cytotoxicity of solid lipid nanospheres (SLN) incorporating doxorubicin or paclitaxel. *International Journal of Pharmaceutics, 210*(1–2), 61–67.

Miller, A. A., Kurschel, E., Osieka, R., & Schmidt, C. G. (1987). Clinical pharmacology of sodium butyrate in patients with acute leukemia. *European Journal of Cancer & Clinical Oncology, 23*(9), 1283–1287.

Miller, S. J. (2004). Cellular and physiological effects of short-chain fatty acids. *Mini Reviews in Medicinal Chemistry, 4*(8), 839–845.

Minotti, G., Menna, P., Salvatorelli, E., Cairo, G., & Gianni, L. (2004). Anthracyclines: Molecular advances and pharmacologic developments in antitumor activity and cardiotoxicity. *Pharmacological Reviews, 56*(2), 185–229.

Minow, R. A., Benjamin, R. S., Lee, E. T., & Gottlieb, J. A. (1977). Adriamycin cardiomyopathy — risk factors. *Cancer, 39*(4), 1397–1402.

Minucci, S., & Pelicci, P. G. (2006). Histone deacetylase inhibitors and the promise of epigenetic (and more) treatments for cancer. *Nature Reviews. Cancer, 6*(1), 38–51.

Mischel, P. S., Shai, R., Shi, T., Horvath, S., Lu, K. V., Choe, G., et al. (2003). Identification of molecular subtypes of glioblastoma by gene expression profiling. *Oncogene, 22*(15), 2361–2373.

Moffat, B. A., Reddy, G. R., McConville, P., Hall, D. E., Chenevert, T. L., Kopelman, R. R., et al. (2003). A novel polyacrylamide magnetic nanoparticle contrast agent for molecular imaging using MRI. *Molecular Imaging, 2*(4), 324–332.

Muldoon, L. L., Sandor, M., Pinkston, K. E., & Neuwelt, E. A. (2005). Imaging, distribution, and toxicity of superparamagnetic iron oxide magnetic resonance nanoparticles in the rat brain and intracerebral tumor. *Neurosurgery, 57*(4), 785–796; discussion 785–796.

Muller, R. H., & Keck, C. M. (2004). Challenges and solutions for the delivery of biotech drugs — a review of drug nanocrystal technology and lipid nanoparticles. *Journal of Biotechnology, 113*(1–3), 151–170.

Muller, R. H., Mader, K., & Gohla, S. (2000). Solid lipid nanoparticles (SLN) for controlled drug delivery — a review of the state of the art. *European Journal of Pharmaceutics and Biopharmaceutics, 50*(1), 161–177.

Muller, R. H., & Olbrich, C. (1999). Solid lipid nanoparticles: Phagocytic uptake, in vitro cytotoxicity and in vitro biodegradation. *Pharmazeutische Industrie, 61*, 564–569.

Nicoloso, M. S., & Calin, G. A. (2008). MicroRNA involvement in brain tumors: From bench to bedside. *Brain Pathology, 18*(1), 122–129.

Niola, F., Evangelisti, C., Campagnolo, L., Massalini, S., Bue, M. C., Mangiola, A., et al. (2006). A plasmid-encoded VEGF siRNA reduces glioblastoma angiogenesis and its combination with interleukin-4 blocks tumor growth in a xenograft mouse model. *Cancer Biology & Therapy, 5*(2), 174–179.

Norden, A. D., Drappatz, J., & Wen, P. Y. (2008a). Antiangiogenic therapy in malignant gliomas. *Current Opinion in Oncology, 20*(6), 652–661.

Norden, A. D., Drappatz, J., & Wen, P. Y. (2008b). Novel anti-angiogenic therapies for malignant gliomas. *Lancet Neurology*, *7*(12), 1152–1160.

Norden, A. D., & Wen, P. Y. (2006). Glioma therapy in adults. *Neurologist*, *12*(6), 279–292.

Nudelman, A., Gnizi, E., Katz, Y., Azulai, R., Cohen-Ohana, M., Zhuk, R., et al. (2001). Prodrugs of butyric acid. Novel derivatives possessing increased aqueous solubility and potential for treating cancer and blood diseases. *European Journal of Medicinal Chemistry*, *36*(1), 63–74.

O'Dwyer, P. J., & Catalano, R. B. (2006). Uridine diphosphate glucuronosyltransferase (UGT) 1A1 and irinotecan: Practical pharmacogenomics arrives in cancer therapy. *Journal of Clinical Oncology*, *24*(28), 4534–4538.

Olbrich, C., Bakowsky, U., Lehr, C. M., Muller, R. H., & Kneuer, C. (2001). Cationic solid-lipid nanoparticles can efficiently bind and transfect plasmid DNA. *Journal of Controlled Release*, *77*(3), 345–355.

Pardridge, W. M. (2007). Drug targeting to the brain. *Pharmaceutical Research*, *24*(9), 1733–1744.

Partlow, K. C., Lanza, G. M., & Wickline, S. A. (2008). Exploiting lipid raft transport with membrane targeted nanoparticles: A strategy for cytosolic drug delivery. *Biomaterials*, *29*(23), 3367–3375.

Parveen, S., & Sahoo, S. K. (2008). Polymeric nanoparticles for cancer therapy. *Journal of Drug Targeting*, *16*(2), 108–123.

Patnaik, A., Rowinsky, E. K., Villalona, M. A., Hammond, L. A., Britten, C. D., Siu, L. L., et al. (2002). A phase I study of pivaloyloxymethyl butyrate, a prodrug of the differentiating agent butyric acid, in patients with advanced solid malignancies. *Clinical Cancer Research*, *8*(7), 2142–2148.

Peart, M. J., Smyth, G. K., van Laar, R. K., Bowtell, D. D., Richon, V. M., Marks, P. A., et al. (2005). Identification and functional significance of genes regulated by structurally different histone deacetylase inhibitors. *Proceedings of the National Academy of Sciences, USA*, *102*(10), 3697–3702.

Peira, E., Marzola, P., Podio, V., Aime, S., Sbarbati, A., & Gasco, M. R. (2003). In vitro and in vivo study of solid lipid nanoparticles loaded with superparamagnetic iron oxide. *Journal of Drug Targeting*, *11*(1), 19–24.

Pellizzaro, C., Coradini, D., Morel, S., Ugazio, E., Gasco, M. R., & Daidone, M. G. (1999). Cholesteryl butyrate in solid lipid nanospheres as an alternative approach for butyric acid delivery. *Anticancer Research*, *19*(5B), 3921–3925.

Peoch, M., Farion, R., Hiou, A., Le Bas, J. F., Pasquier, B., & Remy, C. (2002). Immunohistochemical study of VEGF, angiopoietin 2 and their receptors in the neovascularization following microinjection of C6 glioma cells into rat brain. *Anticancer Research*, *22*(4), 2147–2151.

Petri, B., Bootz, A., Khalansky, A., Hekmatara, T., Muller, R., Uhl, R., et al. (2007). Chemotherapy of brain tumour using doxorubicin bound to surfactant-coated poly(butyl cyanoacrylate) nanoparticles: Revisiting the role of surfactants. *Journal of Controlled Release*, *117*(1), 51–58.

Podio, V., Zara, G. P., Carazzonet, M., Cavalli, R., & Gasco, M. R. (2000a). Biodistribution of stealth and non-stealth solid lipid nanospheres after intravenous administration to rats. *The Journal of Pharmacy and Pharmacology*, *52*(9), 1057–1063.

Podio, V., Zara, G. P., Carazzonet, M., Cavalli, R., & Gasco, M. R. (2000b). Biodistribution of stealth and non-stealth solid lipid nanospheres after intravenous administration to rats. *The Journal of Pharmacy and Pharmacology*, *52*(9), 1057–1063.

Potmesil, M. (1994). Camptothecins: From bench research to hospital wards. *Cancer Research*, *54*(6), 1431–1439.

Pouillart, P. R. (1998). Role of butyric acid and its derivatives in the treatment of colorectal cancer and hemoglobinopathies. *Life Sciences*, *63*(20), 1739–1760.

Qian, Z. M., Li, H., Sun, H., & Ho, K. (2002). Targeted drug delivery via the transferrin receptor-mediated endocytosis pathway. *Pharmacological Reviews*, *54*(4), 561–587.

Reddy, G. R., Bhojani, M. S., McConville, P., Moody, J., Moffat, B. A., Hall, D. E., et al. (2006). Vascular targeted nanoparticles for imaging and treatment of brain tumors. *Clinical Cancer Research*, *12*(22), 6677–6686.

Reid, T., Valone, F., Lipera, W., Irwin, D., Paroly, W., Natale, R., et al. (2004). Phase II trial of the histone deacetylase inhibitor pivaloyloxymethyl butyrate (pivanex, AN-9) in advanced non-small cell lung cancer. *Lung Cancer*, *45*(3), 381–386.

Rejman, J., Oberle, V., Zuhorn, I. S., & Hoekstra, D. (2004). Size-dependent internalization of particles via the pathways of clathrin- and caveolae-mediated endocytosis. *The Biochemical Journal*, *377*(Pt. 1), 159–169.

Rephaeli, A., Entin-Meer, M., Angel, D., Tarasenko, N., Gruss-Fischer, T., Bruachman, I.,et al. (2006). The selectivty and anti-metastatic activity of oral bioavailable butyric acid prodrugs. *Investigational New Drugs*, *24*(5), 383–392.

Rich, J. N., & Bigner, D. D. (2004). Development of novel targeted therapies in the treatment of malignant glioma. *Nature Reviews. Drug Discovery*, *3*(5), 430–446.

Rong, Y., Durden, D. L., Van Meir, E. G., & Brat, D. J. (2006). 'Pseudopalisading' necrosis in glioblastoma: A familiar morphologic feature that links vascular pathology, hypoxia, and angiogenesis. *Journal of Neuropathology Experimental Neurology*, *65*(6), 529–539.

Rosenstein, J. M., & Krum, J. M. (2004a). New roles for VEGF in nervous tissue – beyond blood vessels. *Experimental Neurology*, *187*(2), 246–253.

Rosenstein, J. M., & Krum, J. M. (2004b). New roles for VEGF in nervous tissue – beyond blood vessels. *Experimental Neurology*, *187*(2), 246–253.

Rowinsky, E. K., & Donehower, R. C. (1995). Paclitaxel (taxol). *The New England Journal of Medicine*, *332*(15), 1004–1014.

Saleh, M., Stacker, S. A., & Wilks, A. F. (1996). Inhibition of growth of C6 glioma cells in vivo by expression of antisense vascular endothelial growth factor sequence. *Cancer Research*, *56*(2), 393–401.

Salomone, B., Ponti, R., Gasco, M. R., Ugazio, E., Quaglino, P., Osella-Abate, S., et al. (2000). In vitro effects of cholesteryl butyrate solid lipid nanospheres as a butyric acid prodrug on melanoma cells: Evaluation of antiproliferative activity and apoptosis induction. *Clinical & Experimental Metastasis*, *18*(8), 663–673.

Sanson, M. (2008). Editorial review: Targets for glioma treatment: From bench to bedside. *Current Opinion in Oncology*, *20*(6), 650–651.

Sanson, M., Laigle-Donadey, F., & Benouaich-Amiel, A. (2006). Molecular changes in brain tumors: Prognostic and therapeutic impact. *Current Opinion in Oncology*, *18*(6), 623–630.

Santini, V., Gozzini, A., Scappini, B., Grossi, A., & Rossi Ferrini, P. (2001). Searching for the magic bullet against cancer: The butyrate saga. *Leukemia & Lymphoma*, *42*(3), 275–289.

Sathornsumetee, S., Rich, J. N., & Reardon, D. A. (2007). Diagnosis and treatment of high-grade astrocytoma. *Neurologic Clinics*, *25*(4), 1111–1139, x.

Serganova, I., Doubrovin, M., Vider, J., Ponomarev, V., Soghomonyan, S., Beresten, T., et al. (2004). Molecular imaging of temporal dynamics and spatial heterogeneity of hypoxia-inducible factor-1 signal transduction activity in tumors in living mice. *Cancer Research*, *64*(17), 6101–6108.

Serikawa, T., Kikuchi, A., Sugaya, S., Suzuki, N., Kikuchi, H., & Tanaka, K. (2006). In vitro and in vivo evaluation of novel cationic liposomes utilized for cancer gene therapy. *Journal of Controlled Release*, *113*(3), 255–260.

Serpe, L., Guido, M., Canaparo, R., Muntoni, E., Cavalli, R., Panzanelli, P., et al. (2006). Intracellular accumulation and cytotoxicity of doxorubicin with different pharmaceutical formulations in human cancer cell lines. *Journal of Nanoscience and Nanotechnology*, *6*(9–10), 3062–3069.

Serpe, L., Laurora, S., Pizzimenti, S., Ugazio, E., Ponti, R., Canaparo, R., et al. (2004). Cholesteryl butyrate solid lipid nanoparticles as a butyric acid pro-drug: Effects on cell proliferation, cell-cycle distribution and c-myc expression in human leukemic cells. *Anticancer Drugs*, *15*(5), 525–536.

Shenoy, V. S., Vijay, I. K., & Murthy, R. S. (2005). Tumour targeting: Biological factors and formulation advances in injectable lipid nanoparticles. *The Journal of Pharmacy and Pharmacology*, *57*(4), 411–422.

Shin, J. S., & Abraham, S. N. (2001). Caveolae as portals of entry for microbes. *Microbes and Infection*, *3*(9), 755–761.

Simons, K., & Ikonen, E. (1997). Functional rafts in cell membranes. *Nature*, *387*(6633), 569–572.

Singal, P. K., Li, T., Kumar, D., Danelisen, I., & Iliskovic, N. (2000). Adriamycin-induced heart failure: Mechanism and modulation. *Molecular and Cellular Biochemistry*, *207*(1–2), 77–86.

Soldati, T., & Schliwa, M. (2006). Powering membrane traffic in endocytosis and recycling. *Nature Reviews. Molecular Cell Biology*, *7*(12), 897–908.

Stan, A. C., Casares, S., Radu, D., Walter, G. F., & Brumeanu, T. D. (1999). Doxorubicin-induced cell death in highly invasive human gliomas. *Anticancer Research*, *19*(2A), 941–950.

Steiniger, S. C., Kreuter, J., Khalansky, A. S., Skidan, I. N., Bobruskin, A. I., Smirnova, Z. S., et al. (2004). Chemotherapy of glioblastoma in rats using doxorubicin-loaded nanoparticles. *International Journal of Cancer*, *109*(5), 759–767.

Stevens, P. J., Sekido, M., & Lee, R. J. (2004). A folate receptor-targeted lipid nanoparticle formulation for a lipophilic paclitaxel prodrug. *Pharmaceutical Research*, *21*(12), 2153–2157.

Stommel, J. M., Kimmelman, A. C., Ying, H., Nabioullin, R., Ponugoti, A. H., Wiedemeyer, R., et al. (2007). Coactivation of receptor tyrosine kinases affects the response of tumor cells to targeted therapies. *Science*, *318*(5848), 287–290.

Stupp, R., Mason, W. P., van den Bent, M. J., Weller, M., Fisher, B., Taphoorn, M. J., et al. (2005). Radiotherapy plus concomitant and adjuvant temozolomide for glioblastoma. *The New England Journal of Medicine*, *352*(10), 987–996.

Tan, C., de Noronha, R. G., Roecker, A. J., Pyrzynska, B., Khwaja, F., Zhang, Z., et al. (2005). Identification of a novel small-molecule inhibitor of the hypoxia-inducible factor 1 pathway. *Cancer Research*, *65*(2), 605–612.

Tiwari, S. B., & Amiji, M. M. (2006). A review of nanocarrier-based CNS delivery systems. *Current Drug Delivery*, *3*(2), 219–232.

Toi, M., Matsumoto, T., & Bando, H. (2001). Vascular endothelial growth factor: Its prognostic, predictive, and therapeutic implications. *The Lancet Oncology*, *2*(11), 667–673.

Tondelli, L., Ricca, A., Laus, M., Lelli, M., & Citro, G. (1998). Highly efficient cellular uptake of c-myb antisense oligonucleotides through specifically designed polymeric nanospheres. *Nucleic Acids Research*, *26*(23), 5425–5431.

Tsutsui, Y., Tomizawa, K., Nagita, M., Michiue, H., Nishiki, T., Ohmori, I., et al. (2007). Development of bionanocapsules targeting brain tumors. *Journal of Controlled Release*, *122*(2), 159–164.

Ugazio, E., Marengo, E., Pellizzaro, C., Coradini, D., Peira, E., Daidone, M. G., et al. (2001). The effect of formulation and concentration of cholesteryl butyrate solid lipid nanospheres (SLN) on NIH-H460 cell proliferation. *European Journal of Pharmaceutics and Biopharmaceutics*, *52*(2), 197–202.

Vasir, J. K., & Labhasetwar, V. (2007). Biodegradable nanoparticles for cytosolic delivery of therapeutics. *Advanced Drug Delivery Reviews*, *59*(8), 718–728.

Vijayaraghavalu, S., Raghavan, D., & Labhasetwar, V. (2007). Nanoparticles for delivery of chemotherapeutic agents to tumors. *Current Opinion in Investigational Drugs*, *8*(6), 477–484.

Visone, R., Russo, L., Pallante, P., De Martino, I., Ferraro, A., Leone, V., et al. (2007). MicroRNAs (miR)-221 and miR-222, both overexpressed in human thyroid papillary carcinomas, regulate p27Kip1 protein levels and cell cycle. *Endocrine-Related Cancer*, *14*(3), 791–798.

von Holst, H., Knochenhauer, E., Blomgren, H., Collins, V. P., Ehn, L., Lindquist, M., et al. (1990). Uptake of adriamycin in tumour and surrounding brain tissue in patients with malignant gliomas. *Acta Neurochirurgica (Wien)*, *104*(1–2), 13–16.

Vredenburgh, J. J., Desjardins, A., Herndon, J. E., II, Marcello, J., Reardon, D. A., Quinn, J. A., et al. (2007). Bevacizumab plus irinotecan in recurrent glioblastoma multiforme. *Journal of Clinical Oncology*, *25*(30), 4722–4729.

Wall, M. E., Wani, M. C., Natschke, S. M., & Nicholas, A. W. (1986). Plant antitumor agents. 22. Isolation of 11-hydroxycamptothecin from camptotheca acuminata decne: Total synthesis and biological activity. *Journal of Medicinal Chemistry*, *29*(8), 1553–1555.

Wan, F., You, J., Sun, Y., Zhang, X. G., Cui, F. D., Du, Y. Z., et al. (2008). Studies on PEG-modified SLNs loading vinorelbine bitartrate (I): Preparation and evaluation in vitro. *International Journal of Pharmaceutics*, *359*(1–2), 104–110.

Wang, J. X., Sun, X., & Zhang, Z. R. (2002). Enhanced brain targeting by synthesis of 3',5'-dioctanoyl-5-fluoro-2'-deoxyuridine and incorporation into solid lipid nanoparticles. *European Journal of Pharmaceutics and Biopharmaceutics*, *54*(3), 285–290.

Wang, Y. X., Hussain, S. M., & Krestin, G. P. (2001). Superparamagnetic iron oxide contrast agents: Physicochemical characteristics and applications in MR imaging. *European Radiology*, *11*(11), 2319–2331.

Williams, J., Lansdown, R., Sweitzer, R., Romanowski, M., LaBell, R., Ramaswami, R., et al. (2003). Nanoparticle drug delivery system for intravenous delivery of topoisomerase inhibitors. *Journal of Controlled Release*, *91*(1–2), 167- 172.

Wissing, S. A., Kayser, O., & Muller, R. H. (2004). Solid lipid nanoparticles for parenteral drug delivery. *Advanced Drug Delivery Reviews*, *56*(9), 1257–1272.

Wong, H. L., Bendayan, R., Rauth, A. M., Li, Y., & Wu, X. Y. (2007). Chemotherapy with anticancer drugs encapsulated in solid lipid nanoparticles. *Advanced Drug Delivery Reviews*, *59*(6), 491–504.

Wong, H. L., Bendayan, R., Rauth, A. M., & Wu, X. Y. (2004). Development of solid lipid nanoparticles containing ionically complexed chemotherapeutic drugs and chemosensitizers. *Journal of Pharmaceutical Sciences*, *93*(8), 1993–2008.

Wong, H. L., Bendayan, R., Rauth, A. M., Xue, H. Y., Babakhanian, K., & Wu, X. Y. (2006a). A mechanistic study of enhanced doxorubicin uptake and retention in multidrug resistant breast cancer cells using a polymer-lipid hybrid nanoparticle system. *The Journal of Pharmacology and Experimental Therapeutics*, *317*(3), 1372–1381.

Wong, H. L., Rauth, A. M., Bendayan, R., Manias, J. L., Ramaswamy, M., Liu, Z., et al. (2006b). A new polymer-lipid hybrid nanoparticle system increases cytotoxicity of doxorubicin against multidrug-resistant human breast cancer cells. *Pharmaceutical Research*, *23*(7), 1574–1585.

Xu, H., Lv, M., & Tian, X. (2009). A review on hemisynthesis, biosynthesis, biological activities, mode of action, and structure-activity relationship of podophyllotoxins: 2003–2007. *Current Medicinal Chemistry*, *16*(3), 327–349.

Xu, Z., Gu, W., Huang, J., Sui, H., Zhou, Z., Yang, Y., et al. (2005). In vitro and in vivo evaluation of actively targetable nanoparticles for paclitaxel delivery. *International Journal of Pharmaceutics*, *288*(2), 361–368.

Yang, S. C., Lu, L. F., Cai, Y., Zhu, J. B., Liang, B. W., & Yang, C. Z. (1999). Body distribution in mice of intravenously injected camptothecin solid lipid nanoparticles and targeting effect on brain. *Journal of Controlled Release*, *59*(3), 299–307.

Yang, W., Barth, R. F., Wu, G., Kawabata, S., Sferra, T. J., Bandyopadhyaya, A. K., et al. (2006a). Molecular targeting and treatment of EGFRvIII-positive gliomas using boronated monoclonal antibody L8A4. *Clinical Cancer Research*, *12*(12), 3792–3802.

Yang, W., Barth, R. F., Wu, G., Kawabata, S., Sferra, T. J., Bandyopadhyaya, A. K., et al. (2006b). Molecular targeting and treatment of EGFRvIII-positive gliomas using boronated monoclonal antibody L8A4. *Clinical Cancer Research*, *12*(12), 3792–3802.

Yang, W., Wu, G., Barth, R. F., Swindall, M. R., Bandyopadhyaya, A. K., Tjarks, W., et al. (2008). Molecular targeting and treatment of composite EGFR and EGFRvIII-positive gliomas using boronated monoclonal antibodies. *Clinical Cancer Research*, *14*(3), 883–891.

Yano, A., Shingo, T., Takeuchi, A., Yasuhara, T., Kobayashi, K., Takahashi, K., et al. (2005). Encapsulated vascular endothelial growth factor-secreting cell grafts have neuroprotective and angiogenic effects on focal cerebral ischemia. *Journal of Neurosurgery*, *103*(1), 104–114.

Yasuhara, T., Shingo, T., Muraoka, K., wen Ji, Y., Kameda, M., Takeuchi, A., et al. (2005). The differences between high and low-dose administration of VEGF to dopaminergic neurons of in vitro and in vivo parkinson's disease model. *Brain Research*, *1038*(1), 1–10.

Zara, G. P., Bargoni, A., Cavalli, R., Fundaro, A., Vighetto, D., & Gasco, M. R. (2002a). Pharmacokinetics and tissue distribution of idarubicin-loaded solid lipid nanoparticles after duodenal administration to rats. *Journal of Pharmaceutical Sciences*, *91*(5), 1324–1333.

Zara, G. P., Cavalli, R., Bargoni, A., Fundaro, A., Vighetto, D., & Gasco, M. R. (2002b). Intravenous administration to rabbits of non-stealth and stealth doxorubicin-loaded solid lipid nanoparticles at increasing concentrations of stealth agent: Pharmacokinetics and distribution of doxorubicin in brain and other tissues. *Journal of Drug Targeting*, *10*(4), 327–335.

Zara, G. P., Cavalli, R., Fundaro, A., Bargoni, A., Caputo, O., & Gasco, M. R. (1999). Pharmacokinetics of doxorubicin incorporated in solid lipid nanospheres (SLN). *Pharmacological Research*, *40*(3), 281–286.

Zhang, Y., Jeong Lee, H., Boado, R. J., & Pardridge, W. M. (2002). Receptor-mediated delivery of an antisense gene to human brain cancer cells. *The Journal of Gene Medicine*, *4*(2), 183–194.

Zhang, Y., Zhang, Y. F., Bryant, J., Charles, A., Boado, R. J., & Pardridge, W. M. (2004). Intravenous RNA interference gene therapy targeting the human epidermal growth factor receptor prolongs survival in intracranial brain cancer. *Clinical Cancer Research*, *10*(11), 3667–3677.

Zhou, Y. H., Tan, F., Hess, K. R., & Yung, W. K. (2003). The expression of PAX6, PTEN, vascular endothelial growth factor, and epidermal growth factor receptor in gliomas: Relationship to tumor grade and survival. *Clinical Cancer Research*, *9*(9), 3369–3375.

H.S. Sharma (Ed.)
Progress in Brain Research, Vol. 180
ISSN: 0079-6123

CHAPTER 12

Brain as an HIV sequestered site: use of nanoparticles as a therapeutic option

Christopher J. Destache*

Department of Pharmacy Practice, Creighton University School of Pharmacy & Health Professions, Omaha, NE, USA

Abstract: Patients infected with human immunodeficiency virus (HIV) are at increased risk to develop neurocognitive problems. HIV crosses the blood–brain barrier (BBB) through a variety of means. Once within the brain tissue, HIV stimulates immunoactivation and inflammation that lead to neuronal loss. This review discusses the pathophysiology of HIV within the brain and the treatment modalities used to prevent neurocognitive problems from developing. Modalities that prevent antiretroviral drugs from crossing into the central nervous system and reducing viral load are also discussed. Finally, since drug penetration across the BBB is reduced, the use of nanoparticles as a treatment modality to increase BBB penetration may be an option worth further exploration.

Keywords: HIV-1; antiretroviral nanoparticles; P-glycoprotein; multidrug resistant protein; drug penetration; blood-brain barrier

Introduction

An estimated 33 million people worldwide are infected with the human immunodeficiency virus (HIV) (UNAIDS Epidemic Update, 2008), causing the acquired immunodeficiency syndrome (AIDS). In 2007, the AIDS epidemic has killed approximately 2 million individuals. The central nervous system (CNS) is a major target of HIV and is responsible for a number of neurologic sequelae. HIV infection of the CNS is associated with the development of asymptomatic neurocognitive impairment, HIV-associated mild neurocognitive disorder, and HIV-associated dementia (HAD) that manifests as a clinical syndrome of cognitive, motor, and behavioral dysfunction (Antinori et al., 2007). These neurological complications arise most likely due to the early penetration of HIV-1 into the CNS via infected immune cells such as $CD4^+$ T lymphocytes, dendritic cells, monocytes, and macrophages which are all cellular reservoirs of HIV-1 (Blankson, Persaud, & Siliciano, 2002; McArthur, 2004; Valcour & Paul, 2006). In the late stages of AIDS, HAD is associated with high morbidity and mortality in resource-poor countries where highly active antiretroviral therapy (HAART) is not readily available. However, where there is widespread use of HAART there has been significant reduction in the incidence/severity of HAD (d'Arminio Monforte, 2004). Factors that contribute to the severity of HAD include toxicity, insurgence of drug resistance, poor medication adherence, and sometimes limited access to HAART. Additionally, the constant presence of

* Corresponding author.
Tel.: +402-280-4744; Fax: +402-280-3320;
E-mail: destache@creighton.edu

DOI: 10.1016/S0079-6123(08)80012-X

Table 1. CNS-penetration effectiveness (CPE) ranks

Low	Intermediate	High
Tenofovir	Stavudine (d4T)	Zidvudine (AZT)
Didanosine (ddI)	Lamivudine (3TC)	Abacavir
Nelfinavir	Entricitabine (FTC)	Delavirdine
Ritonavir	Efavirenz	Nevirapine
Saquinavir	Amprenavir	
Enfuvirtide	Fosamprenavir	
Maraviroc	Atazanavir	
Raltegravir	Indinavir	
Darunavir		

Adapted from Letendre et al. (2008).

HIV and the associated immunoactivation and inflammation in CSF is linked to this form of brain injury. Elevated viral loads in the CSF have been reported to predict subsequent neurocognitive impairments (Ellis et al., 2002). Taking these tenets together would suggest that treatment of CNS disease associated with HIV infection may be suboptimal and clinical data has suggested that the CNS effectiveness of anti-HIV drug regimens can be significantly improved by treating with CSF-penetrating antiretroviral drugs in patients with cognitive impairment (Letendre et al., 2004, 2008a; Letendre et al., 2009).

Recently, antiretroviral drugs were ranked based on their ability to penetrate the CNS (Table 1) (Varatharajan & Thomas, 2009). Unfortunately, the majority of antiretroviral compounds have low or moderate CNS penetration. Even when antiretroviral drugs with good CNS penetration are used there are no correlations between their access to the CNS and the slope of CSF viral decline (Eggers, Hertogs, Sturenburg, van Lunzen, & Stellbrink, 2003; Solas et al., 2003). As HIV+ patients who are receiving HAART live longer, an increase in neurocognitive problems could develop from the combined effects of aging and HIV disease, leading to an increase in the incidence and severity of HIV-associated neurocognitive impairment. Additionally, several newer agents have not published CNS penetration. Therefore, the incidence of HIV-induced neurocognitive impairment may increase in the HAART era.

An important long-term goal of antiretroviral compounds is to be able to inhibit HIV-1 virus throughout the body. However, there are difficulties in penetration to areas of the body associated with current antiretroviral drugs (Ghosn et al., 2004). The CNS is one of those areas where antiretrovirals have diminished ability to affect HIV-1 replication and as reported by Smit et al. (2004) to increase the incidence of resistant mutations within the brain. Despite low CNS penetration, Langford et al. (2006) have shown significant reduction in viral loads in brain tissue from HIV+ patients receiving HAART compared to those not receiving HAART. Yet in these patients, the brain tissue viral loads were not reduced to nondetectable ranges after 3 months of HAART. These factors lead investigators to realize that there is more research that is needed to optimize HAART for CNS penetration.

Cells that promote HIV-1 infection of the CNS

HIV-1 penetration through the blood–brain barrier (BBB) occurs without alteration of the BBB permeability. HIV virus penetrates across the BBB by inducing gp120-mediated adsorptive endocytosis, a vesicular mechanism providing a mode of entry into brain microvessel endothelial cells (BMEC) (Banks et al., 2001). HIV-1–proteoglycan interactions are based on electrostatic contacts between basic residuals in gp120 and sulfate groups in proteoglycans, HIV-1 can exploit these interactions to rapidly enter and migrate through the BBB to invade the brain. This is supported by gp120-deficient virus fails to bind and perform transcytosis through human BMEC (Bobardt et al., 2004).

A common feature of brain inflammation due to infection is an increase in BBB permeability. The immune response produces proinflammatory cytokines, chemokines, cellular adhesion molecules, and matrix metalloproteinases at the site of the infection. These mediators alter the structure and function of the BBB as shown by Wolka, Huber, and Davis (2003). The cytokines, namely, tissue necrosis factor-α (TNF-α) and interleukin-1 (IL-1) and metalloproteinases contribute to open the BBB (Petty & Lo, 2002). The increase of BBB permeability has been attributed to a loss of tight junction proteins occluding and ZO-1. Thus HIV-1 takes advantage of the modification to penetrate across the BBB (Fiala et al., 1997).

In early HIV infection, CSF viral loads are thought to result from plasma lymphocytes, with HIV reproduction augmented by continual renewal from the bloodstream. CSF HIV-1 RNA concentrations are maintained by the entry of approximately 10^4–10^6 HIV-1 RNA copies into the CSF daily (Haas et al., 2000). One source associated with HIV-1 viral RNA in brain tissue is $CD4^+$ T lymphocytes in the CNS (Haas et al., 2000, 2003). Resting $CD4^+$ T cells carrying replication-competent HIV-1 have been identified in both chronically infected patients antiretroviral-naïve and those receiving HAART (Chun et al., 1998; Finzi et al., 1997).

Finally, another postulated mechanism is the "Trojan horse" mechanism where infected macrophages cross-activate brain endothelial cells and take up residence in the CNS as infected microglial cells. BMEC and immune cells activated by cytokines overexpress adhesion molecules and ligands which promotes binding of circulating immune cells to brain vasculature. This could facilitate the passage of viral particles between the infected immune cell and the brain endothelial cell, analogous to the transfer of virus between infected immune cells (Liu, Tang, McArthur, Scott, & Gartner, 2000). Macrophages infected by HIV-1 cross the BBB allowing the virus to gain entry into the CNS (Persidsky et al., 1999).

Potential transporters that interfere with antiretroviral CNS penetration

The cerebral capillaries are sites of BBB and CNS penetration across this barrier is restricted by tight junctions between endothelial cells, lack of intracellular fenestrations in brain endothelial cells, as well as by the presence of multiple metabolic enzymes and diverse transport systems that limit antiretroviral CNS penetration (reviewed by Varatharajan & Thomas, 2009). Mechanisms of drug penetration from blood to CSF can occur by a variety of routes including between the cells of endothelium (par cellular movement), directly through the cell wall (active transport, facilitated, and/or passive diffusion), or through vesicular transport (endocytosis).

The two main groups of transporter enzymes that limit CNS drug penetration include the ATP-binding cassette (ABC) transporters and the solute-carrier superfamilies (Loscher & Potschka, 2005).

P-glycoprotein (P-gp) is a 150- to 180-kDa membrane protein encoded by the multidrug resistance gene 1 (MDR-1). It is widely expressed in the liver, kidney, and intestine, as well as the luminal membrane of the BBB. It appears to be important in protecting the brain from hydrophobic molecules and drugs. Uncharged or weakly basic molecules are most efficiently transported by P-gp, but acidic compounds can also be transported. Accumulating evidence shows that protease inhibitors (PIs) are substrates of P-gp and as a consequence limit these drugs from crossing the BBB. As a result of the activity of this efflux transporter, the level of PIs into the CSF is reduced (Bachmeier, Spitzenberger, Elmquist, & Miller, 2005; Eilers, Roy, & Mondal, 2008; Kim et al., 1998; Park & Sinko, 2005; Polli et al., 1999). While P-gp effects on nucleoside reverse transcriptase inhibitors have been studied less, there are reports of reduced CNS penetration of abacavir (Giri et al., 2008). Also, HIV-1-infected T cells and monocytic cell lines have increased P-gp expression which has been shown to reduce the intracellular level of zidovudine (AZT) *in vitro* (Gollapudi & Gupta, 1990). The nonnucleoside reverse transcriptase inhibitors (NNRTIs) have been found not to be a substrate for P-gp; however, all induce the expression and function of this protein with nevirapine the most potent inducer (Stormer, von Moltke, Perloff, & Greenblatt, 2002).

Experiments using primary rat astrocyte cultures have demonstrated that both the expression and transport function of P-gp are downregulated following exposure to HIV gp120. It is thought that these glial cells that harbor the virus within the CNS form a dynamic barrier behind the BBB to further impede the access of antiretroviral drugs to sites of infection within the CNS (Ronaldson & Bendayan, 2006). However, HIV-Tat induces the expression of P-gp in brain endothelial cells which correlated with functional upregulation of the transporter function of P-gp (Hayashi et al., 2005). The HIV PIs increased P-gp activity in a concentration-dependent manner in an *in vitro* model of BBB, raising the possibility that HAART could itself contribute to the brain as a HIV sanctuary site by the induction of drug transporters (Perloff, von Moltke, Fahey, & Greenblatt, 2007). Therefore, it may be appropriate to selectively inhibit P-gp to be able to facilitate the entry of HIV PIs and certain nucleoside reverse transcriptase inhibitors into viral sanctuaries to reduce the development of HAD.

Multidrug resistance proteins (MRP) are ATP-driven efflux transporters localized at the luminal membrane of the brain capillary endothelial cells. These transporters contribute to the nonpermissive nature of the BBB (Dallas, Miller, & Bendayan, 2006). Only the expression of MRP-1, MRP-2, MRP-4, and MRP-5 at the BBB impact drug penetration (Dallas et al., 2006; Eilers et al., 2008; Soontornmalai, Vlaming, & Fritschy, 2006). The HIV PIs saquinavir, ritonavir, and lopinavir have been shown to be substrates of MRP-1 and MRP-2 and these transporters may contribute to the limited penetration of PIs into the brain (Bachmeier et al., 2005; Eilers et al., 2008; Park & Sinko, 2005). A recent *in vitro* study has shown that MRP-1, MRP-2, and MRP-3 were inhibited in a concentration-dependent manner by NNRTIs (delavirdine, efavirenz, and nevirapine), NRTI (abacavir, emtricitabine, lamivudine, and tenofovir). This inhibition was significant for emtricitabine and efavirenz (Weiss, Theile, Ketabi-Kiyanvash, Lindenmaier, & Haefeli, 2007).

Finally, a recent study by Jung et al. (2008) suggested that nelfinavir, ritonavir, saquinavir, and indinavir are possible inhibitors of organic cation transporters (OCTs). Some HIV PIs may be potent inhibitors of cationic drug uptake but poor substrates for human OCT1 (Zhang et al., 2000). Furthermore, lamivudine and zalcitabine are substrates of OCT1 and OCT2 (Jung et al., 2008). Taking this literature together, antiretroviral drugs upregulate ABC and MRP expression to limit CNS penetration, while NRTIs and NNRTIs may affect MRP in the opposite direction.

Colloidal carriers penetration through the BBB

Colloidal drug carriers include micelles, emulsions, liposomes, and nanoparticles (nanospheres and nanocapsules). Only liposomes and nanoparticles have been used for brain drug delivery. The goal of colloidal carriers is to increase the specificity toward cells or tissues, to improve the bioavailability of drugs by increasing their diffusion through biological membranes, and/or protect them against enzyme inactivation. Additionally, colloidal systems may allow access across the BBB of nontransportable drugs by masking their physicochemical characteristics through their encapsulation in these systems.

After intravenous administration, all colloidal systems interact with plasma proteins (immunoglobulins, albumin, complement elements, and fibronectin). This process is known as opsonization is crucial in dictating the subsequent fate of the colloidal particles. Colloidal particles that have hydrophobic surface properties are efficiently coated with plasma components and rapidly removed from the circulation. Colloidal particles that are small and hydrophilic can escape from opsonization process at least partially and remain in circulation for a prolonged period of time. "Steric hindrance" has been applied to avoid the deposition of plasma proteins by adsorbing at the surface of the colloids some surfactant molecules (i.e., copolymers of polyoxyethylene and polyoxyprophylene) or by direct linking of polyethyleneglycol (PEG) at the surface (Peracchia et al., 1998, 1999).

Polymeric micelles as drug delivery systems are formed by amphiphilic copolymers having an A-B di-block structure with A, the hydrophilic (shell) and B, the hydrophobic polymer (core). The polymeric micelles are thermodynamically and kinetically stable in aqueous media. They have a size range of 10–500 nm with narrow distribution. Studies by Kabanov et al. (1989) have shown that poloxamer (Pluronic^{TM}) micelles conjugated with antibodies could improve brain distribution of haloperidol resulting in improved drug efficacy. Pluronic unimers allowed cell penetration in bovine BMEC monolayers by inhibiting P-gp-mediated drug efflux system (Alakhov et al., 1999; Batrakova et al., 2001).

Liposomes are small vesicles composed of unilamellar or multilamellar phospholipids bilayers surrounding aqueous compartments. These are composed of biocompatible and biodegradable lipids similar to biological membranes. Their biophysical properties such as size, surface charge, lipid composition, and concentration of cholesterol are variable and able to control distribution, tissue uptake, and drug delivery. Liposomes have been considered for brain targeting through both intracerebral and intravenous administrations. Most of the studies have focused on tumor therapies to deliver doxorubicin and other antineoplastic agents with the aid of either cationic or pegylated liposomes. These treatments have led to long-term survival and inhibition of tumor growth in patients (Fiorillo et al., 2004; Saito et al., 2004; Siegal, Horowitz, & Gabizon, 1995).

Recent advanced in liposomal formulations include cationic liposomes used to entrap genetic material. Encapsulation of genetic material into cationic liposomes confers protection from the extracellular environment and provides a mechanism for genetic material transfer to target cells. In gene therapy, delivery of plasmid DNA to the endosome would not produce benefits since it is an inappropriate cellular compartment for DNA function. It is critically important that the genetic material escapes from the endosomal compartment into the cytoplasm and is presented in an episomal fashion within the nucleus allowing expression (Davis, 1997). Cationic liposomes have been used for plasmid-mediated transfection of murine brain cells (Roessler & Davidson, 1994). When these liposomes were injected/infused directly into the brain of mice, the expression of transgene could be observed for at least 21 days in the caudate putamen region.

Pegylated liposomes have proven their ability to deliver the drugs owing to their long blood circulating times and their reduced clearance by the reticuloendothelial system (RES). Currently, there are several formulations of pegylated liposomes encapsulating doxorubicin (Caelyx®), amphotericin B, as well as others used in clinical practice (Hau et al., 2004; Kingsley et al., 2006; Koukourakis et al., 2000).

Nanoparticles may be defined as a submicron drug carrier system, generally of polymeric nature with size between 10 and 500 nm. Drugs or other molecules may be dissolved into the nanoparticles, entrapped, encapsulated, and/or adsorbed or attached. The methods of preparation are simple generally and scale-up is relatively easy. The advantage of nanoparticles for drug delivery results from their two properties. First, due to their small size, nanoparticles penetrate into even small capillaries and are taken up within cells allowing efficient drug accumulation at the targeted sites in the body. Secondly, the use of biodegradable materials for nanoparticle preparation, allows sustained drug release at the targeted site over a period of days or even weeks after injection (Vinogradov, Bronich, & Kabanov, 2002).

The use of pegylated-poly(hexadecylcyanoacrylate;PEG-PHDCA) nanoparticles have been investigated for several CNS disease entities including brain tumors (Brigger, Dubernet, & Couvreur, 2002), prion diseases (Calvo et al., 2001a), and experimental autoimmune encephalomyelitis (EAE) (Calvo et al., 2001b). In this technology, the PEG is covalently attached to the hydrophobic block, rather than adsorbed, which seems to be the better choice to avoid the possibility of PEG desorption. After intravenous administration in rats bearing intracerebral gliosarcoma, these particles accumulated preferentially in the tumoral tissue rather than the peritumoral brain tissue or in the healthy contralateral hemisphere. Additionally, the PEG-PHDCA concentrated more in the gliosarcoma than did the nonpegylated counterparts (PHDCA nanoparticles).

Data from my laboratory show that nanoparticles can be fabricated using more than one drug (Destache et al., 2009). We have fabricated poly (D,L-lactide-co-glycolide) (PLGA) polymer with antiretroviral drugs lopinavir, ritonavir, and efavirenz. We have characterized the nanoparticles (size-averaged 320 nm). We have attempted to determine the amount of antiretroviral drug penetration into brain tissue in BALB/c mice compared to free drugs both given intraperitoneally. The dose was 500 mcg of each antiretroviral drug. At specific times, mice ($n = 3$/time point) were euthanized and serum and tissues were harvested. Brain tissue was weighed and placed in methanol for 15–30 min and then homogenized. The samples were assayed by a previously reported high performance liquid chromatographic (HPLC) method adapted to our equipment (Weller, Brundage, Balfour, & Venzina, 2007). These data are presented in Fig. 1. The brain tissue concentration of the antiretroviral drugs is significantly prolonged in the mice that received the nanoparticle antiretroviral drugs. For example, the mean area under the brain concentration–time profile of the three drugs (free) was 1.2 days as compared to 161.8 days from the nanoparticle drugs. Further research is necessary to confirm these results but the conclusion of the research thus far is that the PLGA nanoparticle drugs when given intraperitoneally remain in the body of mice longer than free drugs. This could be advantageous for a number of patient populations affected by HIV.

Overall, the challenge of antiretroviral drug penetration across the BBB has to be overcome to reduce the associated dementia. Our ability to detect HADefore significant disability occurs also has to become more accurate. The use of nanoparticles as a drug delivery system may be able to be used for this purpose. Hopefully in the future, we will have the diagnostic and therapeutic tools to offer HIV patients with dementia with the goal of improved quality of life.

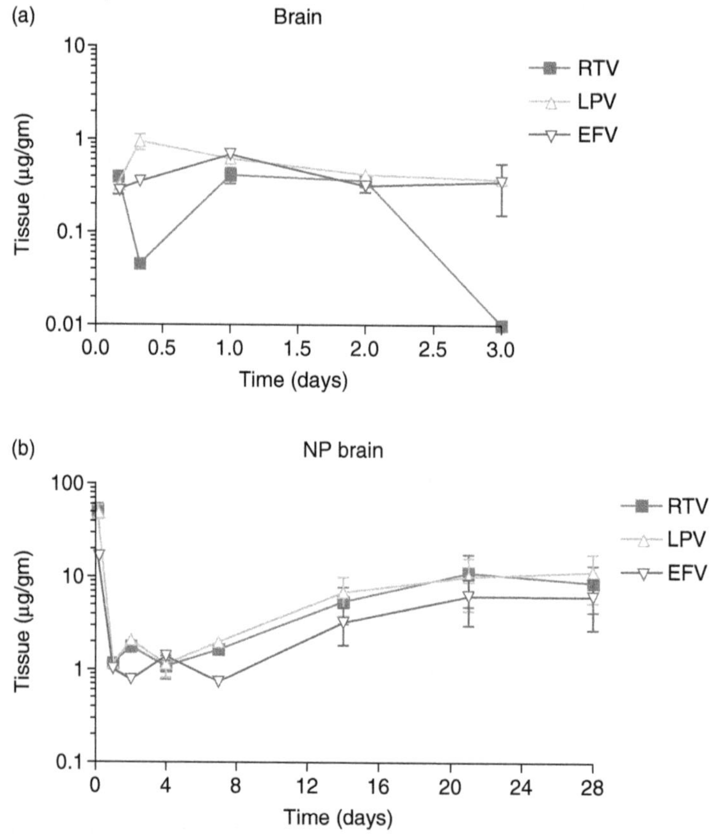

Fig. 1. Antiretroviral pharmacokinetic levels in the brain. Panel a shows the results of the free drugs (ritonavir, lopinavir, and efavirenz) penetration into the brain tissue over time. Panel b shows the results of the PLGA nanoparticle penetration into the brain tissue over time for the same three drugs.

References

Alakhov, V., Klinski, E., Li, S., Pietrzynski, G., Venne, A., Batrakova, E., et al. (1999). Block copolymer-based formulation of doxorubicin from cell screen to clinical trials. *Colloids and Surfaces. B, Biointerfaces*, *16*, 113–134.

Antinori, A., Arendt, G., Becker, J. T., Brew, B. J., Byrd, D. A., Cherner, M., et al. (2007). Updated research nosology for HIV-associated neurocognitive disorders. *Neurology*, *69*, 1789–1799.

Bachmeier, C. J., Spitzenberger, T. J., Elmquist, W. F., & Miller, D. W. (2005). Quantitative assessment of HIV-1 protease inhibitor interactions with drug efflux transporters in the blood-brain barrier. *Pharmaceutical Research*, *22*, 1259–1268.

Banks, W. A., Freed, E. O., Wolf, K. M., Robinson, S. M., Franko, M., & Kumar, V. (2001). Transport of human immunodeficiency virus type 1 pseudoviruses across the blood-brain barrier: Role of envelope proteins and adsorptive endocytosis. *Journal of Virology*, *75*, 4681–4691.

Batrakova, E. V., Li, S., Vinogradov, S. V., Alakhov, V. Y., Miller, D. W., & Kabanov, A. V. (2001). Mechanism of pluronic effect on P-glycoprotein efflux system in blood-brain barrier: Contributions of energy depletion and membrane fluidization. *The Journal of Pharmacology and Experimental Therapeutics*, *299*, 483–493.

Blankson, J. N., Persaud, D., & Siliciano, R. F. (2002). The challenge of viral reservoirs in HIV-1 infection. *Annual Review of Medicine*, *53*, 557–593.

Bobardt, M. D., Salmon, P., Wang, L., Esko, J. D., Gabuzda, D., Fiala, M., et al. (2004). Contribution of proteoglycans to human immunodeficiency virus type 1 brain invasion. *Journal of Virology*, *78*, 6567–6584.

Brigger, I., Dubernet, C., & Couvreur, P. (2002). Nanoparticles in cancer therapy and diagnosis. *Advanced Drug Delivery Reviews*, *54*, 631–651.

Calvo, P., Gouritin, B., Brigger, I., Lasmezas, C., Deslys, J., Williams, A., et al. (2001a). PEGylated polycyanoacrylate nanoparticles as vector for drug delivery in prion diseases. *Journal of Neuroscience Methods*, *111*, 151–155.

Calvo, P., Gouritin, B., Chacun, H., Desmaele, D., D'Angelo, J., Noel, J. P., et al. (2001b). Long-circulating PEGylated polycyanoacrylate nanoparticles as new drug carrier for brain delivery. *Pharmaceutical Research*, *18*, 1157–1166.

Chun, T. W., Engel, D., Berrey, M. M., Shea, T., Corey, L., & Fauci, A. S. (1998). Early establishment of a pool of latently infected resting CD4(+) T cells during primary HIV-1 infection. *Proceedings of the National Academy of Sciences, USA*, *95*, 8869–8873.

Dallas, S., Miller, D. S., & Bendayan, R. (2006). Multi-drug resistance-associated proteins: Expression and function in the central nervous system. *Pharmacological Reviews*, *58*, 140–161.

d'Arminio Monforte, A., Cinque, P., Mocroft, A., Goebel, F. D., Antunes, F., Katlama, C., et al. (2004). Changing incidence of central nervous system diseases in the EuroSIDA cohort. *Annals of Neurology*, *55*, 320–328.

Davis, S. S. (1997). Biomedical applications of nanotechnology – implications for drug targeting and gene therapy. *Trends in Biotechnology*, *15*, 217–224.

Destache, C. J., Belgum, T., Christensen, K., Shibata, A., Sharma, A., & Dash, A. K. (2009). Combination antiretroviral drugs in PLGA nanoparticle for HIV. *BMC Infectious Diseases* (submitted).

Eggers, C., Hertogs, K., Sturenburg, H. J., van Lunzen, J., & Stellbrink, H. J. (2003). Delayed central nervous system virus suppression during highly active antiretroviral therapy is associated with HIV encephalopathy, but not with viral drug resistance or poor central nervous system drug penetration. *AIDS*, *17*, 1897–1906.

Eilers, M., Roy, U., & Mondal, D. (2008). MRP (ABCC) transporters-mediated efflux of anti-HIV drugs, saquinavir and zidovudine, from human endothelial cells. *Experimental Biology and Medicine (Maywood)*, *233*, 1149–1160.

Ellis, R. J., Moore, D. J., Childers, M. E., Letendre, S., McCutchan, J. A., Wolfson, T., et al. (2002). Progression to neuropsychological impairment in human immunodeficiency virus infection predicted by elevated cerebrospinal fluid levels of human immunodeficiency virus RNA. *Archives of Neurology*, *59*(6), 923–928.

Fiala, M., Looney, D. J., Stins, M., Way, D. D., Zhang, L., Gan, X., et al. (1997). TNF-alpha opens a paracellular route for HIV-1 invasion across the blood-brain barrier. *Molecular Medicine*, *3*, 553-564.

Finzi, D., Hermankova, M., Peirson, T., Carruth, L. M., Buck, C., Chaisson, R. E., et al. (1997). Identification of a reservoir for HIV-1 in patients on highly active antiretroviral therapy. *Science*, *278*, 1295–1300.

Fiorillo, A., Maggi, G., Greco, N., Migliorati, R., D'Amico, A., De Caro, M. D., et al. (2004). Second-line chemotherapy with the association of liposomal daunorubicin, carboplatin, and etoposide in children with recurrent malignant brain tumors. *Journal of Neuro-Oncology*, *66*, 179–185.

Ghosn, J., Chaix, M. L., Peytavin, G., Rey, E., Bresson, J. L., Goujard, C., et al. (2004). Penetration of enfuvirtide, tenofovir, efavirenz, and protease inhibitors in the genital tract of HIV-1-infected men. *AIDS*, *18*, 1958–1961.

Giri, N., Shaik, N., Pan, G., Terasaki, T., Mukai, C., Kitagaki, S., et al. (2008). Investigation of the role of breast cancer resistance protein (Bcrp/Abcg2) on pharmacokinetics and central nervous system penetration of abacavir and zidovudine in the mouse. *Drug Metabolism and Disposition*, *36*, 1476–1484.

Gollapudi, S., & Gupta, S. (1990). Human immunodeficiency virus I-induced expression of P-glycoprotein. *Biochemical and Biophysical Research Communications*, *171*, 1002–1007.

Haas, D. W., Clough, L. A., Johnson, B. W., Harris, V. L., Spearman, P., Wilkinson, G. R., et al. (2000). Evidence of a source of HIV type 1 within the central nervous system by ultraintensive sampling of cerebrospinal fluid and plasma. *AIDS Research and Human Retroviruses*, *16*, 1491–1502.

Haas, D. W., Johnson, B. W., Spearman, P., Raffanti, S., Nicotera, J., Schmidt, D., et al. (2003). Two phases of HIV RNA decay in CSF during initial days of multidrug therapy. *Neurology*, *61*, 1391–1396.

Hau, P., Fabel, K., Baumgart, U., Rummele, P., Grauer, O., Bock, A., et al. (2004). Pegylated liposomal doxorubicin-efficacy in patients with recurrent high-grade glioma. *Cancer, 100*, 1199–1207.

Hayashi, K., Pu, H., Tian, J., Andreas, I. E., Lee, Y. W., Hennig, B., et al. (2005). HIV-Tat protein induces P-glycoprotein expression in brain microvascular endothelial cells. *Journal of Neurochemistry, 93*, 1231–1241.

Jung, N., Lehmann, C., Rubbert, A., Knisppel, M., Hartmann, P., van Lunzen, J., et al. (2008). Relevance of the organic cation transporters 1 and 2 for antiretroviral therapy in HIV infection. *Drug Metabolism and Disposition, 36*, 1616–1623.

Kabanov, A. V., et al., Chekhonin, V. P., Alakhov, V. Y., Batrakova, E. V., Lebedev, A. S., Melik-Nubarov, N. S., et al. (1989). The neuroleptic activity of haloperidol increases after its solubilization in surfactant micelles. Micelles as microcontainers for drug targeting. *FEBS Letters, 258*(2), 343–345.

Kim, R. B., Fromm, M. F., Wandel, C., Leake, B., Wood, A. J., Roden, D. M., et al. (1998). The drug transporter P-glycoprotein limits oral absorption and brain entry of HIV-1 protease inhibitors. *Journal of Clinical Investigation, 101*, 289–294.

Kingsley, J. D., Dou, H., Morehead, J., Ravinow, B., Gendelman, H. E., & Destache, C. J. (2006). Nanotechnology: A focus on nanoparticles as a drug delivery system. *Journal of Neuroimmune Pharmacology, 1*, 340–350.

Koukourakis, M. I., Koukouraki, S., Giatromanolaki, A., Kakolyris, S., Georgoulias, V., Velidaki, A., et al. (2000). High intratumoral accumulation of stealth liposomal doxorubicin in sarcomas — rationale for combination with radiotherapy. *Acta Oncologica, 39*, 207–211.

Langford, D., Marquie-Beck, J., deAlmeida, S., Lassaretto, D., Letendre, S., Grant, I., et al. (2006). Relationship of antiretroviral treatment to postmortem brain tissue viral load in human immunodeficiency virus-infected patients. *Journal of Neurovirology, 12*, 100–107.

Letendre, S., Marquie-Beck, J., Capparelli, E., Best, B., Clifford, D., Collier, A. C., et al. (2008). Validation of the CNS penetration-effectiveness rank for quantifying antiretroviral penetration into the central nervous system. *Archives of Neurology, 65*, 65–70.

Letendre S. L., Ellis R. J., Everall, I., Ances, B., Bharti, A., & McCutchan, J. A. (2009). Neurologic complications of HIV disease and their treatment. *Top HIV Med 17*, 46–56.

Letendre, S. L., McCutchan, J. A., Childers, M. D., Woods, S. P., Lazzaretto, D., Heaton, R. K., et al. (2004). Enhancing antiretroviral therapy for human immunodeficiency virus cognitive disorders. *Annals of Neurology, 56*, 416–423.

Liu, Y., Tang, X. P., McArthur, J. C., Scott, J., & Gartner, S. (2000). Analysis of human immunodeficiency virus type 1 gp160 sequences from a patient with HIV dementia: Evidence for monocyte trafficking into brain. *Journal of Neurovirology, 6*, S70–S81.

Loscher, W., & Potschka, H. (2005). Role of drug efflux transporters in the brain for drug disposition and treatment of brain diseases. *Progress in Neurobiology, 76*, 22–76.

McArthur, J. C. (2004). HIV dementia: An evolving disease. *Journal of Neuroimmunology, 157*, 2–10.

Park, S., & Sinko, P. J. (2005). P-glycoprotein and multidrug resistance-associated proteins limit the brain uptake of saquinavir in mice. *The Journal of Pharmacology and Experimental Therapeutics, 312*, 1249–1256.

Peracchia, M. T., Fattal, E., Desmaele, D., Besnard, M., Noel, J. P., Gomis, J. M., et al. (1999). Stealth PEGylated polycyanoacrylate nanoparticles for intravenous administration and splenic targeting. *Journal of Controlled Release, 60*, 121–128.

Peracchia, M. T., Vauthier, C., Desmaele, D., Gulik, A., Dedieu, J. C., Demoy, M., et al. (1998). Pegylated nanoparticles from a novel methoxypolyethylene glycol cyanoacrylate-hexadecyl cyanoacrylate amphiphilic copolymer. *Pharmaceutical Research, 15*, 550–556.

Perloff, M. D., von Moltke, L. L., Fahey, J. M., & Greenblatt, D. J. (2007). Induction of P-glycoprotein expression and activity by ritonavir in bovine brain microvessel endothelial cells. *The Journal of Pharmacy and Pharmacology, 59*, 947–953.

Persidsky, Y., Ghorpade, A., Rasmussen, J., Limoges, J., Liu, X. J., Stins, M., et al. (1999). Microglial and astrocyte chemokines regulate monocyte migration through the blood-brain barrier in human immunodeficiency virus-1 encephalitis. *American Journal of Pathology, 155*, 1599–1611.

Petty, M. A., & Lo, E. H. (2002). Junctional complexes of the blood-brain barrier: Permeability changes in neuroinflammation. *Progress in Neurobiology, 68*, 311–323.

Polli, J. W., Jarrett, J. L., Studenberg, S. D., Humphreys, J. E., Dennis, S. W., Brouwer, K. R., et al. (1999). Role of P-glycoprotein on the CNS disposition of amprenavir (141W94), an HIV protease inhibitor. *Pharmaceutical Research, 16*, 1206–1212.

Roessler, B. J., & Davidson, B. L. (1994). Direct plasmid mediated transfection of adult murine brain cells in vivo using cationic liposomes. *Neuroscience Letters, 167*, 5–10.

Ronaldson, P. T., & Bendayan, R. (2006). HIV-1 viral envelope glycoprotein gp120 triggers an inflammatory response in cultured rat astrocytes and regulates the functional expression of P-glycoprotein. *Molecular Pharmacology, 70*, 1087–1098.

Saito, R., Bringas, J. R., McKnight, T. R., Wendland, M. F., Mamot, C., Drummond, D. C., et al. (2004). Distribution of liposomes into brain and rate tumor models by convection-enhanced delivery monitored with magnetic resonance imaging. *Cancer Research, 64*, 2572–2579.

Siegal, T., Horowitz, A., & Gabizon, A. (1995). Doxorubicin encapsulated in sterically stabilized liposomes for the treatment of a brain tumor model: Biodistribution and therapeutic efficacy. *Journal of Neurosurgery, 83*, 1237.

Smit, T. K., Brew, B. J., Tourtellotte, W., Morgello, S., Gelman, B. B., & Saksena, N. K. (2004). Independent evolution of human immunodeficiency virus (HIV) drug resistance mutations in diverse areas of the brain in HIV-infected patients, with and without dementia, on antiretroviral treatment. *Journal of Virology, 78*, 10133–10148.

Solas, C., Lafeuillade, A., Halfon, P., Chadapaud, S., Hittinger, G., & Lacarelle, B. (2003). Discrepancies between protease inhibitor concentrations and viral load in reservoirs and sanctuary sites in human immunodeficiency virus-infected patients. *Antimicrobial Agents and Chemotheraphy*, *47*, 238–243.

Soontornmalai, A., Vlaming, M. L., & Fritschy, J. M. (2006). Differential, strain-specific cellular and subcellular distribution of multidrug transporters in murine choroid plexus and blood-brain barrier. *Neuroscience*, *138*, 159–169.

Stormer, E., von Moltke, L. L., Perloff, M. D., & Greenblatt, D. J. (2002). Differential modulation of P-glycoprotein expression and activity by non-nucleoside HIV-1 reverse transcriptase inhibitors in cell culture. *Pharmaceutical Research*, *19*, 1038–1045.

Valcour, V., & Paul, R. (2006). HIV infection and dementia in older adults. *Clinical Infectious Disease*, *42*, 1449–1454.

Varatharajan, L., & Thomas, S. A. (2009). The transport of anti-HIV drugs across blood-CNS interfaces: Summary of current knowledge and recommendations for further research. *Antiviral Research*, *82*, A99–A109.

Vinogradov, S. V., Bronich, T. K., & Kabanov, A. V. (2002). Nanosized cationic hydrogels for drug delivery: Preparation, properties and interactions with cells. *Colloids and Surfaces. B, Biointerfaces*, *16*, 291–304.

Weiss, J., Theile, D., Ketabi-Kiyanvash, N., Lindenmaier, H., & Haefeli, W. E. (2007). Inhibition of MRP1/ABCC1, MRP2/ABCC2, and MRP3/ABCC3 by nucleoside, nucleotide, and non-nucleoside reverse transcriptase inhibitors. *Drug Metabolism and Disposition*, *35*, 340–344.

Weller, D. R., Brundage, R. C., Balfour, J. H. H., & Venzina, H. E. (2007). An isocratic liquid chromatography method for determining HIV non-nucleoside reverse transcriptase inhibitor and protease inhibitor concentrations in human plasma. *Journal of Chromatography. B, Analytical Technologies in the Biomedical and Life Sciences*, *848*, 369–373.

Wolka, A. M., Huber, J. D., & Davis, T. P. (2003). Pain and the blood-brain barrier: Obstacles to drug delivery. *Advanced Drug Delivery Reviews*, *55*, 987–1006.

Zhang, L., Gorset, W., Washington, C. B., Blaschke, T. F., Kroet, D. L., & Giacomini, K. M. (2000). Interactions of HIV protease inhibitors with a human organic cation transporter in a mammalian expression system. *Drug Metabolism and Disposition*, *28*, 329–334.

Subject Index

Lightning Source UK Ltd.
Milton Keynes UK
UKHW030002140820
368213UK00001B/16

9 780444 534316